HYPOSPADIAS AND
GENITAL DEVELOPMENT

ADVANCES IN EXPERIMENTAL MEDICINE AND BIOLOGY

Recent Volumes in this Series

A Continuation Order Plan is available for this series. A continuation order will bring delivery of each new volume immediately upon publication. Volumes are billed only upon actual shipment. For further information please contact the publisher.

HYPOSPADIAS AND GENITAL DEVELOPMENT

Edited by

Laurence S. Baskin

University of California, San Francisco
San Francisco, California

Springer Science+Business Media, LLC

Library of Congress Cataloging-in-Publication Data

Hypospadias and Genital Development Symposium (2002: University of California, San Francisco)
    Hypospadias and genital development/edited by Laurence S. Baskin.
        p.   ; cm. — (Advances in experimental medicine and biology, ISSN 0065-2598; v. 545)
    Proceedings of the Hypospadias and Genital Development Symposium, held April 25–26,
2002, at the University of California, San Francisco.
    Includes bibliographical references and index.
    ISBN 978-0-306-48177-2   ISBN 978-1-4419-8995-6 (eBook)
    DOI 10.1007/978-1-4419-8995-6
        1. Hypospadias—Congresses. 2. Hypospadias—Surgery—Congresses. 3. Genitourinary
organs—Surgery—Congresses. I. Baskin, Laurence S. II. Title. III. Series.
    [DNLM]: 1. Hypospadias—etiology—Congresses. 2. Hypospadias—surgery—Congresses.
3. Genitalia, Male—embryology—Congresses. WJ 600 H9976h 2004]
RC896.H976 2002
617.4′63—dc22

                                                                                    2003061972

Proceedings of the Hypospadias and Genital Development Symposium, held April 25–26, 2002, at the
University of California, San Francisco

ISSN 0065-2598

ISBN 978-0-306-48177-2

©2004 Springer Science+Business Media New York
Originally published by Kluwer Academic / Plenum Publishers, New York in 2004

http://www.kluweronline.com

10 9 8 7 6 5 4 3 2 1

A C.I.P. record for this book is available from the Library of Congress

This book is dedicated to my parents, Fred and Cynthia Baskin

PREFACE

The aim of the Hypospadias and Genital Symposium, held at the University of California, San Francisco, was to provide a forum for authoritative investigators who are actively involved in the various disciplines which define the leading edges of hypospadias and genital research. It is important for such investigators to continue to meet for the purpose of discussing the latest developments in their individual fields, to analyze the significance of current research, to discuss new tactics for unresolved problems and to develop new theories and approaches as needed.

The two day conference on hypospadias and genital development research was organized into three sections: 1) Human Studies; 2) Mechanism of Genital Development; and 3) Endocrine Disruptors and Sexual Dimorphism in the Animal Kingdom. Each session was introduced by an expert moderator followed the invited speakers with time for extensive interaction between investigators. This book documents the proceedings of the Hypospadias and Genital Development Symposium.

I would especially like to thank Kari Gaudette for editorial assistants, Cynthia Ashe, Selcuk Yucel, Antonio Souza and the administrative staff in the Department of Urology.

I hope you find this resource useful.

Laurence S. Baskin, M.D.
Program Chair
Chief Pediatric Urology
UCSF

ACKNOWLEDGEMENTS

Support for this symposium is gratefully acknowledged

The National Institute of Health Grant # R13DK-HD5997
UCSF Department of Urology
American Urologic Association

SAN FRANCISCO · Golden Gate Bridge from the Presidio

CONTENTS

SECTION III. ENDOCRINE DISRUPTORS AND SEXUAL

DIMORPHISM IN THE ANIMAL KINGDOM

Section I

Introduction

HYPOSPADIAS

Laurence S. Baskin* M.D., FAAP

1. INTRODUCTION

Hypospadias is one of the most common congenital anomalies occurring in approximately 1:250 newborns or roughly 1 out of 125 live male births (Paulozzi et al., 1997). Hypospadias can be defined as an arrest in normal development of the urethral, foreskin and ventral aspect of the penis.   This results in a wide range of abnormalities with the urethral opening being anywhere along the shaft of the penis, within the scrotum or even in the perineum (Figure 1).

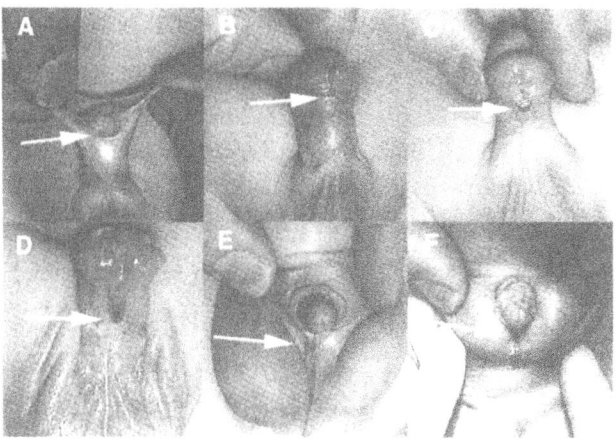

Figure 1.

Variations of hypospadias from mild to severe. A. Mild with the urethral opening on the glans.   B. Mild with the urethral opening at the coronal margin. C. Moderate with the urethral opening on the distal penile shaft. D. Moderate with the urethral opening on the mid penile shaft   E. Severe with the urethral opening at the penoscrotal junction. F. Severe with the urethral opening in the scrotum. (the arrows locate the opening of the hypospadiac urethra meatus) Note that in hypospadias the foreskin is absent on the ventral surface of the penis and excessive on the dorsal aspect. The more severe forms of hypospadias are associated with penile curvature.

*Chief, Pediatric Urology, Associate Professor of Urology and Pediatrics, University of California, San Francisco, San Francisco, CA, 94143-0738, (415) 476-1611, (415) 476-8849 (FAX), lbaskinl@urol.ucsf.edu

Hypospadias and Genital Development, edited by
L. Baskin, Kluwer Academic/Plenum Publishers, 2004

Hypospadias is also associated with penile curvature. Left uncorrected, patients with severe hypospadias may need to sit down to void and tend to shun intimate relationships because of the fears related to abnormal sexuality. Babies born with severe hypospadias and penile curvature may have "ambiguous genitalia" in the newborn period, making an immediate and accurate sex assignment difficult.

Hypospadias is classified by the location of the urethral meatus (Figure 2).

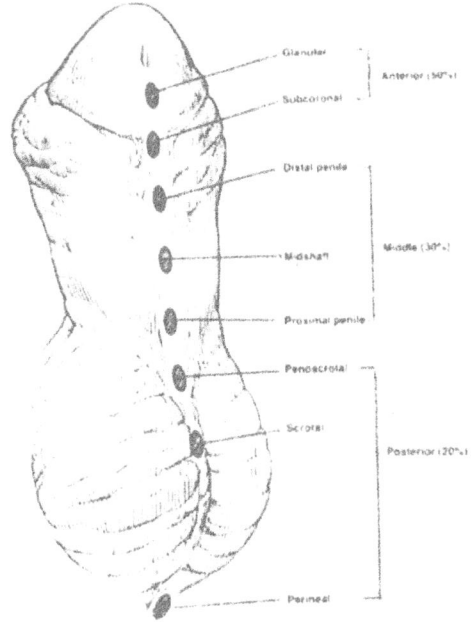

Figure 2. Classification and Incidence of Hypospadias: Anterior, Middle and Posterior.

Anterior hypospadias is described as glandular (meatus on the ventral surface of the glans penis); coronal (meatus in the balanopenile furrow); or distal (in the distal third of the penile shaft). Middle hypospadias is along the middle third of the penile shaft. Posterior hypospadias extends through the proximal third of the penile shaft to the perineum and is described as posterior penile (at the base of the shaft); penoscrotal (at the base of the shaft in front of the scrotum); scrotal (on the scrotum or between the genital swellings): or perineal (behind the scrotum or behind the genital swellings). As noted, chordee or penile curvature is a downward curvature of the penis that typically accompanies the more severe forms of hypospadias. Standard classification of hypospadias does not take into account the associated penile curvature. In reality a patient with severe curvature and an anterior urethral meatus may in fact require a more extensive surgery to correct both the curvature and the abnormal urethra.

2. Historical Notes

Throughout Greek culture, there was high appreciation for the goddess Hermaphrodite, half man, half woman. Many statues reflect hypospadiac genitalia, perhaps indicative of admiration for this condition. It is, therefore, understandable why it was not until the first and second centuries A.D. that the Alexandrian surgeons Heliodorus and Antyllus are given credit for the first attempted correction of this anomaly by amputation of the distal curved portion (Rogers, 1973). Sexually, the dystopia of the meatus may cause *impotentia generandi,* which is illustrated from the following historic note concerning Henry II of France. Henry II was known to have hypospadias, as recorded by his physician Fernal. His marriage with Catherine the Medici was infertile until Fenral "advised his patient that in such cases *coitus more ferarum* permitted him to overcome the difficulty" (Ombredanne, as quoted by Van der Muelen, 1964). Henry II then proceeded to sire three kings of France, along with seven other children.

3. Embryology: Development of the Male External Urogenital System

Formation of the external male genitalia is a complex developmental process involving genetic programming, cell differentiation, hormonal signaling, enzyme activity, and tissue remodeling. By the end of the first month of gestation, the hindgut and future urogenital system reach the ventral surface of the embryo at the cloacal membrane. The cloacal membrane divides the urorectal septum into a posterior, or anal half and an anterior half, the urogenital membrane. Three protuberances appear around the latter. The most cephalad is the genital tubercle. The other two, the genital swellings, flank the urogenital membrane on each side. Up to this point, the male and female genitalia are essentially indistinguishable. Under the influence of testosterone in response to a surge of luteinizing hormone from the pituitary, masculinization of the external genitalia takes place. One of the first signs of masculinization being an increase in the distance between the anus and the genital structures, followed by elongation of the phallus, formation of the penile urethra from the urethral groove and development of the prepuce (Jirasek et al., 1968; Hinman, 1993).

At eight weeks gestation the external genitalia remain in the indifferent stage (Figure 3A).

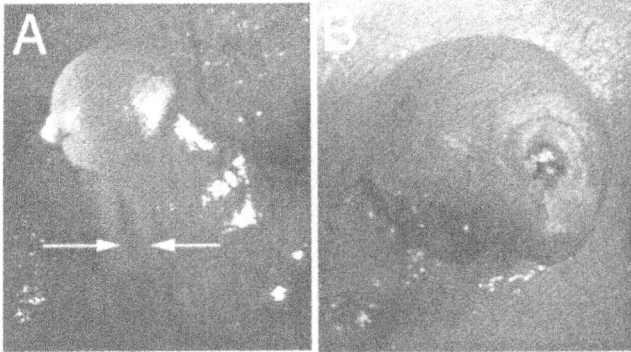

Figure 3.
Normal Male Genitalia Development: A. 10 weeks gestation. Note the open urethra, prominent urethra folds (arrows) and the glandular epithelial skin tag. B. 16 weeks gestation. Note that penile and urethra development are complete.

The urethral groove on the ventral surface of the phallus is between the paired urethral folds (Baskin et al, 2001). The penile urethral forms as a result of fusion of the medial edges of the endodermal urethral folds. The ectodermal edges of the urethral groove fuse to form the median raphe. By 12 weeks the coronal sulcus separates the glans from the shaft of the penis. The urethral folds have completely fused in the midline on the ventrum of the penile shaft. During the 16th week of gestation the glandular urethral appears. The mechanism of the glandular urethral formation remains controversial. Evidence suggest two possible explanations; 1) endodermal cellular differentiation (new theory) or 2) primary intrusion of the ectodermal tissue from the glans (old theory) (Figure 4).

Anatomical and immunohistochemical studies advocate the new theory of endodermal differentiation which shows that epithelium of the entire urethra is of urogenital sinus origin (Kurzrock et al., 1999). The entire male urethra, including the glandular urethra, is formed by dorsal growth of the urethral plate into the genital tubercle and ventral growth and fusion of the urethral folds. Under proper mesenchymal induction, urothelium has the ability to differentiate into a stratified squamous phenotype with characteristic keratin staining thereby explaining the cell type of the glans penis (Kurzrock et al., 1999). There is no evidence of an ectodermal ingrowth or a solid ectodermal cord filling the glans as was historically proposed (old theory) (Glenister, 1954).

The future prepuce is forming at the same time as the urethra and is dependent on normal urethral development. At about eight weeks gestation, low preputial folds appear on both sides of the penile shaft which join dorsally to form a flat ridge at the proximal edge of the corona. The ridge does not entirely encircle the glans because it is blocked on the ventrum by incomplete

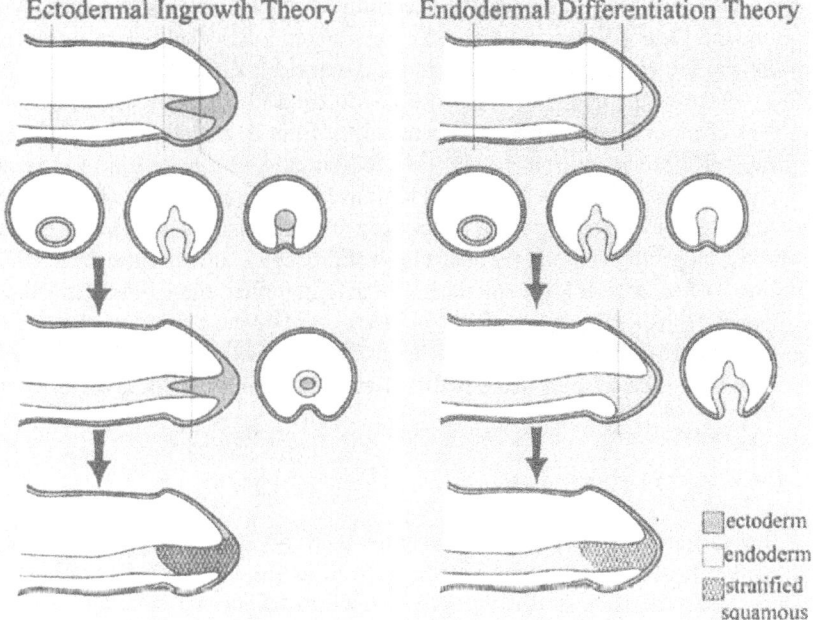

Figure 4.
Two theories of urethral development. The older theory of ectodermal intrusion and the newer theory of endodermal differentiation. (Used with permission from Dr. Kurzrock)

development of the glandular urethra. Thus, the preputial fold is transported distally by active growth of the mesenchyme between it and the glanular lamella. The process continues until the preputial fold (foreskin) covers all of the glans. (Figure 3B) The fusion is usually present at birth, but subsequent desquamation of the epithelial fusion allows the prepuce to retract. If the genital folds fail to fuse, the preputial tissues do not form ventrally; consequently, in hypospadias, preputial tissue is absent on the ventrum, and excessive dorsally (Figure 1).

The chronology of penile differentiation commences when "passive", or default, female differentiation is interrupted by androgenic hormones triggered by male genes (Conte and Grumbach, 1995). At the molecular level testosterone must be converted to 5α-dihydrotestosterone (DHT) by the microsomal enzyme, type 2 5α-reductase, for complete differentiation of the penis with a male-type urethra and glans (Wilson et al., 1993). Testosterone dissociates from its carrier proteins in the plasma and enters cells via passive diffusion (Kelce and Wilson, 1997). Once in the cell, testosterone binds to the androgen receptor (AR) and induces changes in conformation, protecting it from degradation by proteolytic enzymes (Kelce and Wilson, 1997). This conformational change is also required for AR dimerization, DNA binding, and transcriptional activation, all necessary for testosterone to be expressed. Androgen binding also displaces heat shock proteins, possibly relieving constraints on receptor dimerization or DNA binding. After entering the nucleus the AR complex then binds androgen response element DNA regulatory sequences within the androgen responsive genes and activates them. DHT also binds the AR, with enhanced androgenic activity, in part because of its slow dissociation rate from the AR (Griffin et al., 1995).

4. Incidence of Hypospadias

Seven European countries, including Norway, Sweden, England and Wales, Hungary, Denmark, Italy, and France, as well as the United States published independent reports of increasing rates of hypospadias during the 1960s, 1970s, and 1980s (Paulozzi, 1999). Recently, data from the International Clearinghouse for Birth Defects Monitoring Systems (ICBDMS) was analyzed to determine if these increases were worldwide, continuing, and whether they revealed any geographic patterns. The data from ICBDM suggest that increases in hypospadias are not a worldwide trend. Increases were most notable in the United States, Norway, and Denmark (Paulozzi, 1999). Also, it was determined that increases were not seen in the less affluent and less industrialized nations (gross domestic product, GDP, was used as a marker of affluence and industrialization) for which data were available. Increasing trends in England, Canada, and Northern Netherlands appeared to be leveling off since 1985 (Paulozzi,, 1999).

It is difficult to draw conclusions from international monitoring of birth defects. Different registries have different reporting methodologies and diagnostic criteria as well as varying degrees of physician compliance with reporting. However, as noted above, two independent surveillance systems in the United States with consistent and unchanging diagnostic criteria have also reported significant increases in hypospadias. While these surveillance programs are not flawless the data warrant particular mention (Paulozzi et al., 1997).

Data from the Metropolitan Atlanta Congenital Defects Program (MACDP), a population-based registry that uses active case ascertainment in 22 hospitals and clinics in the Atlanta, Georgia area, indicated that the total hypospadias rate almost doubled from 1968 to 1993 ($p < 10^{-6}$) at an annual rate of increase of 2.9% (Paulozzi et al., 1997). No one hospital in the Atlanta metropolitan region was responsible for the observed increases. Of particular significance, is that the incidence of severe hypospadias increased. Severe cases involve the urethral opening on the shaft of the penis or on the scrotum or perineum. Between 1968 and 1990 the severe cases increased from 1.1 to 2.7 per 10,000 live births and by 1993 to 5.5 per 10,000 births per year ($p < 10^{-6}$) (Paulozzi et al., 1997).

The Birth Defects Monitoring Program (BDMP), a program which gathers diagnoses recorded on newborn discharge summaries from hospitals nationwide, also reported an increase in hypospadias. Incidences increased from 20.2 per 10,000 live births (includes both male and female births) in 1970 to 39.7 per 10,000 live births in 1993 ($p < 10^{-6}$) (Paulozzi et al., 1997).

Both independent surveillance programs indicate a near doubling in reported rates of hypospadias. It is unlikely that this increase is due to flaws in the surveillance programs. For example, if increased sensitivity were to account for the observed increases, the surveillance programs would initially have had to record fewer than half of all defects. This is unlikely as no major changes in case ascertainment has occurred in the MACDP or the BDMP during that time period. It is possible that physicians' reporting habits of hypospadias have changed over time—particularly an increased reporting of mild hypospadias. This is not consistent, however, with reports from the MACDP that indicate that the ratio of mild to severe hypospadias decreased from 4.2 from 1968 to 1982 to 2.6 from 1983 to 1993 (Paulozzi et al., 1997). It is important to note, however, that as the ratio of mild to severe hypospadias decreased, the number of unclassified hypospadias also decreased making this data more difficult to interpret. This raises the question whether the mild cases are under-reported. Nonetheless, these longitudinal studies support an increasing incidence of hypospadias in the United States.

5. Associated Anomalies

Undescended testis and inguinal hernia are the most common associated anomalies with hypospadias. In one series, 9.3% of hypospadias patients had an undescended testis (Khuri et al., 1981). Posterior hypospadias had a 32% incidence, middle 6%, and anterior 5%. Khuri and associates also found the overall incidence of inguinal hernia to be 9%, with 17% associated with posterior hypospadias. Ross and colleagues (Ross et al., 1959) reported a similar incidence of either cryptorchidism or hernia in 16%, and Cerasaro and co-workers (Cesasaro et al., 1986) found 18% of 301 patients had these anomalies. A utriculus masculinus (utricle) is found in a high percentage of the more severe cases. For example, Devine and associates (Devine et al., 1980) found that of 44 patients whose meatus was posterior, 6 (14%) had a utricle. Of seven with perineal hypospadias, four (57%) had a utricle, compared with 2 of 20 (10%) with a penoscrotal meatus. None of the 17 patients with penile hypospadias had a utricle. Shima and colleagues (Shima et al., 1979) found a utricle in 1 of 29 glandular, 3 of 74 penile, 11 of 122 penoscrotal, and 10 of 29 scrotal-perineal hypospadias cases. Thus the incidence is 14% in 151 penoscrotal or perineal hypospadias cases. If both studies are combined, there is an 11% incidence of a utricle in severe hypospadias. Usually there are no complications from the presence of a utricle except for infection and difficulty passing a catheter (Ritchey et al., 1988).

It is not surprising that urinary tract anomalies are infrequent, because the external genitalia are formed much later than the supravesical portion of the urinary tract. McArdle and Lebowitz found only 6 genitourinary anomalies among 200 patients with hypospadias (3%) (McArdle and Lebowitz, 1975). Cerasaro and associates found 1.7% (4 of 233) patients with significant anomalies (Cerasaro et al., 1986). Avellan found only 1% renal anomalies in his Swedish patients with hypospadias, whereas he found 7% of patients with other systems involved (Avellan, 1975). In a review of 169 patients, Shelton and Noe do not recommend routine urinary tract evaluation (Shelton and Noe, 1985). Khuri and co-workers reviewed 1076 patients. Urinary tract anomalies considered to be significant were ureteropelvic junction obstruction, severe reflux, renal agenesis, Wilms' tumor, pelvic kidney, crossed renal ectopia, and horseshoe kidney. They conclude that patients with hypospadias and an associated inguinal hernia or undescended testis need not have further urinary tract evaluation. However, patients who have hypospadias in association with other organ system anomalies should undergo an upper urinary tract screening with an abdominal ultrasound.

Henderson and associates (1976) studied urogenital abnormalities in 500 sons of women treated with diethylstilbestrol (DES). He found 2% had hypospadias, and 4.4% had abnormalities of the penile urethra (Goldman, 1977). Gupta and Goldman (1986) implicated metabolic disturbances of the arachidonic acid cascade in midline fusion defects, such as cleft palate and hypospadias (Gupta and Goldman, 1986). Hypospadias occurs in association with a number of well-known syndromes as well as with Wilms' tumor.

6. Etiology of Hypospadias

Reports of increasing incidences of hypospadias have raised questions concerning etiology, treatment, and prevention. To date there is no sound understanding of the etiology of hypospadias that can inform primary prevention efforts and improve therapeutics. For example, in a recent study, 33 patients with severe (scrotal or penoscrotal) hypospadias were evaluated with a range of diagnostic techniques including clinical assessment, ultrasonography, karyotyping, endocrine evaluation, and molecular genetic analysis of the AR gene and the 5α-reductase gene to classify and determine the cause of the hypospadias. In 12 patients, diagnoses were determined. The remaining 64% of patients were classified as hypospadias of unknown etiology (Albers et al., 1997). Other investigators have also attempted to link abnormalities in androgen metobolism and/or receptor to hypospadias. For example, Gearheart and associates could not find any deficiencies in either androgen receptor levels or 5-_ reductase in their study of preputial skin from hypospadias boys (Gearhart et al., 1988). Allera et al. analyzed 9 patients with severe hypospadias and found a defect in open-reading frame of the androgen receptor in only one patient (Allera et al., 1995). Sutherland et al. also concluded that the mutations in the androgen receptor gene are rarely associated with hypospadias (Sutherland et al., 1996). Using single strand conformational polymorphism analysis they found a missense mutation of exon 2 of the androgen receptor gene in 1 of 40 patients with distal hypospadias. Molecular biology techniques have demonstrated that defects in the androgen receptor gene are definitely associated with isolated hypospadias. However, the frequency of these genetic defects accounts for an extremely small subset of cases implying that other factors are responsible for hypospadias.

7.   Genetic Impairment

Uncertainty about the etiology of hypospadias is in part a reflection of the complicated and incompletely understood molecular and cellular basis for normal phallus and urethral development (Wilson et al., 1981). It is well established that the urethral folds fuse during the formation of the penile urethra in the seventh to twelfth week after ovulation. Fusion of the folds eventually leads to a fully formed penis that is dependent on the synthesis of testosterone by the fetal testis. Adequate virilization of the urogenital sinus and external genitalia during embryogenesis is dependent on the conversion of testosterone to DHT by 5α-reductase. Genetic defects anywhere along this pathway can interfere with proper fusion of the urethral folds and result in hypospadias (Aaronson et al., 1997).

Increasingly, researchers are examining the role of cellular signals other than testosterone and DHT in the morphogenesis of the phallus and the etiology of hypospadias. Normal embryogenesis of the urogenital system depends on epithelial-mesenchymal interactions and it has been hypothesized that aberrant signaling between epithelium and mesenchyme could lead to hypospadias (Kurzrock et al., 1999). For example, prostate development requires "testosterone-dependent *Sonic hedgehog* (Shh) expression in the epithelium of the urogenital sinus" (Podlasek et al., 1999). It is likely that researchers may find similar genetic, signaling molecules involved in epithelial-mesenchymal interactions in the phallus that play a role in its development.

Another area of investigation with respect to the etiology of hypospadias is the expression and regulation of homeobox (Hox) genes. These genes are transcriptional regulators that play an essential role in directing embryonic development. Genes of the Hox A and Hox D clusters are expressed in regionalized domains along the axis of the urogenital tract. Transgenic mice with loss of function of single Hox A or Hox D genes exhibit homeotic transformations and impaired morphogenesis of the urogenital tract (Dolle et al., 1991; Benson et al., 1996; Hsieh-Li et al., 1995; Podlasek et al., 1997). Human males with hand-foot-genital (HFG) syndrome, an autosomal dominant disorder characterized by a mutation in HOXA13, exhibit hypospadias of variable severity, suggesting that HOXA13 may be important in the normal patterning of the penis (Mortlock and Innis, 1997; Donnenfeld et al., 1992; Fryns et al., 1993). Furthermore, recent research has shown that the embryonic expression of certain Hox genes is regulated by hormonal factors (Ma et al., 1998). Estrogen and the synthetic estrogen diethylstilbestrol (DES) for example, inhibit Hoxa-9, Hoxa-10, Hoxa-11, and Hoxa-13 genes in mice. Thus, in addition to defects in Hox genes, it is possible that improper regulation or expression of hormonal factors during embryogenesis could disrupt normal expression of Hox genes and lead to reproductive tract anomalies.

8. Hormone Receptor Impairment

As noted above, studies examining the androgen receptor (AR) have yielded even less insight into the etiology of hypospadias. In fact, a number of studies have concluded that defects in the androgen receptor or mutations in the androgen receptor coding sequence are rare in patients with hypospadias (McPhaul et al., 1993; Hiort et al., 1994). In addition, Bentvelsen et al. measured AR expression in the foreskins of boys with hypospadias and age-matched controls and found no significant difference in mean AR content (Bentvelsen et al., 1995). However, Bentvelsen et al. did not measure the mean AR content in the foreskin during gestational development.

9. Enzyme Impairment

Despite the central role that testosterone plays, attempts to ascribe all hypospadias to an underlying genetic defect in this pathway have been only partly successful. Aaronson et al. determined the incidence of defects in three major enzymes in the biosynthetic pathway of testosterone—3β-hydroxysteroid dehydrogenase, 17α-hydroxylase, and 17,20-lyase—in 30 boys with fully descended testes but with penoscrotal or proximal shaft hypospadias (Aaronson et al., 1997). A total of 15 boys had evidence of impaired function of one or more of these enzymes suggesting that half the boys had an underlying defect in the biosynthesis of testosterone. This still left 50% of the cases unexplained (Aaronson et al., 1997).

10. Environmental Factors and Endocrine Disruptors

In the past, environmental factors were generally ruled out as causes for hypospadias (Harris, 1990; Stoll et al., 1990). More recently, multi-causality models include environmental contaminants to determine the risk of developing a given phenotype. For example, familial clustering of hypospadias among first-degree relatives has been perceived as under the influence of a strong genetic and heritable component, but there have been many exceptions where genetics were ruled out. More recently it has been suggested that environmental influences should be considered as well, taking into consideration that families share similar exposure. And, especially, in those cases where the effects are the most profound, genetic predisposition exacerbated by environmental exposure should be considered (Fritz and Czeizel, 1996).

Attempts to determine risk factors for hypospadias have yielded a number of maternal and paternal risk factors. Among traditional studies of maternal risk factors for congenital anomalies, maternal age and primiparity are significantly associated with hypospadias, although some studies have contested the maternal age effect (Harris, 1990). Paternal risk factors associated with hypospadias include abnormalities of the fathers' scrotum or testes (Sweet et al., 1974) and low spermatozoa motility and abnormal sperm morphology (Fritz and Czeizel, 1996). It has been suggested that the recent increase in hypospadias reflects the improvement in fertility treatment, contributing to more subfertile men fathering children. As the authors state, this ". . relaxed-selection hypothesis, which states that there is a redistribution in the number of children born to fertile and infertile (subfertile) couples, may account for the increasing number of other defects and cancers of male genitalia observed today and the fall in sperm counts" (Fritz and Czeizel, 1996).

In addition to parental risk factors there is strong consensus in the literature that boys with hypospadias have lower birth weight (Fredell et al., 1998). Fredell et al. examined hypospadias in discordant monozygotic twins and found that the twin with hypospadias weighed 78% of the twin without hypospadias. The birth weight difference was still significant when compared with birth weight difference between healthy monozygotic twins. Another study found that boys with hypospadias have a lower placental weight than control boys (Stoll et al., 1990). A 1995 meta-analysis of first trimester exposure to progestins and oral contraceptives, did not indicate an increased risk of hypospadias (Raman-Wilms et al., 1995). Exposure to DES was excluded in this study. However, a number of studies do list gestational exposure to progestins as a causal agent. Another recent study links a maternal vegetarian diet in pregnancy with an increase in the incidence of hypospadias (North and Golding, 2000). This study looked at 51 boys with hypospadias from a group of 7928 boys born to mothers taking part in the Avon Longitudinal Study of Pregnancy and Childhood. The authors hypothesize that vegetarians have a greater exposure to phytoestrogens than omnivores. The phytoestrogens may come in the from of soya which is high in isoflavones or related to endocrine disrupters in pesticides and fertilizers (Price and Fenwick, 1985).

While these risk factors do not reveal direct information about the etiology of hypospadias, they provide additional information that may reveal a common developmental pathway and can inform future research. For example, there is growing evidence that androgens play a central role in the lower birth weight of girls compared to boys (de Zegher et al., 1998). Androgens are also crucial to the development of the male reproductive tract. Thus, perhaps exposure to an agent that compromises the weight-gaining advantage of androgen during gestation could play a role in the development of hypospadias and lowered birth weight.

Increasing rates of hypospadias have paralleled reports of other untoward endpoints related to male reproductive health, including increases in testicular cancer, (Bergstrom et al., 1996) increasing incidences of cryptorchidism, and decreasing semen and sperm quality.[54] In another study, 8% of patients diagnosed (n = 252) with undescended testes had urogenital anomalies and over 50% of those had hypospadias (Cheng et al., 1996). The increasing incidence over the past 50 years of multiple endpoints co-occurring with increasing production and use of synthetic chemicals has raised concerns that environmental factors may play a role in the etiology of these problems (Toppari et al., 1996). It is now well documented from wildlife studies and accompanying laboratory data, that a number of synthetic and natural chemicals commonly found in the environment can mimic or antagonize hormones or otherwise

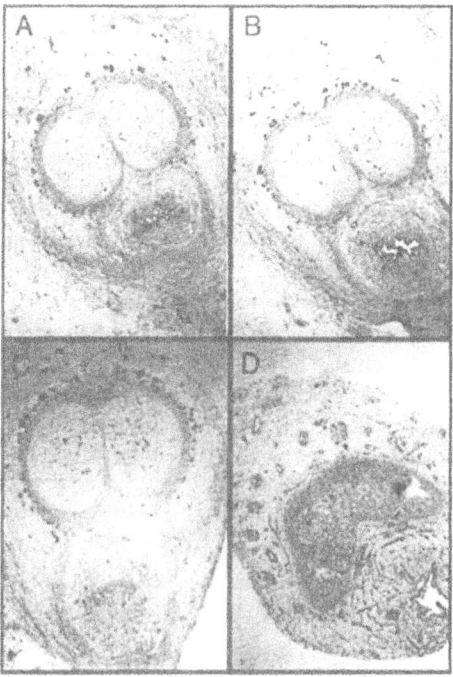

Figure 5A-D. Normal Human fetal penis, 25 weeks. A-H Transverse sections distal to proximal immunostained with neuronal marker S-100 (25X). Note localization of S-100 nerve marker in brown completely surrounding the cavernous bodies up to the junction with the urethral spongiosum along the penile shaft except at the 12 o'clock position (A-D).

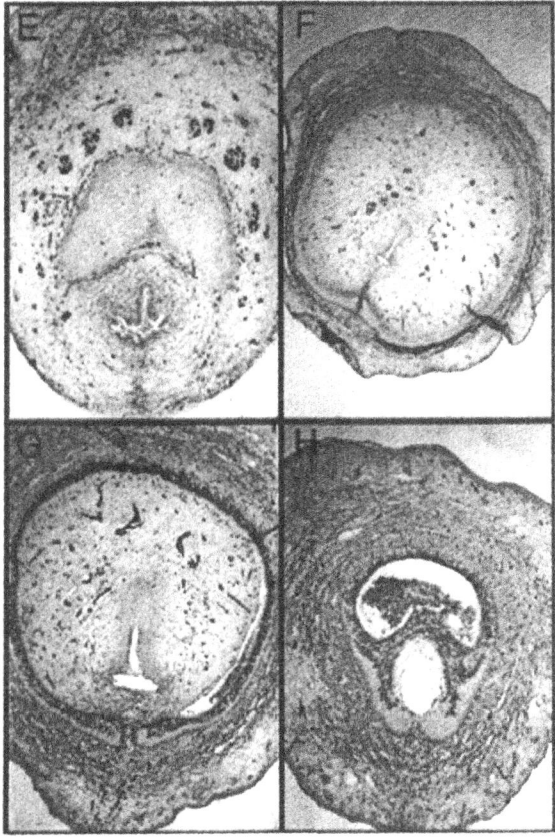

Figure 5E-H. On the proximal penis at the point where the corporal bodies split into two (E) and continue in a lateral fashion inferior an adjacent to the pubic rami the nerves localize to an imaginary triangular area at the 11 and 1 o'clock position. At this point (E) the nerves reach there furthest vertical distance from the corporeal body (~ one half the diameter of the corporeal body) and continue (F-G) in a tighter formation at the 11 and 1 o'clock position well away from the urethra.

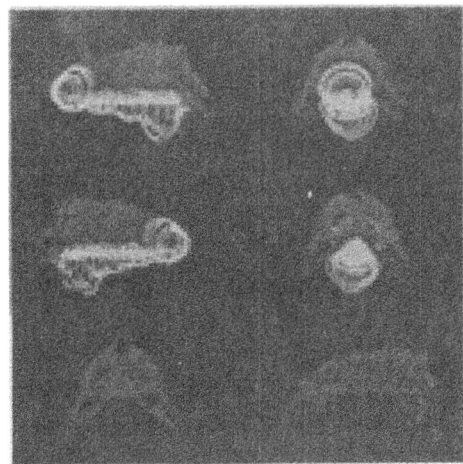

Figure 6.
Normal Human fetal penis, 25 weeks' gestation
Four views of a computer-generated three-dimensional reconstruction (A, Side; B, Front; C, Side; D, Back E, Front (without urethral); F, Side (without urethral)). Note the nerves in red and their absence at the 12 o'clock position. The tunica is represented in blue, the urethral lumen in yellow and the urethral spongiosum and prepuce in lime.

interfere with the development and function of the endocrine and reproductive systems (Rolland et al., 1995; Colborn et al., 1999). Whether endocrine disruptors are having an impact on male reproductive health and on hypospadias in particular, is difficult to determine (Shakkebaek et al., 1998). Regardless, public health agencies world-wide are increasingly concerned about endocrine disruption and it remains an active area of research (Jensen, 1998; Groshart et al., 1999; EPA, 1998; Baskin et al., 2001).

11. Penile Anatomy

Surgical repair of hypospadias requires an expert understanding of the normal anatomy of the penis, as well as an understanding of the anatomy of the hypospadiac penis. The human penis consists of paired corpora cavernosa covered by a thick, elastic tunica albuginea, with a midline septum (Figure 5) (Hinman, 1993).

The urethral spongiosum lies in a ventral position, intimately engaged between the two corporal bodies. Buck's fascia surrounds the corporal cavernosa and splits to contain the corpus spongiosum in a separate compartment. Recent work has shown that the neurovascular bundle lies deep to Buck's fascia and where the two crural bodies join to form the corporal bodies, the neurovascular bundle completely fans out around the corporal cavernosa, all the way to the junction of the corporal spongiosum (Figure 5 and 6).

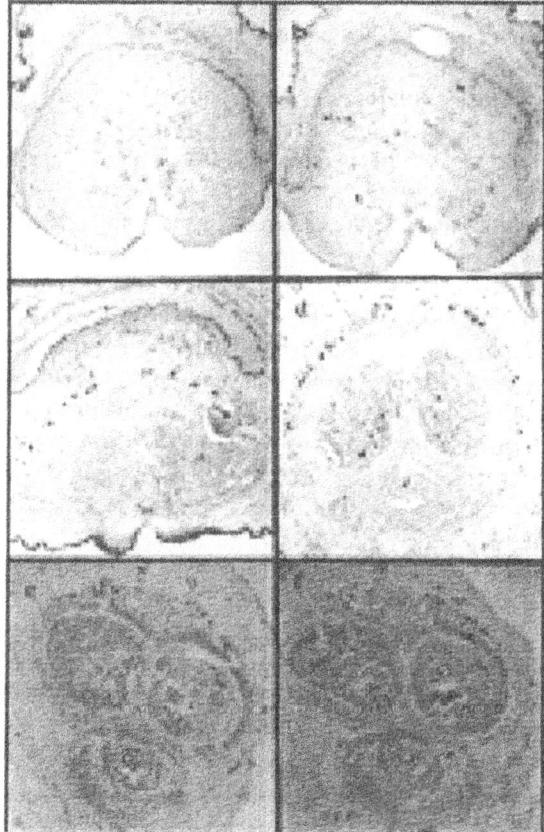

Figure 7.
Hypospadiac Penis, 33 weeks' gestation   A-H Transverse sections distal to proximal immunostained
with neuronal marker S-100 (20X).   Note the anatomy of the hypospadias penis is the same as the
normal penis except for the abnormal formation of the distal urethra and glans (A-C). The nerves
are black staining in Figures A-D and brown staining in E and F.

(Baskin et al., 1997). This concept disagrees with the classic dogma that the neurovascular bundle lies in the 11 and 1 o'clock position. Superior to Buck's fascia is the dartos fascia which lies immediately beneath the skin. This fascia contains the blood supply to the prepuce. The prepuce is supplied by two branches of the inferior external pudendal arteries, the superficial penile arteries (Hinman, 1991). These arteries divide into the anterolateral and posterolateral branches. The island flap is typically based on the anterolateral superficial vessels. The onlay island flap and tubularized island flap are dependent on careful preservation of these blood vessels. In hypospadias surgery, the outer skin survives from remaining subcutaneous vessels.

12. Neuro and Vascular Anatomy of Hypospadias

The normal anatomy of the penis is to be compared to the anatomy of the hypospadiac penis (Figure 7) (Baskin et al., 1998). Except at the region of the abnormal urethral spongiosum and glans, the hypospadiac and normal penises show no difference in neuronal innervation, corporal cavernosa and tunica albuginea architecture and blood supply. The nerves in both the normal and hypospadiac penises start as two well-defined bundles superior and slightly lateral to the urethra. As the two crural bodies converge into the corporal cavernosal bodies, the nerves diverge spreading around the cavernosal bodies up to the junction with the urethral spongiosum, not limiting themselves to the 11 and 1 o'clock position (Figure 7 and 8). The 12 o'clock position in the hypospadiac penis is spared of any neuronal structures just as in a normal penis. The most striking difference between the normal penis and the hypospadiac penis is a difference in vascularity (Figure 7). The hypospadiac penis has huge endothelial-lined vascular channels filled with red blood cells. In contrast, the normal penis has well defined, small capillaries around the urethra, fanning into the glans.

Anatomical studies of the urethral plate do not show any evidence of fibrosis or scarring (Erol et al., 2000). The urethral plate is well vascularized, has a rich nerve supply and an extensive muscular and connective tissue backing (Figure 9). These features may explain the successful use of incorporating the urethral plate into hypospadias reconstruction.

13. Surgical Techniques

The goal of hypospadias surgery is to correct the penile curvature, reposition the meatus to allow for intercourse and proper delivery of semen and to reconstruct a forward directed stream. Each hypospadiac penis is different and therefore one operation will not solve all the reconstructive problems. When planning hypospadias surgery, the issues at hand are reconstruction of a new urethra, correction of penile curvature, creation of a new meatus, skin coverage, and finally correction of any penile scrotal transposition and/or bifid scrotum.

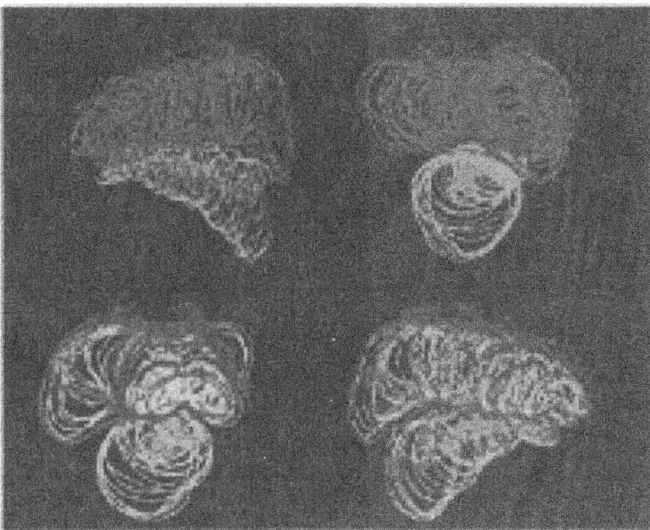

Figure 8.
Hypospadiac Penis, 33 weeks' gestation   Four views of a computer-generated three-dimensional reconstruction. (A, Side; B, Back; C, Front; D, Side). Note the nerves in red and their absence at the 12 o'clock position. The tunica of the corporal bodies is represented in blue in A and B and blue and yellow in C and D, the urethral lumen in orange and the urethral spongiosum in lime. The hypospadiac penis has the same innervation and anatomy as the normal penis except the abnormal urethral spongiosum.

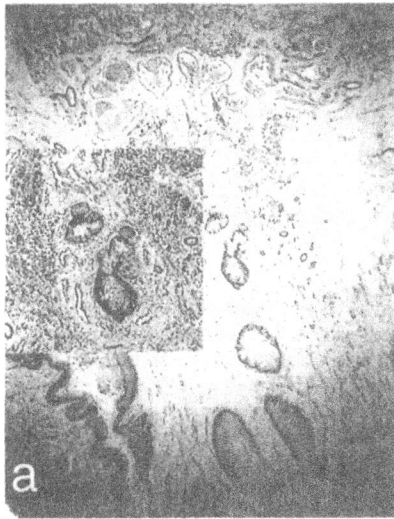

Figure 9.
The urethral plate in a newborn human penis with proximal hypospadias, (25x, α-actin immunostaining)   The urethral plate is well vascularized, without any evidence of fibrosis or scarring. Glans or the abortive attempt at the formation of the urethra are seen within the plate (see insert).

These issues must be assessed prior to surgery, but flexibility needs to be maintained in that the quality of the hypospadiac urethra may be difficult to assess until the patient is under anesthesia. An anterior hypospadias may therefore turn into a proximal hypospadias, especially after the correction of penile curvature and resection of abortive or inadequate urethral tissue. A surgeon's approach to hypospadias depends on his personal preference, skill and experience. The father of modern hypospadias surgery, Dr. John W. Duckett, coined the term "hypospadiology" building his knowledge and technique on an extensive familiarity with those who have come before him. The reader is refered to multiple chapters and articles on specific surgical techniques (Baskin, 2001; Baskin et al., 1996; Duckett and Baskin, 1996; Snodgrass et al., 1996; Baskin and Duckett, 1998).

REFERENCES

Aaronson, I.A., Cakmak, M.A. and Key, L.L., 1997,.Defects of the testosterone biosynthetic pathway in boys with hypospadias, *J Urol.* **157(5):** p. 1884-8.

Albers, N., et al., 1997, Etiologic classification of severe hypospadias: implications for prognosis and management [see comments], *J Pediatr.* **131(3):** p. 386-92.

Allera, A., et al., 1995, Mutations of the androgen receptor coding sequence are infrequent in patients with isolated hypospadias, *J Clin Endocrinol Metab.* **80(9):** p. 2697-9.

Avellan, L., 1975, The incidence of hypospadias in Sweden, *Scand J Plast Reconstr Surg.* **9:** p. 129-138.

Baskin, L.S., Duckett, J.W. and Lue, T.F., 1996, Penile curvature, *Urology.* **48(3):** p. 347-56.

Baskin, L.S., Lee, Y.T. and Cunha, G.R., 1997, Neuroanatomical ontogeny of the human fetal penis, *Br J Urol.* **79(4):** p. 628-40.

Baskin, L. and Duckett, J., 1998, Hypospadias: Long-Term Outcomes, in Pediatric Surgery and Urology: Long-Term Outcome, M. Pierre D.E. Mouriquand, Editor. W.B. Saunders: London. p. 559-567.

Baskin, L.S., et al., 1998, Anatomical studies of hypospadias, *J Urol.* **160(3 Pt 2):** p. 1108-15; discussion 1137.

Baskin, L.S., Colborn, T. and Himes, K., 2001, Hypospadias and endocrine disruption: is there a connection? *Environmental Health Perspectives.* **109:** p. 1175-1183.

Baskin, L.S., et al., 2001, Urethral Seam Formation and Hypospadias, *Cell Tissue Res.* **305:** p. 379-387.

Baskin, L.S., 2001, **in press.**, Hypospadias: a critical analysis of cosmetic outcomes using photography, *British Journal of Urology.*

Benson, G.V., et al., 1996, Mechanisms of reduced fertility in Hoxa-10 mutant mice: uterine homeosis and loss of maternal Hoxa-10 expression, *Development.* **122(9):** p. 2687-96.

Bentvelsen, F.M., et al., 1995, Decreased immunoreactive androgen receptor levels are not the cause of isolated hypospadias, *Br J Urol.* **76(3):** p. 384-8.

Bergstrom, R., et al., 1996, Increase in testicular cancer incidence in six European countries: a birth cohort phenomenon, *J Natl Cancer Inst.* **88(11):** p. 727-33.

Carlsen, E., et al., 1995, Declining semen quality and increasing incidence of testicular cancer: is there a common cause? *Environ Health Perspect.* **103 Suppl 7:** p. 137-9.

Cerasaro, T.S., Brock, W.A. and Kaplan, G.W., 1986, Upper urinary tract anomalies associated with congenital hypospadias: is screening necessary? *J Urol.* **135(3):** p. 537-8.

Cheng, W., Mya, G.H. and Saing, H., 1996, Associated anomalies in patients with undescended testes, *J Trop Pediatr.* **42(4):** p. 204-6.

Colborn, T., Short, P. and Gilbertson, M., 1999, Health effects of contemporary-use pesticides: The wildlife/human connection, in *Toxicology and Industrial Health*.

Conte, F. and Grumbach, M., 1995, Pathogenisis, classification, diagnosis and treatment of anomalies of sex, in *Endocrinology*. DeGroot, Editor. Saunders.

Devine, C., et al., 1980, Utricular configuration in hypospadias and intersex, *J Urol*. **123:** p. 407-411.

de Zegher, F., et al., 1998, Androgens and fetal growth, *Horm Res*. **50(4):** p. 243-4.

Dolle, P., et al., 1991, HOX-4 genes and the morphogenesis of mammalian genitalia, *Genes Dev*. **5(10):** p. 1767-7.

Donnenfeld, A.E., Schrager, D.S. and Corson, S.L., 1992, Update on a family with hand-foot-genital syndrome: hypospadias and urinary tract abnormalities in two boys from the fourth generation, *Am J Med Genet*. **44(4):** p. 482-4.

Duckett, J. and Baskin, L., 1996, Hypospadias, in *Adult and Pediatric Urology*, J. Gillenwater, et al., Editors. Mosby: St. Louis.

EPA, 1998, Endocrine Disruptor Screening and Testing Advisory Committee (EDSTAC). Final Report. United States Environemental Protection Agency.

Erol, A., et al., 2000, Anatomical studies of the urethral plate: why preservation of the urethral plate is important in hypospadias repair, *BJU Int*. **85(6):** p. 728-34.

Fredell, L., et al., 1998, Hypospadias is related to birth weight in discordant monozygotic twins, *J Urol*. **160(6 Pt 1):** p. 2197-9.

Fritz, G. and Czeizel, A.E., 1996, Abnormal sperm morphology and function in the fathers of hypospadiacs, *J Reprod Fertil*. **106(1):** p. 63-6.

Fryns, J.P., et al., 1993, The hand-foot-genital syndrome: on the variable expression in affected males, *Clin Genet*. **43(5):** p. 232-4.

Gearhart, J.P., et al., 1988, Androgen receptor levels and 5 alpha-reductase activities in preputial skin and chordee tissue of boys with isolated hypospadias, *J Urol*. **140:** p. 1243-6.

Glenister, T., 1954, The origin and fate of the urethral plate in man, *J Anat*. **288:** p. 413-418.

Griffin, J., et al., 1995, The Androgen Resistance Sydromes: Steroid 5 alpha-Reductase Deficiency, Testicular Feminization and Related Disorders, in *The Metabolic and Molecular Bases of Inherited Disease*., C. Scriver, Editor., McGraw-Hill,: Philadelphia. p. 2967-2998.

Goldman, A., 1977, Abnormal organogenesis in the reproductive system, in *Handbook of Teratology*, J. Wilson and F. Fraser, Editors., Plenum Press: New York. p. 391-419.

Groshart, C., Okkerman, P. and Folkers, G., 1999, Report to the European Commission: Identification of Endocrine Disruptors, First Draft Interim Report, Delft, The Netherlands: BKH Consulting Engineers.

Gupta, C. and Goldman, A., 1986, Arachidonic acid cascade is involved in the masculinization of embryonic external genitalia in mice, *Proc Natl Acad Sci*. **83:** p. 4346-4349.

Harris, E.L., 1990, Genetic epidemiology of hypospadias, *Epidemiol Rev*. **12(29):** p. 29-40.

Hinman, F., Jr., 1991, The blood supply to preputial island flaps, *J Urol*. **145(6):** p. 1232-5.

Hinman, F.J., 1993, Penis and male urethra, in *UroSurgical Anatomy*, F.J. Hinman, Editor. WB Saunders Co.: Philadelphia. p. 418-470.

Hiort, O., et al., 1994, Molecular characterization of the androgen receptor gene in boys with hypospadias, *Eur J Pediatr*. **153(5):** p. 317-21.

Hsieh-Li, H.M., et al., 1995, Hoxa 11 structure, extensive antisense transcription, and function in male and female fertility, *Development*. **121(5):** p. 1373-85.

Jensen, K., 1998, European Parliament's Committee on the Environment, Public Health and Consumer Protection. Report on Endocrine Disrupting Chemicals. Rapporteur.

Jirasek, J., Raboch, J. and Uher, J., 1968, The relationship between the development of gonads and external genitals in human fetuses, *Am J Obstet Gynecol*. **101:** p. 830.

Kelce, W.R. and Wilson, E.M., 1997, Environmental antiandrogens: developmental effects, molecular mechanisms, and clinical implications, *J Mol Med*. **75(3):** p. 198-207.

Khuri, F., Hardy, B. and Churchill, B., 1981, Urologic anomalies associated with hypospadias, *Urol Clin North Am*. **8:** p. 565-571.

Kurzrock, E., Baskin, L. and Cunha, G., 1999, Ontogeny of the Male Urethra: Theory of endodermal differentiation, *Differentation*. **64:** p. 115-122.

Kurzrock, E., et al., 1999, Epithelial-mesenchymal interactions in development of the mouse fetal genital tubercle, *Cells Tissues and Organs*. **164:** p. 1015-1020.

Ma, L., et al., 1998, Abdominal B (AbdB) Hoxa genes: regulation in adult uterus by estrogen and progesterone and repression in mullerian duct by the synthetic estrogen diethylstilbestrol (DES). *Dev Biol.* **197(2):** p. 141-54.

McArdle, F. and Lebowitz, R., 1975, Uncomplicated hypospadias and anomalies of upper urinary tract. Need for screening? *Urology.* **5:** p. 712-716.

McPhaul, M.J., et al., 1993, Genetic basis of endocrine disease. 4. The spectrum of mutations in the androgen receptor gene that causes androgen resistance, *J Clin Endocrinol Metab.* **76(1):** p.17-23.

Mortlock, D.P. and Innis, J.W., 1997, Mutation of HOXA13 in hand-foot-genital syndrome [see comments], *Nat Genet.* **15(2):** p. 179-80.

North, K. and Golding, J. 2000, A maternal vegetarian diet in pregnancy is associated with hypospadias. The ALSPAC Study Team. Avon Longitudinal Study of Pregnancy and Childhood. *BJU Int.* **85(1):** p. 107-13.

Paulozzi, L., Erickson, D. and Jackson, R., 1997, Hypospadias Trends in Two US Surveillance Systems, *Pediatrics.* **100(5):** p. 831-834.

Paulozzi, L.J., 1999, International trends in rates of hypospadias and cryptorchidism, *Environ Health Perspect.* **107(4):** p. 297-302.

Podlasek, C.A., Duboule, D. and Bushman, W., 1997, Male accessory sex organ morphogenesis is altered by loss of function of Hoxd-13, *Dev Dyn.* **208(4):** p. 454-65.

Podlasek, C.A., et al., 1999, Prostate development requires Sonic hedgehog expressed by the urogenital sinus epithelium, *Dev Biol.* **209(1):** p. 28-39.

Price, K. and Fenwick, G., 1985, Naturally occuring oestrogens in foods- a review, *Food Add Contam,.* **2:** p. 73-106.

Raman-Wilms, L., et al., 1995, Fetal genital effects of first-trimester sex hormone exposure: a meta-analysis, *Obstet Gynecol.* **85(1):** p. 141-9.

Ritchey, M.L., et al., 1988, Management of mullerian duct remnants in the male patient, *J Urol.* **140(4):** p. 795-9.

Rogers, B., 1973, History of external genital surgery, in *Plastic and Reconstructive Surgery of the Genital Area,* C. Horton, Editor., Little, Brown & Co. Boston. p. 3-47.

Rolland, R., Gilbertson, M. and Colborn, T., 1995, Environmentally induced alternations in development: A focus on wildlife, *Environ Health Perspect.* **103(suppl 4)**.

Ross, F., Farmer, A. and. Lindsay, W., 1959, Hypospadias--a review of 230 cases, *Plast Reconstr Surg.* **24:** p. 357-368.

Shelton, T.B. and Noe, H.N., 1985, The role of excretory urography in patients with hypospadias, *J Urol.* **134(1):** p. 97-9.

Shima, H., et al., 1979, Developmental anomalies associated with hypospadias, *J Urol.* **122:** p. 619-621.

Skakkebaek, N.E., et al., 1998, Germ cell cancer and disorders of spermatogenesis: an environmental connection? *Apmis.* **106(1):** p. 3-11; discussion 12.

Snodgrass, W., et al., 1996, Tubularized incised plate hypospadias repair: results of a multicenter experience, *J Urol.* **156(2 Pt 2):** p. 839-41.

Stoll, C., et al., 1990, Genetic and environmental factors in hypospadias, *J Med Genet.* **27(9):** p. 559-63.

Sutherland, R.W., et al., 1996, Androgen receptor gene mutations are rarely associated with isolated penile hypospadias, *J Urol.* **156(2 Pt 2):** p. 828-31.

Sweet, R., et al., 1974, Study of the incidence of hypospadias in Rochester, Minnesota 1940-1970, and a case-control comparison of possible etiologic factors, *Mayo Clin Proc.* **49:** p. 52-58.

Toppari, J., et al., 1996, Male reproductive health and environmental xenoestrogens [see comments], *Environ Health Perspect.* **104 Suppl 4:** p. 741-803.

Van der Meulen, J., 1964, in *Hypospadias Monograph..* Leiden, The Netherlands: AG Stenfert, Kroese, NV.

Wilson, J.D., George, F.W. and Griffin, J.E., 1981, The hormonal control of sexual development, *Science.* **211(4488):** p. 1278-84.

Wilson, J.D., Griffin, J.E. and Russell, D.W., 1993, Steroid 5 alpha-reductase 2 deficiency, *Endocr Rev.* **14(5):** p. 577-93.

Section II

Human Studies

EPIDEMIOLOGY OF HYPOSPADIAS

Adapted from a presentation by J. David Erickson, DDS, Ph.D. from the
Center for Disease Control

1. INTRODUCTION

·Birth defects affect about 3% of babies and some are born with major structural
chromosomal abnormalities. Birth defects are a major cause of infant mortality. As other
causes of infant mortality have been controlled, birth defects have risen in importance. It
is surprising how very little we know about the cause of birth defects and how to prevent
them. What is known is that prescription medications cause birth defects. The Center
for Disease Control (CDC) efforts in the prevention of birth defects and the efforts of
epidemiologists worldwide started as a result of the realization that exposure to environmental agents is deleterious. Specifically, thalidomide, the drug that caused an epidemic of limb reduction deformities in Europe in the late 1950s and early 1960s, spurred
on the idea that synthetic compounds may be teratogenic (Hamilton and Poswillo, 1972).

From the 1970's until the 1990's a national program from the CDC recorded the number of cases of hypospadias (Paulozzi, Erickson et al. 1997). Over the last 20 years this
program reported a rise in hypospadias in the United States from about two per thousand
to about four per thousand (Paulozzi, Erickson et al. 1997; Paulozzi 1999). It is clear that
there has been an increase in reporting in this country. This was documented in the
national program and in the program that the CDC has run in metropolitan Atlanta since
1968.

Why the increase in incidence of hypospadias? Is the increase a result changes in
incidence or prevalence at birth, is it a result of an increased awareness, or an increase in
desires to do surgery that result in the recognition? No clear answer exist. One certainly
cannot rule out that there is a real increase in incidence or prevalence at birth.

2. CDC Surveillance Program

The program at the CDC has the goals of improving surveillance for birth defects and
attempting to understand the causes. The CDC is particularly interested in identifying
environmental causes of birth defects that might lead to prevention strategies. Unfortunately, there are not many that have been discovered. The CDC is interested in pursuing
policy change that might result in better birth outcomes, and running prevention activities. This consist of surveillance systems or monitoring, to keep track of the frequency
for which hypospadias occurs. Epidemiological studies are initiated to identify risk fac-

Hypospadias and Genital Development, edited by
L. Baskin, Kluwer Academic/Plenum Publishers, 2004

tors or protective factors for the birth defect in question. There are many times the public is concerned, particularly on a local level, when an increase in incidence of hypospadias is reported.

Presently, the CDC participates in three surveillance systems. There is a guidance system, that was initiated 1968. This is a highly aggressive system where the CDC staff directly interacts with the local medical care community trying to ascertain cases of birth defects or hypospadias. The CDC has a consortium of approximately 30 state health departments, that are interested in the prevention of birth defects and in many ways is the result of the increasing importance of birth defects as a cause of infant mortality. Finally, The CDC participates in the International Clearinghouse for Birth Defects, which is an organization that has been around for about 30 years and was formed in the aftermath of the thalidomide tragedy.

In Atlanta, the CDC established, in conjunction with Emory and the Georgia State Health Department, continuous data collection since late 1967. This consortium operates in five counties in Georgia and the five major counties of the metropolitan Atlanta area. This consortium has under surveillance about 47,000 births per year that yields approximately 1,000 to 1,500 cases of all types of major congenital structural malformations each year. This group keeps track of what is reported on birth and death certificates and data within these counties, data from the Emory Genetics Group and 17 different obstetric and pediatric hospitals. Currently, in the surveillance program there are not any outpatient facilities that are not hospital-connected. Outpatient diagnoses that are made outside of a hospital and treated outside of a hospital may not be reported in the surveillance system. Is this possible with hypospadias? Most likely and this may be a source of missed cases.

3. Folic acid neural tube defects

The data on hypospadias is not as strong and the possible risk factors not as well defined as the role of folic acid in neural tube defects.. Folic acid has been shown to reduce the risk of neural tube defects, anencephaly and spina bifida (De Walle et al., 2003; Berry et al., 1999; Guney et al., 2003). There has been a myriad of research projects beginning in 1980 with a seminal study by Dick Smithells from Leeds, England, who did a nonrandomized but controlled trial of vitamin supplementation in women who had already had a pregnancy affected by a neural tube defect (Smithells, 1996). He found quite a dramatic decrease in risk for women who had folic acid during pregnancy. This study generated more research but did not change public policy.

In the early 1990s two well done randomized control trials conclusively demonstrated that folic acid could reduce the risk of neural tube defect by 50 to 75% (Kirke et al., 1992; Werler et al., 1993). The CDC did a similar study in the 1990s in the Peoples' Republic of China. This pattern of data and the conclusiveness and the soundness of the research resulted in major changes in public policy in this country. That major change in public policy was a fortification of our flour supply with folic acid. In 1998 the Food and Drug Administration mandated that all enriched cereal grain products be fortified with folic acid, especially and solely for the purpose of preventing spina bifida and anencephaly (Honein et al., 2001; AAP, 1999).

There have been dramatic changes in the rates of anencephaly and spina bifida over a long period of time that are well documented. In contrast, the changes in the reported rates of hypospadias have been from a relatively recent report from the CDC. The CDC reported on data that was gathered up to 1995 showing a constant change in the rate of hypospadias (Paulozzi, Erickson et al. 1997). Since that time the rate has not changed much, but there has been a change of about two per thousand to three and a half to four per thousand (Paulozzi 1999). There has also been a dramatic increase in rates of reported sex organ defects in total that remains unexplained (Toppari, Larsen et al. 1996).

4. Hypospadias Risks

Hypospadias data is also available from a number of surveillance programs and birth defects monitoring programs from around the world (Landrigan, Carlson et al. 1998; Aho, Koivisto et al. 2000; Toppari, Kaleva et al. 2001). There is a tremendous heterogeneity of the rates. In Australia; the reported rate is somewhere around one or two per ten thousand. In contrast, the rate in New Zealand is reported to be ten times higher, raising questions of accurate reporting. It is doubtful that these sorts of differences reflect real differences in incidence. The rate in South Africa, based on two years of data documents an intermediate rate between that of South Africa and New Zealand. The incidence in Alberta, Canada has shown a higher rate than Canada as a whole, apparently increasing at one time and now decreasing. Norway has also reported some ups and downs, but is more or less stable. England and Wales have intermediate rates. There are tremendous differences in the rates within Italy, which could reflect real differences in incidence, or just reporting variations.

In a review article by Baskin et.al. a number of risk factors were identified, however, the retrospective nature of the reviewed data and the small numbers leave the interpretation of the data in question (Baskin, Colborn et al. 2001). A report in Lancet commented on the apparent transgenerational effects of DES-increased rate of hypospadias in the offspring of the women who were themselves exposed to DES in utero (Klip et al., 2002)). A study of DDE emanating out of the collaborative perinatal study, which was a prospective study of approximately 50,000 pregnancies in this country 30 or 40 years ago, showed a possible increase in hypospadias (Henderson et al., 1976). The Avon longitudinal study of pregnancy and child development in Great Britain showed a significant increased risk of hypospadias among women who were vegetarians (North and Golding 2000).

There are some risk factors associated with hypospadias from the CDC data that have not been published but are supported by the Atlanta case-control study. This study mostly focused on folic acid. All types of hypospadias were reported in this case control study, including mild and moderate not just "severe" types. The data revealed a highly statistically significant association of maternal asthma and hypospadias risks. Hypospadias was also associated in a highly statistically significant way with morning sickness and the use of medications used to counteract the symptoms. Finally, a relatively low odds ratio but nevertheless, significant was an increase risk of hypospadias if the mother's occupation was a nurse.

There are some new opportunities for collecting more information on identifying risk factors that cause hypospadias. Specifically factors which can be modified from the environmental situation to prevent hypospadias. The CDC has a consortium of seven state-university alliances, all participating in a common case control study known as the National Birth Defects Prevention Study. In addition, the CDC in Nevada participates, so there are eight centers. The information is being collected on 30 major birth defects including so-called second and third degree hypospadias. Presently, 12,000 interviews have been completed with 16,000 planned in the upcoming years. One of the problems with this study, so far as hypospadias is concerned is the tremendous variation in the frequency of cases of hypospadias that are entered into this study by the eight different centers, for reasons we that remain obscure.

There are currently either ongoing or proposed studies in the context of this eight-center study to look at androgen receptors, pesticide exposures, and maternal nutrition. Of particular interest is the issue of vegetarian diets and the issue of low fat intake. The Smith-Lemli-Opitz syndrome in which patients have hypospadias is associated with altered cholesterol metabolism is the catalyst for exploring dietary intake.

5. Conclusion

In the United States the Department of Health and Human Services and the Environmental Protection Agency are working together to promote the idea of doing another collaborative perinatal study. This study would involve a larger number of preganancies and use modern technology such as DNA samples, environmental samples and gene array technology. Another goal of the study is to standardize the examination of hypospadias so that reporting variations can be minimized. This study will be an important opportunity to learn more about the risk factors for hypospadias as well as many other kinds of birth defects and developmental disabilities.

References

Aho, M., A. M. Koivisto, et al. 2000, Is the incidence of hypospadias increasing? Analysis of Finnish hospital discharge data 1970-1994, *Environ Health Perspect* 108(5): 463-5.
American Academy of Pediatrics, Folic acid for the prevention of neural tube defects, Committee on Genetics, 1999 Aug, 104 (2 Pt 1):325-7.
Baskin, L. S., T. Colborn, et al. 2001, Hypospadias and endocrine disruption: is there a connection? *Environmental Health Perspectives* 109: 1175-1183.
Berry, R.J., Li, Z.,Erickson, J.D., Li, S., Moore, C.A., Wang, H., Mulinare, J., Zhao, P., Wong, L.Y., Gindler, J., Hong, S.X., Correa, A., 1999, Prevention of neural-tube defects with folic acid in China. China-U.S. Collaborative Project for Neural Tube Defect Prevention [corrected; erratum to be published], *N Engl J Med*, Nov 11, 341(20): 1485-90.
de Walle, H.E., Reefhuis, J., Cornel, M.C., 2003, Folic acid prevents more than neural tube defects: a registry-based study in the northern Netherlands, *Eur J Epidemiol.*, 18(3):279-80.
Guney, O., Canbilen, A., Konak, A., Acar, O., 2003, The effects of folic acid in the prevention of neural tube development defects caused by phenytoin in early chick embryos, *Spine*, Mar 1, 28(5):442-5.
Hamilton, W.J., Poswillo, D. E., 1972, Limb reduction anomalies induced in the mamoset by thalidomide, *J Anat.*, Apr, 111(3):505-6.
Henderson, B.E., Benton, B., Cosgrove, M., Baptista, J., Aldrich, J. Townsend, D., Hart, W., Mack, T.M., 1976, Urogenital tract abnormalities in sons of women treated with diethylstillbestrol, *Pediatrics*, Oct, 58(4):505-7.
Honein, M.A., Paulozzi, L.J., Mathews, T.J. Erickson, J.D., Wong, L.Y., 2001, Impact of folic acid fortification of the US food sulpply on the occurrence of neural tube defects. *JAMA*, Jun 20, 285(23):2981-6. Erratum in: JAMA 2001 Nov 14, 286(18):2236.
Kirke, P.N., Daly, L.E., Elwood, J.H., 1992, A randomized trial of low dose folic acid to prevent neural tube defects. The Irish Vitamin Study Group, *Arch Dis Child*, Dec, 67(12):1442-6.
Klip, H., Verloop, J., van Gool, J.D., Koster, M.E., Burger, C.W., van Leeuwen, F.E., 2000, OMEGA Project Group, Hypospadias in sons of women exposed to diethylstillbestrol in utero: a cohort study, *Lancet*, Mar 30, 359(9312):1102-7.
Landrigan, P. J., J. E. Carlson, et al. 1998, Children's health and the environment: a new agenda for prevention research, *Environ Health Perspect* 106 Suppl 3: 787-94.

North, K. and J. Golding 2000, A maternal vegetarian diet in pregnancy is associated with hypospadias. The ALSPAC Study Team. Avon Longitudinal Study of Pregnancy and Childhood, *BJU Int* **85**(1): 107-13.

Paulozzi, L., D. Erickson, et al. 1997, Hypospadias Trends in Two US Surveillance Systems, *Pediatrics* **100**(5): 831-834.

Paulozzi, L. J. 1999, International trends in rates of hypospadias and cryptorchidism, *Environ Health Perspect* **107**(4): 297-302.

Toppari, J., M. Kaleva, et al. 2001, Trends in the incidence of cryptorchidism and hypospadias, and methodo-logical limitations of registry-based data, *Hum Reprod Update* **7**(3): 282-6.

Toppari, J., J. C. Larsen, et al. 1996, Male reproductive health and environmental xenoestrogens, *Environ Health Perspect* **104 Suppl 4**: 741-803.

Werler, M.M., Shapiro, S., Mitchell, A.A., 1993, Periconceptional floic acid exposure and risk of occurrent neural tube defects, *JAMA*, Mar 10, **269**(10):1257-61.

ENDOCRINE EVALUATION OF HYPOSPADIAS

Grace Hyun, M.D. and Thomas F. Kolon*, M.D. F.A.A.P.

Division of Urology
The Children's Hospital of Philadelphia
Department of Urology
University of Pennsylvania School of Medicine

1. INTRODUCTION

Hypospadias is a congenital defect of the penis resulting in an incomplete development of the anterior urethra (Duckett, 1998). The abnormal urethral opening may be any place along the shaft of the penis or open onto the scrotum or perineum. By the end of the first month of gestation, the hindgut reaches the surface of the embryo at the cloacal membrane on the ventral surface. The cloacal membrane is then divided by the urorectal septum into the anal and urogenital membrane. The genital tubercle and genital swellings appear around the urogenital membrane. A surge of luteinizing hormone from the pituitary gland leads to an increase of testosterone. Testosterone biosynthesis occurs at about 9 weeks gestation and hCG-LH receptors are present on fetal Leydig cells by at least 12 weeks of gestation. Androgens are critical for the induction, growth and differentiation of the genital tubercle into the male phenotype via the androgen receptor (AR), a nuclear transcription factor. Dihydrotestosterone, a metabolite of testosterone, mediates the action of androgens on the external genitalia and is known to bind to the AR with a higher affinity than testosterone. Masculinization of the external genitalia, namely elongation of the phallus, formation of the penile urethra from the urethral groove and development of the prepuce, depends on adequate testosterone synthesis, adequacy of the enzymatic pathway mediating the cascade and responsive end organs. During the third fetal month, the wolffian ducts complete their development and involution of the müllerian system occurs under the influence of müllerian inhibiting substance (MIS or AMH). A deficiency in the formation or delay in the maturation of the enzymes in the testosterone steroid cascade during these critical embryogenesis periods may be responsible for incomplete development.

Hypospadias is a common anomaly occurring in approximately 1/250 to 1/300 live births. In the 1970s and 1980s, birth defect surveillance systems reported transient 1.5 to 2-fold increases in the birth prevalence of hypospadias in Norway (Bjerkedal and

---

*Assistant Professor of Urology, The Children's Hospital of Philadelphia, 34th Street and Civic Center Boulevard, Wood Center, 3rd Floor, Philadelphia, PA 19104, 215-590-4690, 215-590-3985 (fax), kolon@email.chop.edu

Hypospadias and Genital Development, edited by
L. Baskin, Kluwer Academic/Plenum Publishers, 2004

Bakketeig, 1975), Sweden (Kallen and Winberg, 1982), Denmark (Kallen et al., 1986), England (Matlai and Beral, 1985), and Hungary (Czeizel, 1985). A hypospadias prevalence rate of 4.1 per 1000 births was found in a longitudinal study at the Mayo Clinic from 1940 to 1970, and rates did not significantly very over time (Sweet et al., 1974). In 1997, the CDC studied the birth prevalence of hypospadias in the United States and found the total hypospadias rate (mild and severe) doubled in two birth defects surveillance systems (Metropolitan Atlanta Congenital defects Program and Birth Defects Monitoring Program). For severe hypospadias, the rate increased 3- to 5-fold (0.11 to 0.55/1000 births) a statistically significant change (Paulozzi et al., 1997). Concern was raised involving exposure to environmental risk factors with either estrogenic or anti-androgenic effects as the cause of increased hypospadias rates in the United States (Dolk, 1998).

2. Inheritance

There is a well-recognized familial clustering of hypospadias, and male relatives of boys with hypospadias are more likely to have this condition compared to the general population. In a study by Sorenson (Sorenson, 1953), male relatives of 103 index cases with non-syndromic hypospadias were evaluated. Of the 103 index cases, 28% had at least one other family member with hypospadias. The more severe the hypospadias in the index case, the higher the incidence of hypospadias in a 1st degree relative. With 1st degree hypospadias, 3.5 of relatives were affected; with 2nd degree, 9.1%, and with 3rd degree, 16.7%.

In a description by Bauer et al. of 307 cases of familial hypospadias, 25% of families had a second family member, in addition to the index child, with this anomaly and 7% had 3 affected members (Bauer et al., 1981). The risk of a second male sibling being born with hypospadias when this anomaly occurred for the first time in the index child was 11%. No boy with mild or distal hypospadias had a brother with this anomaly. When the index child had a moderate or severe (proximal) hypospadias, a 12% to 19% chance existed, respectively, for hypospadias to occur in an additional male sibling. An increased risk of hypospadias among twins has also been described (Kallen et al., 1986). The prevalence of hypospadias is higher among members of male-male pairs and lower among males in male-female pairs. Concordance among twins of the same sex was 18% for both mild and severe forms.

Family data do not suggest a Mendelian pattern of inheritance, and a multifactorial pattern is the most consistent explanation for familial clustering of severe hypospadias (Fredell et al., 2002). Allelic variants in genes involved in androgen production and metabolism may individually produce small risks that are not in themselves sufficient to produce hypospadias. When genetic susceptibility is combined with exposure to anti-androgenic agents, however, gene and environment risk factors may interact to surpass a threshold, resulting in occurrence of this birth defect.

3. Androgens

Although common, the exact etiology of hypospadias is unknown. Testicular function in boys with hypospadias has been widely evaluated during the last decades. Most series revealed a poor testicular response to human chorionic gonadotropin (hCG) stimulation during childhood (Knorr, 1979; Nonomura, et al., 1984; Shima et al., 1986; Shima et al.,

1992). Allen and Griffin suggested that hypospadias is not only a dysmorphic phenomenon, but also a local manifestation of systemic endocrinopathy. Of 15 patients with severe hypospadias in their series, 11 had an endocrine abnormality. Nearly half of their patients had a substandard response to hCG, although 4 eventually had an improved response (Allen and Griffin, 1984).

Since the development of external male genitalia is androgen dependent multiple studies have been performed evaluating the role of endocrine factors, both exogenous and endogenous, in hypospadias. Via the AR, testosterone stabilizes and induces the internal wolffian structures to develop into the epididymis, seminal vesicles and vas deferens. In the embryonic urogenital tract, T is converted to DHT via the enzyme steroid 5 alpha-reductase type 2 (SRD5A2). DHT binds to the AR and drives differentiation of the external genitalia, penis and scrotum, and prostate gland ((Siiteri and Wilson, 1974; Russell and Wilson, 1999). A deficiency or inhibition of 5α reductase type 2, which converts testosterone(T) to dihydrotestosterone(DHT), feminizes the genitalia. Similarly the administration of androgens to females masculinizes the phallus.

The relative importance of T and DHT in genital tract development is illustrated in men with male pseudohermaphroditism due to congenital SRD5A enzyme deficiency (Griffin and Wilson, 1989; Imperato-McGinley et al., 1991). They have a normal XY karyotype but ambiguous external genitalia, leading to a female sex of rearing in many cases. T and estrogen (E) levels are normal to elevated, but DHT levels are reduced. These individuals have normal T-dependent wolffian duct derivatives--epididymis, seminal vesicle, vas deferens--and lack müllerian duct derivatives. The DHT-dependent external genitalia are abnormal, however, and a small phallus, bifid scrotum, blind vaginal pouch, and varying degrees of hypospadias are present. At puberty, male pseudohermaphrodites virilize with male muscular development, deepening of the voice, enlargement of the phallus and production of semen, but have a hypoplastic prostate and scanty beard. Women with this genetic disorder are phenotypically normal and have normal reproductive function.

Thus, normal penile development is dependent on 3 factors: testosterone, its conversion via SRD5A2 dihydrotestosterone, and a functional AR. Kim et al investigated the distribution of AR and 5α reductase type 2 in the developing human fetal external genitalia with special emphasis on urethra formation. Twenty fetal genital specimens from normal human males (12-20 weeks gestation) were sectioned serially. Stained sections throughout male genital development documented the expression of AR and 5 alpha-reductase type 2 in the phallus. Between 12 and 14 weeks of gestation, AR was localized to epithelial cells of the urethral plate in the glans, the tubular urethra of the penile shaft, and stromal tissue surrounding the urethral epithelium. In the fetal penis between 16 and 20 weeks gestation, the density of AR expression was greatest in urethral epithelial cells versus the surrounding stromal tissues. There was a characteristic pattern of AR expression in the glanular urethral epithelium between 16 and 20 weeks gestation. AR expression was greater along the ventral aspect of the glanular urethra than along the dorsal aspect of the urethral epithelium. The expression of 5 alpha-reductase type 2 was localized to the stroma surrounding the urethra, especially along the urethral seam area in the ventral portion of the remodeling urethra. These anatomical studies support the hypothesis that androgens are essential for the formation of the ventral portion of the urethra and that abnormalities in either the AR or 5 alpha-reductase type 2 may explain the occurrence of hypospadias (Kim et al., 2001).

Mutations in the AR gene in males can cause a spectrum of androgen insensitivity syndromes. Phenotypes vary from individuals with female external genitalia (complete androgen insensitivity) through patients with genital ambiguity (partial androgen insensitivity) to individuals with normal male genitalia but infertility (minimum androgen insensitivity). Most reports of AR gene mutations in individuals with hypospadias have included patients with other genitourinary malformations, and so the disorder analyzed was partial androgen insensitivity syndrome rather than hypospadias. When patients with isolated hypospadias have been studied, mutations in the AR gene have occasionally been found. There has been an extensive search for abnormalities in the AR in patients with hypospadias. Gearhart et al did not note any deficiencies in AR levels or 5α-reductase in their study of preputial skin from boys with hypospadias (Gearhart et al., 1988). Allera et al evaluated 9 patients with severe hypospadias and detected a defect in open reading frame of the AR gene in only one patient Allera et al., 1995). Sutherland et al also concluded that mutations in the AR gene are rarely associated with hypospadias. Using single strand conformational polymorphism and sequence analysis they discovered a missense mutation of exon 2 of the AR gene in 1 of 40 patients with distal hypospadias (Sutherland et al., 1996). Whereas most earlier studies have not fully evaluated exon 1 which contains the transcription activation domain of the AR gene, Kolon et al examined 39 consecutive boys with isolated hypospadias and found no mutations of exons 1-8 (Kolon et al., 2001).

AR genes with different CAG (glutamine) repeat lengths in exon 1 represent polymorphisms in the AR gene. A reduction in the number of glutamine repeats results in increased transcription-factor activity of the AR while an increase in the number of glutamine repeats is associated with decreases in AR activity (Hakimi et al., 1996; MacLean et al., 1995). Cassella et al investigated the CAG repeat length in boys with hypospadias and/or cryptorchidism. Although increased repeat lengths were seen in hypospadiac patients, only those with both hypospadias and cryptorchidism had long polymorphic repeats that were statistically significant compared to normal controls (Casella et al., 2001). Polymorphism in the AR gene may modify effects due to allelic variants in the SRD5A2 gene and/or anti-androgenic exposures. Molecular biology techniques have demonstrated that defects in the AR gene are definitely associated with isolated hypospadias. However, the frequency of these genetic defects accounts for an extremely small subset of cases, implying that other factors are responsible for hypospadias.

4. Endocrine profiles

Since hypospadias is an example of incomplete virilization, multiple studies have evaluated the endocrine and genetic profiles of patients with hypospadias to determine the defects, if any, in the testosterone pathway (Figure 1). Aaronson et al (Aaronson et al., 1997) evaluated 30 boys with 46XY karyotype, fully descended testes and penoscrotal or proximal shaft hypospadias to determine the incidence of defects in 3 enzymes (3β hydroxysteroid dehydrogenase, 17α hydroxylase and 17,20 lyase) in the testosterone biosynthetic pathway (Table 1).

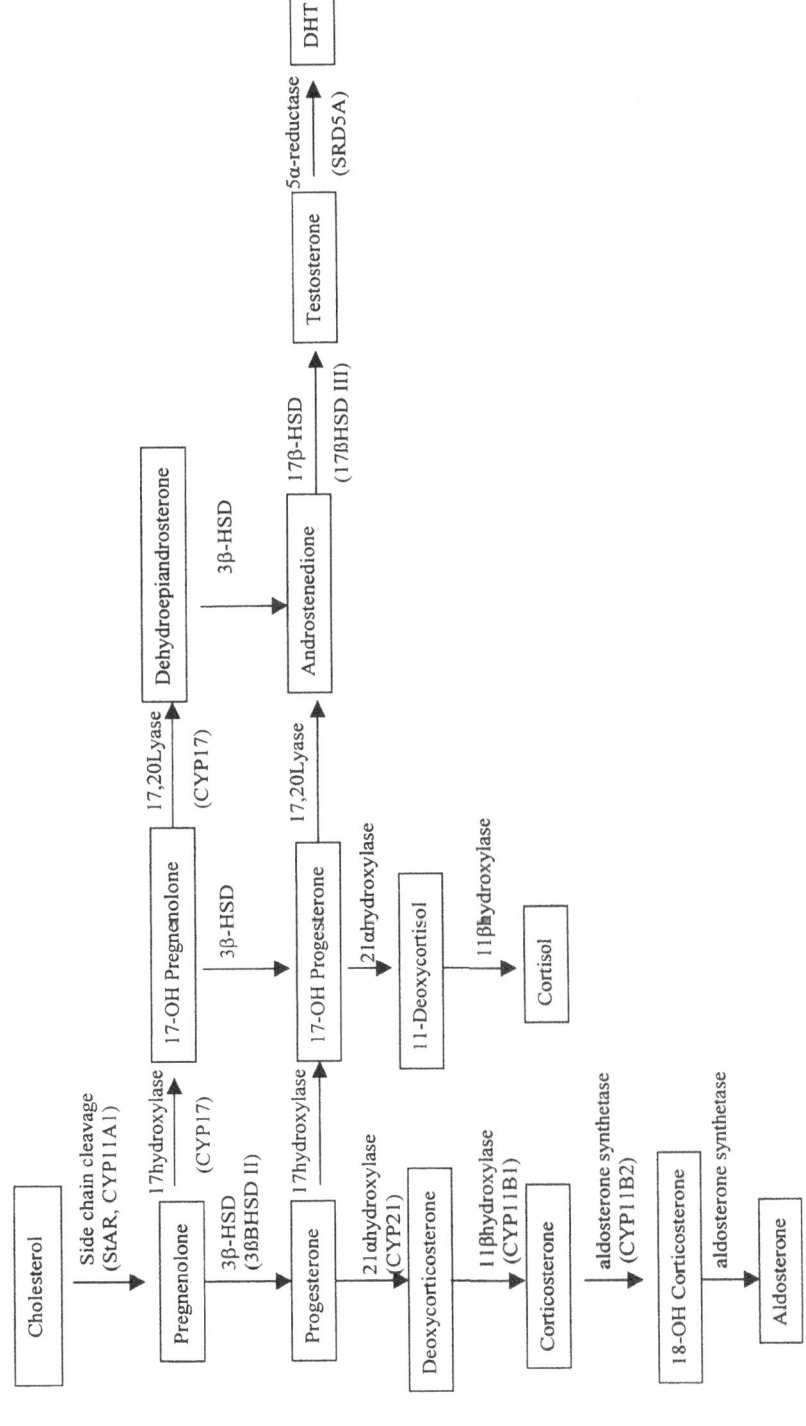

Figure 1: The steroid biosynthetic pathways with the responsible enzymes and genes (parentheses).

TABLE 1: Testosterone biosynthesis defect.

Hormone level (ratio)	Enzyme deficiency
↑17OHP	21- or 11-hydroxylase
↑DHEA:andro	3ß HSD
↑P:17OHP	17 alpha hydroxylase
↑17OHP:andro	17,20 lyase
↑Andro:T	17ß HSD
↑T:DHT	5 alpha reductase

17OHP= 17-OH progesterone, DHEA=dehydroepiandrosterone, andro=androstenedione, T=testosterone,

HSD=hydroxysteroid dehydrogenase

Serum concentrations of the metabolites mediated by these enzymes were measured from which the precursor to product ratios were calculated. Seven patients underwent adrenocorticotropic hormone stimulation. A total of 11 boys had evidence of impaired function of 3β hydroxysteroid dehydrogenase alone or in combination with impaired 17,20 lyase or 17 alpha hydroxylase activity. An additional 4 boys had evidence of isolated 17,20 lyase deficiency. Thus, of the 30 boys studied, 15 (50%) had evidence of a testosterone biosynthetic defect. The effect of adrenocorticotropic hormone stimulation varied widely. With the exception of 2 patients, serum cortisol levels were normal. Similarly, basal testosterone levels were low but with the normal prepubertal range.

Kolon et al. studied patients presenting with hypospadias and cryptorchidism in order to classify them as part of the intersex spectrum (Kolon et al., 2001). An extensive hormonal evaluation involving all enzymes of the testosterone cascade showed 16% of 105 consecutive boys had a testosterone synthesis defect and a further 9.5% had a gonadotropin/testis axis defect. None exhibited severe cortisol or aldosterone deficiency. Of particular interest is the finding that physical exam is not predictive of the hormonal or intersex etiology.

Holmes et al. (Holmes et al., 2001), however, studied the enzymatic function of 3β-hydroxysteroid dehydrogenase, p450c17 (17,20 lyase, 17α hydroxylase) and 17β-hydroxysteroid dehydrogenase of 75 patients and found that the enzymatic activity of the androgen pathway was varied and differences based on age or the severity of hypospadias did not reach statistical significance. Therefore, no abnormalities in the androgen pathway as a cause of hypospadias could be documented. They concluded that the enzymatic dysfunction of these enzymes was a rare cause of hypospadias.

5.   Steroid 5 α reductase type II

Animal and human studies indicate that SRD5A enzyme activity is critical in formation of the external male genital tract, and DHT produced by this enzyme is essential for normal development of the penis, scrotum and prostate. Some consider SRD5A2 a primary hypospadias susceptibility gene because it is in the direct causal pathway for production of DHT. Finasteride is a 5α reductase inhibitor that blocks the conversion of testosterone to DHT. In rats, in utero exposure to low doses of finasteride results in decreased anogenital distance, transient production of nipples, hypospadias and decreased prostate weight in male offspring (Clark et al., 1990; Clark et al., 1993). The critical period for these effects was gestation day 16-17 in the rat, following initiation of T synthesis by the fetal testes on Gestation Day 15. Additional studies have been carried out in rhesus monkeys (Prahalada et al., 1997). Oral administration of the drug from gestation day 20 to 100 resulted in male genital tract abnormalities consisting of hypospadias, preputial adhesions to the glans, underdeveloped scrotum and small penis. No abnormalities were seen in female fetuses.

Male pseudohermaphrodites have deficiencies only in the SRD5A2 enzyme, and the gene is located on the short arm of chromosome 2 (2p23). The gene for SRD5A1 is located on the short arm of chromosome 5 (5p15), and is normal in male pseudohermaphrodites (Thigpen et al., 1992). SRD5A2 enzyme activity alone has been found in human fetal genital tissue; SRD5A1 activity has not been detected in any fetal tissues and is first expressed at birth in the liver and in non-genital skin. In children, SRD5A1 expression is localized to the sebaceous gland in non-genital skin and is markedly elevated at puberty (Thigpen et al., 1993). The virilization of male pseudohermaphrodites which occurs at puberty is likely to be influenced by synthesis of DHT in non-genital skin via SRD5A1 activity. SRD5A2 activity is considered responsible for formation of male external genitalia and prostate in embryonic life, and SRD5A1 activity does not play a role in this process. The SRD5A2 gene is an important susceptibility gene for hypospadias, and allelic variants of this gene may predispose fetuses to developing hypospadias.

In children, T: DHT ratios have been measured in serum after hCG stimulation in order to detect SRD5A2 deficiency. These parameters are highly variable and may be difficult to interpret in prepubertal children. T/DHT ratios are generally increased only in the most severely affected patients (Hiort et al., 1996; Sinnecker et al., 1996). Direct evaluation of allelic variants in the SRD5A2 gene has also been used to diagnose SRD5A2 deficiency. Phenotypes vary widely in patients with SRD5A2 allelic variants, spanning the range from completely feminized to normally masculinized external genitalia. Recommendations have been made that SRD5A2 deficiency be considered not only in patients with female or ambiguous phenotypes, but also in those with mild symptoms of under-masculinization as encountered in patients with non-syndromic hypospadias (Sinnecker et al., 1996).

Studies have revealed deletions, missense and nonsense mutations, splicing defects, and frameshift mutations individually distributed in all 5 exons of the SRD5A2 gene. Approximately two-thirds of male pseudohermaphrodites are homozygous, and one-third compound heterozygous for these mutations (Nonomura et al., 1984; Thigpen et al., 1992). Despite extensive characterization of individual mutations, it has not been possible to correlate severity of male pseudohermaphroditism with specific mutations in the SRD5A2 gene inherited by an affected individual. Silver and Russell identified SRD5A2 gene mutations in 8.6% of 81 boys with isolated hypospadias. These mutations were seen in both distal and proximal hypospadias, although hormonal studies were not performed to see if the genetic alterations translated to actual hormonal deficiencies (Silver and Russell, 1999).

6.   17 β hydroxysteroid dehydrogenase

The HSD17B isozymes catalyze the final steps in androgen and estrogen biosynthesis. The HSD17B3 isozyme is found in the microsomes of the testes where it reduces the weak androgen androstenedione into T (Geissler et al., 1994; Andersson, 1995). Familial male pseudohermaphroditism due to a deficiency in the HSD17B3 isozyme has been reported (Castro-Magana, 1993). The phenotype caused by deficits in HSD17B3 isozyme is similar to that of SRD5A2 deficiency. The external genitalia are usually female at birth, although mild to moderate ambiguity with hypospadias is occasionally present. Although typically raised as females, all affected individuals develop male body habitus and normal male secondary sexual characteristics at puberty. Gonads are palpable in the inguinal canals or labial folds, and normal wolffian structures develop. The metabolism of androstenedione to T, although deficient, may produce enough T for stabilization of the wolffian ducts but not enough T to serve as a precursor for DHT, resulting in inadequate masculinization of the external genitalia. Alternatively, T produced by extra-gonadal tissues may be sufficient to stabilize the wolffian ducts (Rosler, 1992). Deficiency of HSD17B3 is rarely diagnosed at birth and is usually discovered in 46XY females due to the presence of severe gynecomastia, amenorrhea and hirsutism at puberty. Diagnosis of HSD17B3 deficiency is made on the basis of an abnormally high androstenedione to T ratio. These ratios are typically altered at puberty but are not a sensitive indicator of the deficiency in infants (Kohn et al., 1985).

Mutations in the HSD17B3 gene have been found in consanguinous families or small case series of individuals with a family history of genital ambiguity. Male pseudohermaphrodites with either homozygous SRD5A2 or HSD17B3 gene defects were found in this Turkish kindred, while other affected males were genetically more complex (homozygous for the SRD5A2 defect and heterozygous for the HSD17B3 defect, homozygous for the SRD5A2 defect and heterozygous for the HSD17B3 defect, or homozygous for the HSD17B3 defect and heterozygous for the SRD5A2 defect). Male pseudohermaphrodites with homozygous SRD5A2 or HSD17B3 gene defects were phenotypically distinguishable from the heterozygotes due to gynecomastia (Can et al., 1998). Mutations in the HSD17B3 gene have not yet been examined in large populations of infants with non-syndromic hypospadias. Allelic variants in the HSD17B3 gene may accompany variants in the SRD5A2 gene, as demonstrated in the aforementioned Turkish kindred, and modify effects due to DHT depletion from this source as well as from anti-androgenic exposure.

Feyaerts et al. prospectively studied 32 consecutive 46,XY boys with hypospadias (Feyaerts et al., 2002). Endocrine evaluation consisted of measuring luteinizing hormone, follicle-stimulating hormone, anti-müllerian hormone, testosterone, dihydrotestosterone, progesterone, 17α-hydroxypregnenolone, 17α-hydroxyprogesterone, dehydroepiandrosterone sulfate and androstenedione. In all but 3 patients gonadal stimulation with 1,500 IU hCG every other day for 12 days was performed and steroid concentrations were reassessed after the test. An adrenocorticotropic hormone (ACTH) test was performed in 2 patients and molecular study of the AR gene was performed in 28. An increase to 37.37 nmol./l. progesterone (normal 0.1 to 0.5) and 17α-hydroxyprogesterone to 25.48 nmol./l. (normal 1.18 ± 0.66) before hCG stimulation was noted in 1 patient. These abnormal results were not found after hCG stimulation but reappeared after the ACTH test. This result might be related to a partial mix of 17α-hydroxylase/17,20-lyase deficiency but no mutation was found after complete sequencing of gene CYP17. Of the 32 patients, 4 had an insufficient response to hCG stimulation (testosterone less than 10 nmol./l.), including 1 with a low AMH level of 180 pmol./l. (normal 451 ± 198) and an increased dehydroepiandrosterone sulfate level of 1,995 nmol./l. (normal 59 ± 41) before hCG stimulation. Partial androgen insensitivity was suspected in 1 patient because he had a high testosterone response (29.96 nmol./l.) after hCG stimulation but no mutation of the gene of the AR gene was detected. Two patients with proximal hypospadias had isolated decreased AMH levels, consistent with Sertoli cell insufficiency. They concluded that although the patients demonstrated several abnormal endocrine screenings, no significant endocrine defects were seen on a genetic level.

7.   Müllerian inhibiting substance

In another recent study, Austin et al measured the serum levels of müllerian inhibiting substance, testosterone and androstenedione in 29 boys with distal hypospadias (midshaft or less) and 21 normal boys undergoing circumcision (Austin et al., 1999). Within each respective age group, MIS levels in boys with hypospadias were significantly higher compared with those of normal boys. Serum MIS levels were not statistically different when normal boys or boys with hypospadias were compared across age groups. In boys age 12 months or younger, serum testosterone levels in boys with hypospadias were not significantly different from those of normal boys but serum androstenedione levels in boys with hypospadias were significantly lower compared with those of normal boys of similar age. At 13 to 24 months of age, testosterone levels were higher in normal boys whereas androstenedione levels were not different between hypospadiac and normal boys. These results suggest that MIS may have some role in the development of hypospadias, possibly mediated by MIS inhibition of CYP17 gene expression.

8. In vitro fertilization

Silver et al. demonstrated a 5-fold increased risk for hypospadias after in vitro fertilization (IVF) in the Greater Baltimore Medical Center compared to statewide incidence figures in Maryland (Silver et al., 1999). All women who had IVF procedures also received progesterone to support the pregnancy early after embryo transfer in the 1st trimester. These results may indicate that hypospadias is more common in couples with a history of infertility or that progestins administered as part of an IVF protocol may disturb the fetal endocrine environment.

Some allelic variants such as SRD5A2 polymorphism or AR gene CAG repeats may produce a modest decrease in DHT action that is not sufficient to produce hypospadias. If an infant carrying these variant alleles is exposed in utero to an anti-androgen during critical periods of development, the combined risk from these factors may be sufficient to reduce DHT levels below a threshold and result in hypospadias. Anti-androgens of concern in this scenario may be inhibitors of the SRD5A2 enzyme; a wide range of steroidal and non-steroidal agents have been found to inhibit this enzyme (Nonomura et al., 1984). Other anti-androgens of concern are progestins administered for pregnancy support or birth control.

Experimental studies have shown that progestins administered to laboratory animals during pregnancy can cause hypospadias (Goldman and Bongiovanni, 1967). Progestins can act as substrates for SRD5A2 and reduce the conversion of T to DHT by competitive inhibition (Dean and Winter, 1984). However, the association between increased risk for hypospadias and 1st trimester exposure to oral contraceptives, hormones for pregnancy support, and hormone-based pregnancy tests is controversial (Sweet et al., 1974; Calzolari et al., 1986; Czeizel et al., 1979; Monteleone et al., 1981). Mau (Mau, 1981) prospectively examined the risk for hypospadias associated with 1st trimester exposure to progestins. He found that 1.2% of the 408 male infants from pregnancies exposed to progestins for pregnancy support had hypospadias, and 2% of 151 male infants exposed via hormonal pregnancy tests, compared to 0.8% of 3043 unexposed male infants. The relative risks of 1.5 and 2.4 were not significantly different from 1.0. Aarskog (Aarskog, 1979) reported similar results. A total of 1.4% of 441 male infants exposed to progestins in utero had hypospadias compared to 0.7% of 25,101 unexposed male infants for a relative risk of 1.9, which was not statistically significantly different. When exposure to oral contraceptives was analyzed separately, a significant relative risk of 3.1 was obtained.

A study by Angerpointer (Angerpointer, 1984) reported a positive relationship between the severity of 515 cases of hypospadias and in utero exposure to progestins for threatened abortion. Progestins were taken by 3% of mothers of the mildest cases, 9% of mothers of cases with penile hypospadias, 13% of mothers of cases with scrotal hypospadias and 18% of mothers of perineal cases. A recent meta-analysis of human studies, however, found no association between progestin exposures and external genital anomalies in male infants (Raman-Wilms, 1995). Epidemiologic studies suggest there may be approximately a 2-fold relative risk for hypospadias with 1st trimester exposure to progestins. However due to the relative rarity of the exposure and the outcome, it has been difficult to conclusively demonstrate such an association.

Exogenous hormone exposure is another possible etiology in hypospadias, specifically environmental contamination. In this regard it is well established that humans continually ingest substances with known estrogenic activity, such as insecticides used in crop production, natural plant estrogens, byproducts of plastic production, pharmaceuticals and so forth. North et al reported that mothers who were vegetarian in pregnancy had a higher chance of giving birth to a boy with hypospadias, compared with omnivores who did not supplement their diet with iron. As vegetarians have a greater exposure to phytoestrogens than do omnivores, these results support the possibility that phytoestrogens have a deleterious effect on the developing male reproductive system (North and Golding, 2000).

8.   Conclusion

In summary, male urethral development is under hormonal control.   Hypospadias appears to have a multifactorial etiology. Mild defects in the testosterone biosynthesis pathway may be causative for isolated hypospadias either exclusively or, more likely, in concert with decreased androgen receptor activity, parental exposure to anti-androgens during fetal development, or other unidentified developmental genes.

REFERENCES

Aaronson, I.A., Cakmak, M.A., Key, L.L., 1997, Defects of the testosterone biosynthetic pathway in boys with hypospadias. *J Urol*,157(5):1884-1888.

Aarskog D., 1979, Current concepts: maternal progestins as a possible cause of hypospadias. *N Engl J Med*, **300**:75-78.

Allen, T.D. and Griffin, J.E., 1984, Endocrine studies in patients with advanced hypospadias. *J Urol*, **131**:310-314.

Allera, A., Herbst, M.A., Griffin, J.E. et al., 1995, Mutations of the AR gene coding sequence are infrequent in patients with isolated hypospadias. *J Clin Endocrinol Metab*, **80**:2697-9.

Andersson S., 1995, 17 beta hydroxysteroid dehydrogenase: isozymes and mutations. *J Endocrinol*, **146**:197-200.

Angerpointer T., 1984, Hypospadias- Genetics, epidemiology and other possible aetiological influences. *Z Kinderchir*, **39**:112-118.

Austin, P.F., Siow, Y., Fallat, M.E., et al., 2002, The Relationship Between Müllerian Inhibiting Substance and Androgens in Boys with Hypospadias. *J Urol*, **168**:1784-1788.

Bauer, S.B., Retik, A.B., and Colodny, A.H., 1981, Genetic aspects of hypospadias. *Urol Clin North Am*, **8**:559-564.

Bjerkedal, T. and Bakketeig, L.S., 1975, Surveillance of congenital malformations and other conditions of the new born. *Int J Epidemiol*, **4**:31-36.

Calzolari, E., Contrero, M.R., Roncarati, E., et al., 1986, Aetiological factors in hypospadias. *J Med Genet*, **23**:333-337.

Can, S., Zhu ,Y-S, Cai, L-Q, et al., 1998, The identification of 5_ reductase 2 and 17_ hydroxysteroid dehydrogenase 3 gene defects in male pseudohermaphrodites from a Turkish kindred. *J Clin Endocrinol Metab*, **83**:560-569.

Casella, R., Kolon ,T.F., Maduro, M., et al., CAG repeat length in exon 1 of the androgen receptor gene of patients with cryptorchidism and/or hypospadias. Presented at annual meeting of Section on Urology, American Academy of Pediatrics, San Francisco, California, October 20–22, 2001.

Castro-Magana, M., Angulo, M., and Uly, J., 1993, Male hypogonadism with gynecomastia caused by late-onset deficiency of testicular 17-ketosteroid reductase. *N Engl J Med*, **328**:1297-1301.

Clark, R., Antonello, J., Grossman, S., et al., 1990, External genitalia abnormalities in male rats exposed in utero to finasteride, a 5α reductase inhibitor. *Teratology;* **42**:91-100.

Clark, R., Antonello, J., Grossman, S., et al., 1993, Critical developmental periods for effects on male rat genitalia induced by finasteride, a 5α reductase inhibitor. *Toxicol Appl Pharmacol*, **119**:34-40.

Czeizel, A., Toth, J. and Erodi, E., 1979, Aetiological studies of hypospadias in Hungary. *Hum Hered*, **38**:45-50.

Czeizel, A., 1985, Increasing trends in congenital malformations of male external genitalia. *Lancet*, 1:462-463.

Dean, H.J. and Winter, J.S., 1984, The effect of five synthetic progestational compounds on 5 alpha reductase activity in genital skin fibroblast monolaYERS. *Steroids*, 43:13-17.

Dolk, H., 1998, Rise in the prevalence of hypospadias. *Lancet*, 351:770.

Duckett, J.W., 1998, Hypospadias. In: Walsh PC, Retik AB, Vaughan ED, Wein AJ, eds. Campbell's Urology, 7th ed. Philadelphia, PA: W. B. Saunders Co., vol. 2, chapt. 68,2093-2119

Feyaerts, A., Forest, M.G., Morel ,Y., et al., 2002, Endocrine Screening in 32 Consecutive Patients with Hypospadias. *J Urol*, 168:720-725.

Fredell, L., Iselius ,L., Collins, A. et al., 2002, Complex segregation analysis of hypospadias. *Hum Genet*, 111:231-234.

Gearhart, J.P., Linhard, H. R., Berkovitz, G.D. et al., 1988, Androgen receptor levels and 5 alpha-reductase activities in preputial skin and chordee tissue of boys with isolated hypospadias. *J Urol*, 140(5 pt 2):1243-1246.

Geissler, W.M., Davis, D.L., Wu ,L., et al., 1994, Male pseudohermaphroditism caused by mutations of testicular 17 beta hydroxysteroid dehydrogenase3. *Nat Genet*, 7:34-39.

Griffin, J. and Wilson, J. The androgen resistance syndromes: 5α reductase deficiency, testicular feminization and related disorders. In The metabolic Basis of Inherited Disease, C Scriver, A Beaudet, W Sly, D Valle, eds., 6th Ed., Vol II, McGraw-Hill, NY, 1989: 1919-1944.

Goldman, A.S. and Bongiovanni, A.M.., 1967, Induced genital anomalies. *Ann NY Acad Sci*, 142:755-767.

Hakimi, J.M., Rondinelli, R.H., Schenberg, M.P., et al., 1996, Androgen receptor gene structure and function in prostate cancer. *World J Urol*, 14:329-337.

Hiort, O., Willenbring, H., Albers, N., et al., 1996, Molecular genetic analysis and human chorionic gonadotropin stimulation tests in the diagnosis of prepubertal patients with partial 5 alpha reductase deficiency. *Eur j Pediatr*, 155:445-451.

Holmes, N.M., Miller, W.L. and Baskin, L.S., Defects in androgen production as a role in the etiology of hypospadias. Presented at annual meeting of Section on Urology, American Academy of Pediatrics, San Francisco, California, October 20–22, 2001.

Imperato-McGinley ,J., Miller, M., Wilson, J., et al., 1991, A cluster of male pseudohermaphrodites with 5α reductase deficiency in Papua New Guinea. *Clin Endocrinol*, 34:293-298.

Kallen, B. and Winberg, J., 1982, An epidemiological study of hypospadias in Sweden. *Acta Paediatr Scand Suppl*, 293:3-21.

Kallen, B., Bertolini, R., Castilla, A., et al., 1986, A joint international study on the epidemiology of hypospadias. *Acta Paediatr Scand Suppl*, 324:1-52.

Kim, K.S., Liu ,W., Cunha, G.R., et al., 2002, *Cell Tissue Res.*, 307(2):145-53. Epub 2001 Nov 27.

Knorr, D., 1979, Plasma testosterone in male puberty. II. hCG stimulation test in boys with hypospadias. *Acta Endocrinol* (Copenh), 90:365.

Kohn, G., Lasch, E.E., El-Shawwa, R., et al., 1985, Male pseudohermaphroditism due to 17-beta hydroxysteroid dehydrogenase deficiency in a large Arab kinship: studies on the natural history of the defect. *J Pediatr Endocr*, 1:29-37.

Kolon, T.F., Murthy, L., Gonzales, E.T., Jr., et al., Prevalence of intersex in hypospadias and cryptorchidism. Presented at annual meeting of Section on Urology, American Academy of Pediatrics, San Francisco, California, October 20–22, 2001.

MacLean, H.E., Warne, G.L., and Zajac, J.D., 1995, Defects of androgen receptor function: from sex reversal to motor neurone disease. *Molec Cell Endocrinol*, 112:133-141.

Matlai, P. and Beral, V., 1985, Trends in congenital malformations of external genitalia. *Lancet*, 1:108.

Mau, G., 1981, Progestins during pregnancy and hypospadias. *Teratology*, 24:285-287.

Monteleone, N.R., Castilla, E.F. and Paz, J.E., 1981, Hypospadias: An epidemiologic study in Latin America. *Am J Med Genet*, 10:5-19.

Nonomura, K., Fujieda, K., Sakakibara, N. et al., 1984, Pituitary and gonadal function in prepubertal boys with hypospadias. *J Urol*, 132:595-598.

North, K. and Golding, J., 2000, A maternal vegetarian diet in pregnancy is associated with hypospadias. The ALSPAC Study Team. Avon Longitudinal Study of Pregnancy and Childhood. *BJU Int.*, 85(1):107-13.

Paulozzi, L.J., Erickson, D.J., and Jackson, R.J., 1997, Hypospadias trends in two American surveillance systems. *Pediatrics*, 100:831-834.

Prahalada, S., Tarantal, A.F., Harris, G.S., et al., 1997 , Effects of finasteride, a type 2 5-alpha reductase inhibitor, on fetal development in the rhesus monkey (Macaca mulatta). *Teratology*. Feb 55(2):119-31.

Raman-Wilms, L., Tseng, A., Wighardt, S., et al., 1995, Fetal genital effects of first tri85 imester sex hormone exposure: a meta-analysis. *Obstet Gynecol*, 141-149.

Rosler, A., 1992, Steroid 17 beta hydroxysteroid dehydrogenase in man: an inherited form of male pseudohermaphroditism. *J Steroid Biochem Mol Biol*,**43**:989-1002.

Russell, D.W. and Wilson, J.D., 1994, Steroid 5 alpha reductase: two genes/two enzymes. *Ann Rev Biochem*, **63**:25-61.

Shima, H., Ikoma, F., Yabumoto, H., et al., 1986, Gonadotropin and testosterone response in prepubertal boys with hypospadias. *J Urol*, **135**:539-542.

Shima, H., Yabumoto, H., Okamoto, E., et al., 1992, Testicular function in patients with hypospadias associated with enlarged prostatic utricle. *Br J Urol*, **69**:192-195.

Siiteri, P. and Wilson, J., 1974, Testosterone formation and metabolism during male sexual differentiation in the human embryo. *J Clin Endocrinol Metab*, **38**:113-124.

Silver, R.I. and Russell, D.W., 1999, 5-reductase type 2 mutations are present in some boys with isolated hypospadias. *J Urol*, **162**:1142-1145.

Silver, R.I., Rodriguez ,R., Chang, T.S., et al., 1999, In vitro fertilization is associated with an increased risk of hypospadias. *J Urol*, **161**:1954-1957.

Sinnecker, G.H., Hiort, O., Dibbelt, L., et al., 1996, Phenotypic classification of male pseudohermaphroditism due to steroid 5 alpha reductase 2 deficiency. *Am J Med Genet*, **63**:223-230.

Sorenson, H.R. 1953, Hypospadias with special reference to aetiology. Copenhagen, *Munksgaard*, 94.

Sutherland, R.W., Weiner, J.S., Hicks, J.P., et al., 1996, Androgen receptor gene mutations are rarely associated with isolated penile hypospadias. *J Urol*, **156**:828-831.

Sweet, Ra, Schrott, H.G., Kurland, R., et al., 1974, Study of the incidence of hypospadias in Rochester, Minnesota 1940-1970 and a case control comparison of possible etiologic factors. *Mayo Clin Proc*, **49**:52-57.

Thigpen, A., Davis, D., and Milatovich, A., 1992, Molecular genetics of steroid 5 alpha reductase deficiency. *J Clin Invest*, **90**:799-809.

Thigpen, A., Silver ,R., Guileyardo, J., et al., 1993, Tissue distribution and ontogeny of steroid 5 alpha reductase isozyme expression. *J Clin Invest*, **92**:903-910.

ENDOCRINE ABNORMALITIES IN BOYS WITH HYPOSPADIAS

Richard I. Silver*, M.D., FAAP, FACS

Assistant Professor of Urology and Pediatrics
Albert Einstein College of Medicine

Attending Pediatric Urologist
Schneider Children's Hospital
Long Island Jewish Medical Center
New Hyde Park, NY

1. INTRODUCTION

Hypospadias is one of the most common problems treated by the pediatric urologist, with an incidence of about 2 - 4 cases per 1,000 male births, or about 1 in 335 boys in modern series (Belman, 2002). Although the problem is common and the surgical management of hypospadias has made tremendous advances, with details provided elsewhere in this textbook, the etiology of most cases of hypospadias is often unclear. Since the development of the urethra is a hormone dependent event, logic suggests - and most authorities consider - that the development of hypospadias is in some way related to a relative deficiency of androgen action at the time of genital development. Given the recent advances in cellular and molecular biology, new theories of urethral development based on mesenchymal-epithelial interaction – and related derangements that might lead to hypospadias - are also emerging (Baskin, 2000). Research into the etiology of hypospadias continues to shed light on the endocrine, genetic, environmental, and other miscellaneous factors that may contribute to the occurrence of this specific birth defect. The recent developments in this field are the topic of this review.

2. Embryology

Normal development of the human male urethra is a process that is hormone dependent and occurs between weeks 8 and 14 of gestation. Although development of the internal

---

*19 Flower Lane, Manhasset, NY 11030, Tel: 516-869-6564, Fax:516-869-6459,
Email: rsilver@kids-urology.com

Hypospadias and Genital Development, edited by
L. Baskin, Kluwer Academic/Plenum Publishers, 2004

genitalia is dependent on testosterone (T), development of the external genitalia, including the urethra, is dependent on dihydrotestosterone (DHT). Since the conversion of T to DHT in the genitalia is catalyzed by the enzyme 5α-reductase type 2, (Figure 1), coded for by the gene at locus 2p23, and the proper action of DHT on external genital tissues requires an androgen receptor, coded for by the gene at locus Xq11-12, these additional biological components are critical to the process of normal genital development – and especially development of the male urethra.

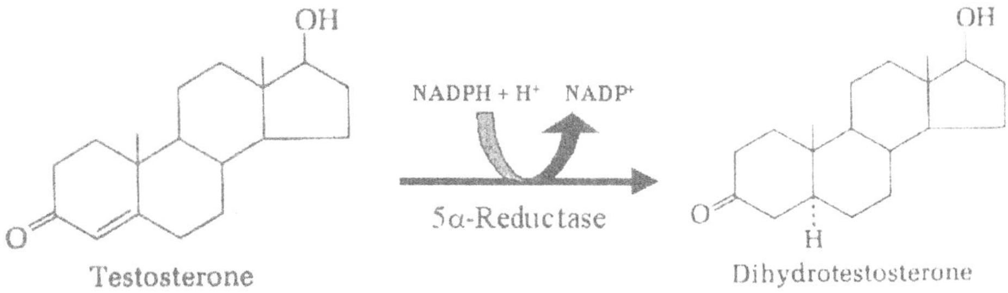

Figure 1. The Steroid 5α-reductase reaction. Testosterone is converted to the more potent dihydrotestosterone using NADPH as a cofactor.

3. Androgens and Gonadotropins

Numerous reports published in the literature, for at least three decades, have suggested an abnormal androgen milieu in boys with hypospadias. A number of studies have suggested either hypergonadotropic hypogonadism (due to impaired testicular testosterone production) or hypogonadotropic hypogonadism (due to impaired pituitary gonadotropin production) in boys with hypospadias. (1) In 1984, Allen and Griffin evaluated 15 children with hypospadias and found that 7 of the 15 children showed a poor testosterone response to HCG stimulation, suggesting that hypospadias is a local manifestation of a systemic endocrinopathy (Allen and Griffin, 1984). (2) In 1984, Nonomura et al. studied the endocrine abnormalities in boys with hypospadias and found that the basal LH levels and the LH response to LHRH stimulation was higher than that in normal controls, and that the T response to HCG stimulation was lower than that for normal controls. In addition, the basal LH levels in proximal hypospadias were higher than that for distal hypospadias, suggesting, perhaps quite logically, more severe hypogonadism in patients with more severe hypospadias (Nonomura et al., 1984), (3) In 1986, Shima et al. found that boys with hypospadias had an impaired response of LH to LHRH stimulation, a low basal level of LH, and a T response to HCG stimulation that was impaired in direct proportion to the degree of hypospadias (Shima et al., 1986). (4) Another report by Shima et al., in 1992, found low testosterone levels after HCG stimulation in boys with enlarged prostatic utricles, a condition that is often associated with severe hypospadias (Shima et al., 1992a). (5) In 1997, Aaronson et al. reported data that 50% of boys with penoscrotal or proximal shaft hypospadias have a defect in testicular testosterone biosynthesis, related to 3 specific enzymatic defects (Aaronson, 1997). (6) In 2002, Austin et al. reported that boys with hypospadias age 1-12 months have normal testosterone but low androstenedione levels, whereas boys with hypospadias ages 13-24 months have low testosterone but normal androstenedione levels (the exact opposite results in similar boys just one year older), and that the mean Müllerian Inhibitory Substance (MIS) levels in boys with hypospadias, in both age groups, are higher than normal (Austin et al., 2002).

Hypogonadism has also been identified in older patients with hypospadias. (7) In 1976, Raboch et al. studied a series of 42 peri-pubertal boys and young men with severe hypospadias and found reduced testosterone levels (Raboch et al., 1976). (8) In 1990, Gearhart et al. studied 16 adult patients with "mild" hypospadias (distal to the penoscrotal junction) and found a pattern of hypergonadotropic hypogonadism, with elevated LH levels, but normal T levels (Gearhart et al., 1990). Therefore, both of these reports show that systemic hypogonadism in boys with hypospadias can persist, even after childhood.

However, some studies are unable to show any endocrine abnormalities in boys with hypospadias, even when such hypospadias is severe. (1) In 1976, Walsh et al. investigated the testosterone response to HCG stimulation in boys with cryptorchidism and hypospadias (without a control group); the T response to HCG stimulation was uniform across these groups, and those with hypospadias had the highest incremental increase in testosterone, and the highest testosterone levels, after HCG stimulation (Walsh et al., 1976). (2) Feyaerts et al., in 2002, studied 32 consecutive boys with hypospadias who underwent thorough endocrine screening, and found no significant endocrine defects (Feyaerts et al., 2002). (3) A study by Boehmer et al., in 2001, examined 63 unselected cases of severe hypospadias and identified an etiology in only 31% of cases, including 17% due to complex genetic syndromes, 9.5% due to chromosomal anomalies, and 1 patient each due to the vanishing testis syndrome, the androgen insensitivity syndrome, and 5α-reductase type 2 deficiency (Boehmer et al., 2001). Therefore, even in cases of severe hypospadias, in which specific endocrine derangements might be more frequent and easier to find, only 3 / 63 patients (5%) with severe hypospadias, as reported by Boehmer et al., had such a condition. (4) It is also interesting to note that, in 1992, Shima et al. found that 16 boys with chordee, but without hypospadias, had no evidence of hypogonadism, with a normal T and LH response to HCG and GnRH, respectively (Shima et al., 1992b). Since boys with just chordee do not seem to have the hypogonadism that is commonly seen in boys with hypospadias, it is possible that chordee without hypospadias could have a different embryologic etiology than that for hypospadias; alternatively; it may just represent the mildest form in the spectrum of the condition.

Taken collectively, these reports offer strong, although sometimes inconsistent, evidence regarding the androgen-related endocrine derangements that may be found in boys with isolated hypospadias. The preponderance of the evidence suggests that a significant subset of boys with hypospadias have either hypergonadotropic or hypogonadotropic hypogonadism. In addition, elevated MIS levels may play some role in the development of hypospadias (Austin et al., 2002), perhaps as an endogenous inhibitor of testosterone synthesis (Teixeira et al., 1999). However, since some boys show no evidence of hypogonadism, the etiology of hypospadias is almost certainly multifactorial, with an abnormal androgen milieu only one of the potential causes of this problem.

4. 5α-Reductase Type 2

The enzyme 5 α-reductase converts testosterone (T) to dihydrotestosterone (DHT), a much more potent androgen required for normal urethral development (Figure 1). Two isoenzymes of 5_-reductase exist, named type 1 and type 2 for the order in which they were discovered, and are expressed differently throughout the various tissues of the body. In the human male genitalia, the 5α-reductase type 2 enzyme - abbreviated by geneticists as SRD5A2 (the gene is italicized, the enzyme is not) - predominates and is responsible for the T to DHT conversion in these tissues.

Studies on the male fetal rat involving mRNA in situ hybridization, with attention to the urogenital tract, show 5α-reductase type 2 in the mesenchyme of the urogenital sinus and seminal vesicle, and the type 1 isoenzyme in the epithelium of the urogenital sinus, the bladder, and the urethral bulb (Berman et al., 1995). Further mRNA in situ hybridization studies on the male fetal rat, with attention to the genital tubercle, has confirmed the presence of the type 2 mRNA message in the mesenchyme around the urethra (tissue analogous to the corpus spongiosum) and the type 1 mRNA message in the urethral epithelium ((Tian and Russell, 1997). These studies suggest that, in the rat, the mesenchyme around the urethra may be more critical than the epithelium for the normal development of this organ.

Experiments in human male fetal tissues have revealed similar results. Immunohistochemistry has shown that the 5α-reductase type 2 enzyme is present in the stroma of the corpus spongiosum but not in the epithelium of the pendulous urethra (Levine et al., 1996). Recent studies have confirmed these findings and revealed the enzyme in the stroma along the ventral urethral seam in the human male fetus (Figure 2) (Kim et al., 2002).

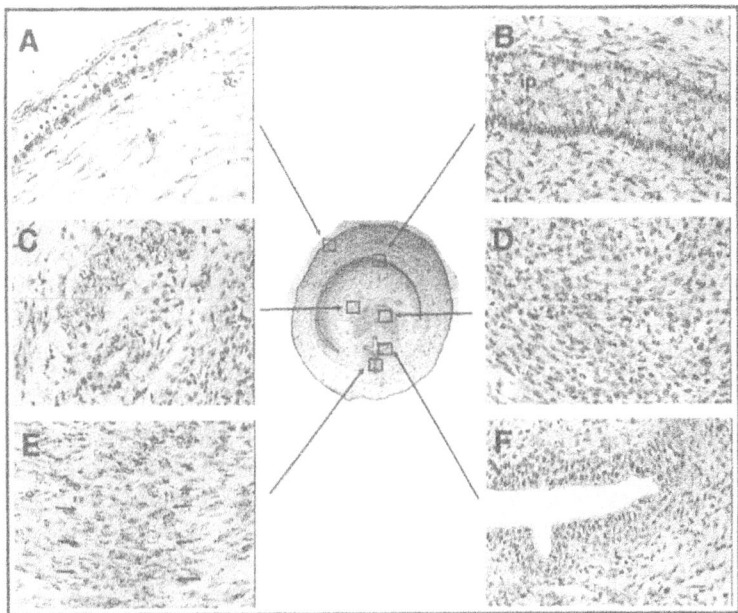

Figure 2. Immunohistochemical detection of 5α-reductase type 2 in the fetal external genitalia at 16.5 weeks of gestation. Note the 5α-reductase expression (brown stain) in the skin (A), inner prepuce (B), glans (C), and corpus cavernosum (D). In the urethra, note strongly positive stain in the stroma, the faint positive stain in the luminal epithelial cells, and the lack of staining in the basal cells (F). The stroma of the urethral seam area (multiple arrows) is strongly positive (E). From Kim et al., 2002.

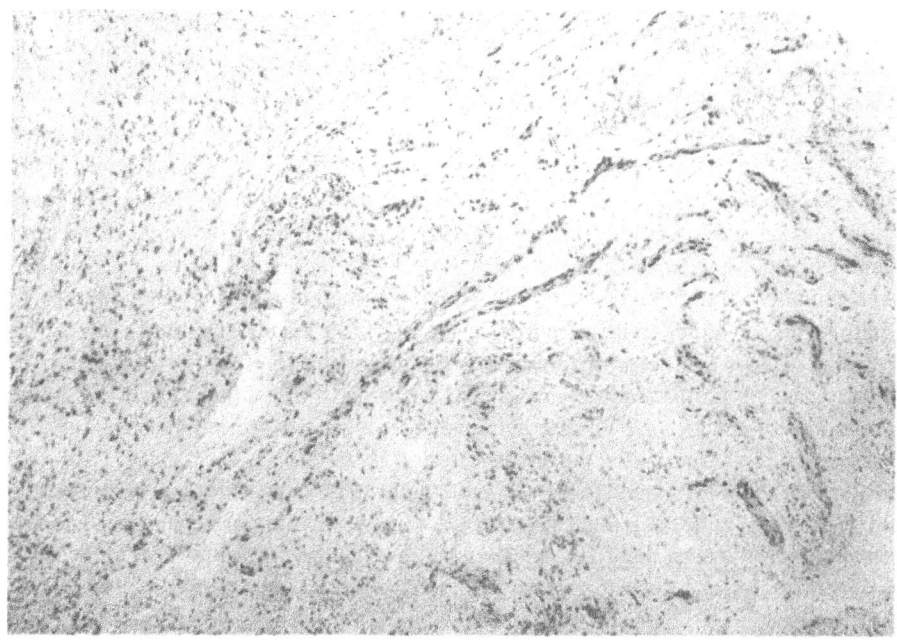

Figure 3a
Immunohistochemistry for 5α-reductase type 2 in (a) normal human foreskin, and (b) in foreskin from a boy with hypospadias. Note that the staining pattern and intensity is similar.

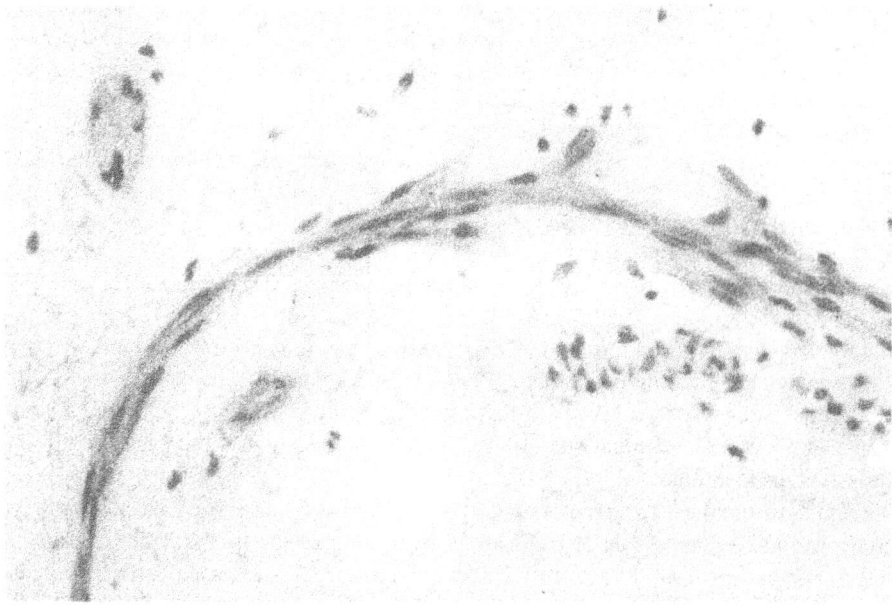

Figure 3b

Other immunohistochemistry experiments using mature genital tissues taken from boys at surgery for circumcision and hypospadias repair have revealed the 5α-reductase type 2 enzyme in the foreskin from boys both with and without hypospadias, with similar patterns and intensity of staining (Figure 3, a and b) (R. I. Silver, previously unpublished data).

From a genetic perspective, mutations in both alleles of the 5 α-reductase type 2 gene, in 46, XY genetic males, usually result in 5α-reductase deficiency and male pseudohermaphroditism with hypospadias. The fact that males with 5α-reductase deficiency have hypospadias led to the idea that mutations in 5α-reductase type 2 might be present in boys with isolated hypospadias (a male phenotype with descended testes, no signs of an intersex condition, and no evidence of endocrine abnormality). This is despite a report by Gearhart et al., in 1988, that 5α-reductase <u>activity</u> (based on quantitative T to DHT conversion) in boys with hypospadias is normal (Gearhart et al., 1988). However, it is important to note that this study involved only 10 patients, a relatively small sample size, and could have missed a 5α-reductase abnormality if the incidence in hypospadias patients is relatively rare, and especially if the incidence is less than 10%.

A follow-up study by Silver and Russell, reported in 1999, with nearly 100 patients and genetic evaluation by SSCP (single strand conformation polymorphism) analysis and DNA sequencing, indicates that about 10% of boys with isolated hypospadias have a mutation in the 5α-reductase type 2 gene (Silver and Russell, 1999). In this study, 9 / 93 boys with isolated hypospadias (9.7%) were found to have at least one allele with a mutation in their 5α-reductase type 2 genes. Three different mutations were identified: A49T, L113V, and H231R. The A49T genotype was the most common mutation, and is generally present (at least in heterozygotes) with less severe forms of hypospadias (Figure 4). The A49T mutations were present in 7 / 9 patients with mutations (78%) and in all patients with hypospadias with a native urethral meatus distal to the penoscrotal junction. The A49T and L113V mutations had not been previously reported in a case of 5α-reductase deficiency, but may ultimately prove to be common in boys with hypospadias. Interestingly, the family history data from this study were not helpful to predict which patients might harbor a 5α-reductase mutation (Silver and Russell, 1999).

A subsequent report by Zhou et al. from China, in 1999, found that 3 / 23 Chinese patients with hypospadias (13%) also had a 5α-reductase type 2 mutation (Zhou et al., 1999). The incidence of this abnormality in the Chinese population is similar to that in the American study. However, this study identified a different set of mutations in the Chinese *SRD5A2* genes; 2 of the 3 patients had a single R227Q mutation, and the third patient was a compound heterozygote with an R227Q in one allele and an F186L mutation in the other allele. Therefore, a total of 5 different 5α-reductase type 2 mutations have been found in boys with isolated hypospadias in the two studies reported so far, and with different mutations among the different populations (Figure 5 and Table 1).

It will be interesting to see if boys with hypospadias in other countries will have a similar incidence of 5α-reductase mutations (10-13%), and whether or not the same or different mutations will be identified.

The findings from these two reports suggest that a <u>partial</u> deficiency of 5 α-reductase type 2 activity and inadequate levels of DHT in the fetal urethra may be sufficient to cause the phenotype of hypospadias without other clinical features of 5α-reductase deficiency. Since the 5α-reductase type 2 enzyme is present in the stroma around the male fetal urethra

Hypospadias Data - Mutant Genes (n=9)

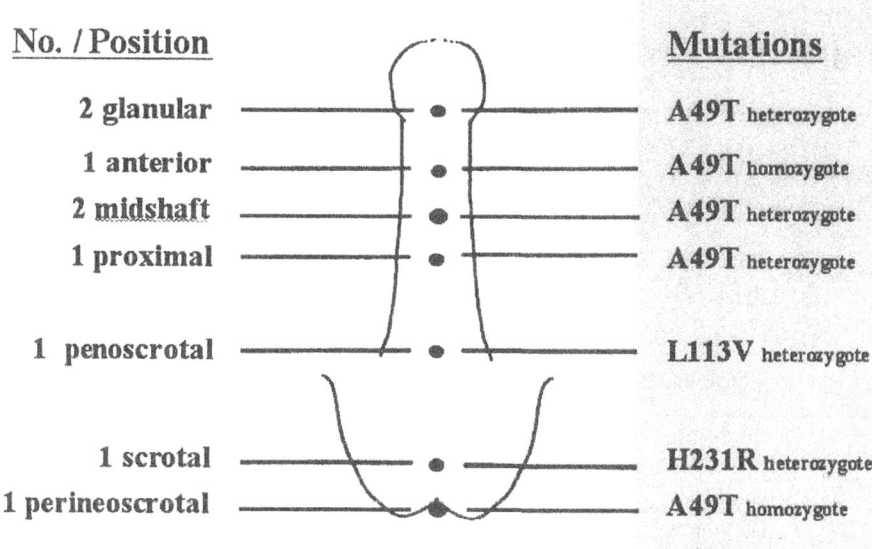

No. / Position		Mutations
2 glanular		A49T heterozygote
1 anterior		A49T homozygote
2 midshaft		A49T heterozygote
1 proximal		A49T heterozygote
1 penoscrotal		L113V heterozygote
1 scrotal		H231R heterozygote
1 perineoscrotal		A49T homozygote

Figure 4. Schematic showing the phenotype and *SRD5A2* genotype of 9 of 93 consecutive subjects with isolated hypospadias and later identified with 5α-reductase type 2 mutations. Data from Silver and Russell, 1999.

Table 1

5α-Reductase Type 2 Mutations in Boys with Isolated Hypospadias

From Silver and Russell, 1999
A49T   Alanine (GCC) replaced by Threonine (ACC) at amino acid position 49
H231R   Histidine (CAC) replaced by Arginine (CGC) at amino acid position 231
L113V   Leucine (CTC) replaced by Valine (GTC) at amino acid position 113

From Zhou, et al., 1999
R227Q   Arginine (CGA) replaced by Glutamine (CAA) at amino acid position 227
F186L   Phenylalanine (TTT) replaced by Leucine (CTT) at amino acid position 186

Mutation	Normal Amino Acid	Amino Acid Position	Mutant Amino Acid	Reference
A49T	Alanine	49	Threonine	Silver/Russell
H231R	Histidine	231	Arginine	Silver/Russell
L113V	Leucine	113	Valine	Silver/Russell
R227Q	Arginine	227	Glutamine	Zhou
F186L	Phenylalanine	186	Leucine	Zhou

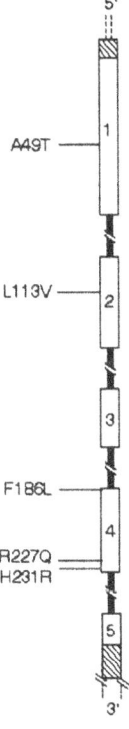

Figure 5. Schematic showing the structure of the 5α-reductase type 2 gene and the mutations identified so far in boys with isolated hypospadias. Figure courtesy of Dr. David Russell.

(Figure 2) (Kim et al., 2002), it is reasonable to conclude that a male fetus that lacks functionally active 5α-reductase type 2 adjacent to the urethra at the time of its normal development – either because of a genetic defect, impaired transcription and/or translation, biochemical inhibition, faulty expression, or any other cause for its proper and timely action – may demonstrate some of the phenotypic features of 5α-reductase deficiency, such as genital ambiguity and/or hypospadias.

It also must be considered that other causes of 5 α-reductase type 2 underexpression may contribute to the development of hypospadias. With this in mind, it is important to recognize that both of the 5α-reductase isoenzymes demonstrate a variable expression pattern during an individual lifetime, a phenomenon known as ontogeny. The expression of 5α-reductase type 2 varies depending on the specific tissue, and specific time, when it is examined. For example, although the liver and extragenital skin begin to express the type 2 enzyme after birth, there is no such expression in those organs in the fetus. In contrast, the type 2 enzyme is expressed in male fetal genital tissues within the first trimester, with gradually increasing levels of expression up until the time of birth, and then steady expression for the remainder of life (Figure 6) (Thigpen et al., 1993).

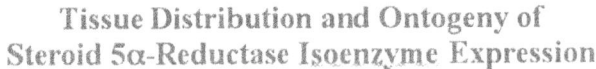

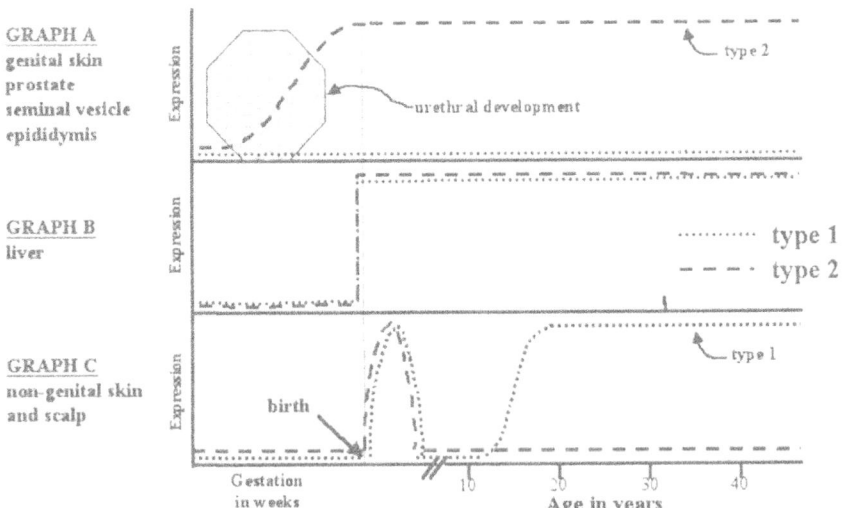

Figure 6. Tissue Distribution and Ontogeny of Steroid 5α-Reductase Isoenzyme Expression. Temporal expression of 5α-reductase isoenzymes in various tissues as determined by qualitative immunoblotting. Arbitrary (qualitative) expression is indicated on the ordinate and age is indicated on the abscissa. Note the three different expression patterns for different tissue types.
Graph A. The male genital tissues express only the type 2 isoenzyme, with gradually increasing expression during gestation. The shaded area is expanded for analysis in figure 7.
Graph B. The liver expresses both the type 1 and 2 isoenzyme, but only after birth.
Graph C. The non-genital skin and scalp expresses the enzymes in a unique and interesting pattern, or ontogeny. There is no expression of either enzyme prior to birth, with a burst of expression by both isoenzymes developing in early childhood, with cessation prior to puberty. After puberty, the type 1 isoenzyme is expressed in these tissues, whereas the type 2 enzyme is not.
Data extracted and modified from Thigpen et al., 1993.

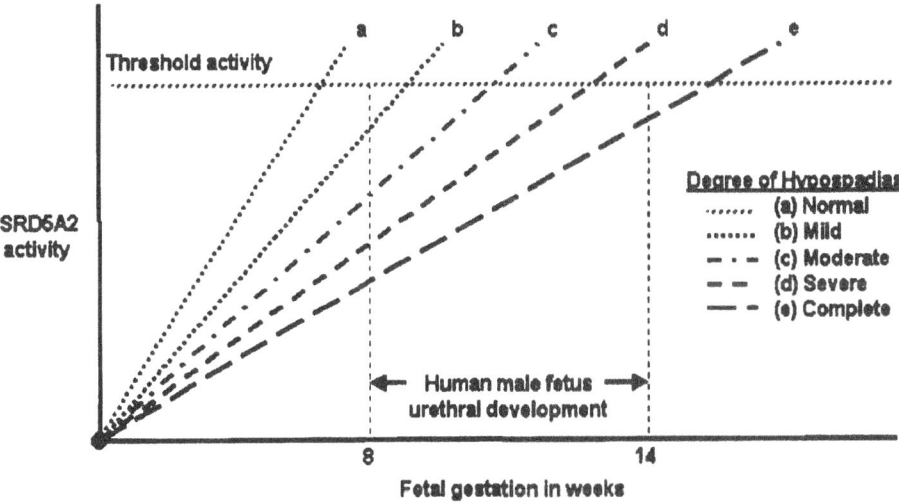

Figure 7. Graphic representation for a theory regarding how a delayed or reduced expression level of the 5α-reductase type 2 enzyme could result in the development of hypospadias. The model presumes a threshold level of 5α-reductase activity necessary to catalyze sufficient conversion of T to DHT to provide normal embryologic development of the male genitalia. The threshold level of activity must be reached by the onset of the critical period of urethral development (gestation week 8), and maintained throughout this period (until gestation week 14), to stimulate complete closure of the urethral folds and the development of a properly formed male urethra. The model also presumes that the slope of the curve, indicating a gradual rise in the SRD5A2 expression level, may vary between individuals. As the slope for expression decreases, the overall delivery of DHT to the urethral folds decreases, and results in increasingly more severe forms of hypospadias. However, as long as the SRD5A2 expression reaches a normal level prior to birth, there will be no postnatal evidence of a fetal endocrine abnormality other than the hypospadias.

The   gradually increasing expression of 5   α-reductase type   2 in the male fetal genitalia provides another possible explanation for the embryologic development of isolated hypospadias - in the absence of any detectable postnatal endocrinopathy. It is theoretically possible and scientifically rational that a <u>temporally delayed onset of expression</u>, or a <u>reduced level of expression,</u> of 5α-reductase type 2 during the critical period of urethral development could result in variable phenotypic forms of isolated hypospadias. Since the human male fetal urethra develops between weeks 8 and 14 of gestation, a delay of 5α-reductase type 2 expression in the genitalia by even a few days could significantly decrease the delivery of DHT to the urethral folds, the precursor of the urethra. This alone might be sufficient to cause a phenotype of mild hypospadias. More prolonged delays of expression could result in increasingly severe forms of hypospadias and ultimately approach the phenotype of 5α-reductase deficiency, even with the presence of normal *SRD5A2* genes (Figure 7). This model allows for a continuous spectrum of hypospadias phenotypes and is also consistent with the observation that most patients with hypospadias do not have clinically apparent endocrine abnormalities. Confirmation of this theory will require periodic monitoring of fetal 5α-reductase type 2 activity in the genitalia, and later correlation with the postnatal phenotype. Unfortunately, the technology to perform this work, even in an animal model, is not yet available.

5.   Androgen Receptor

A functional androgen receptor (AR) is necessary for a circulating androgen signal to reach the cell nucleus in target tissues and trigger the hormonally sensitive cell functions. Since the androgen receptor is an X-linked gene, mutations are transmitted as an X-linked recessive trait from maternal carriers to their children. Statistically, a woman with an AR mutation in one allele will transmit this mutation to one-half of her children; of those children who inherit the abnormal gene, the boys will be clinically affected and the girls will be clinically silent carriers.

Males with an AR mutation and "androgen resistance" generally have a 46, XY karyotype, inguinal or abdominal testes, an external female phenotype, and various degrees of genital ambiguity - depending on the specific androgen receptor mutation. In some cases of mild androgen resistance, the genital ambiguity may be quite mild and be limited to minor abnormalities of external genital development, such as hypospadias.

Animal experiments using in situ   mRNA hybridization in the fetal mouse have revealed the spatial and temporal expression of the AR mRNA transcripts in the genital tubercle, the labioscrotal folds, and the urethral folds in the developing mouse external genitalia (Crocoll et al., 1998). Other studies using immunohistochemistry and adult rats have identified the AR protein in the reproductive tract tissues in rats in both sexes (Pelletier et al., 2000).

To evaluate AR expression in and around the human urethra,   Kalloo et al., in 1993, used immunohistochemistry to identify AR expression in human male fetal genitalia in subjects 18 - 22 weeks of gestation. Fetal AR protein expression was identified in the corpora cavernosa, the stroma of the inner prepuce, the periphery of the glans, and the periurethral mesenchyme (the primordium of the corpus spongiosum). However, AR protein expression was not seen in the epithelium of the preputial or penile shaft skin, or the urethral plate in the distal glans. Interestingly, canalization of the urethral plate began where it joined the patent distal penile urethra, where AR expression could be identified. Also, interestingly, this pattern of AR expression did not appear to be sexually dimorphic, with a similar pattern of expression in the female human fetus (Kalloo et al., 1993).

More recently, Kim et al., in 2002, used   immunohistochemistry to examine AR protein expression in human fetal genital tissues at different time points during gestation. From 12 - 14 weeks of gestation, AR is expressed in the epithelium of the urethral plate in the glans, the tubular urethra of the penile shaft, and the stromal tissue surrounding the urethral epithelium. From 16 - 20 weeks gestation, the density of AR expression is greater in urethral epithelium than in the surrounding stromal tissues. Also from 16 - 20 weeks gestation, a noteworthy pattern of AR expression in the glandular urethral epithelium is present, with greater expression along the ventral half of the urethra than along the dorsal half of the urethra (Figure 8) (Kim et al., 2002). As it relates to a discussion of hypospadias, this has obvious clinical significance.

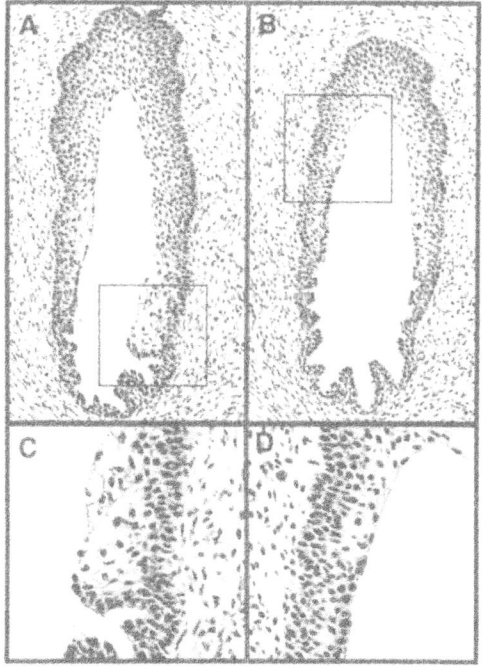

Figure 8. Immunohistochemical detection of AR in the mid to proximal glans. The density of AR expression is greater in the ventral portion than in the dorsal portion of the urethral epithelium. Note the intense staining in the luminal epithelial cells. Arrows indicate the upper margin of the increased AR expression (A and B). Note the luminal cuboidal epithelial cells in rows showing strong AR expression (C and D). From Kim et al., 2002.

Given this background, a number of studies over the last three decades have looked at the androgen receptor density, integrity, and function in boys with isolated hypospadias. The methods of analysis vary from study to study, making pooling and analysis of the data difficult. Results are generally reported as AR tissue concentration - often expressed as ligand (androgen) binding capacity, or total binding capacity; affinity – often expressed as the dissociation constant, Kd; in vitro biochemical behavior in the laboratory; and, DNA mutation analysis. A summary of the data from these studies is listed in Table 2.

Early studies before routine PCR (polymerase chain reaction) gene amplification and DNA mutation analysis used indirect evidence to evaluate the quantity {concentration, or total binding capacity (TBC)} and quality (affinity, or Kd) of the AR. A study by Svensson and Snochowski, in 1979, showed that the mean androgen receptor concentration in the foreskin of 11 boys with hypospadias was lower than that of controls (Svensson and Snochowski, 1979). However, a separate study by Coulam et al., in 1983, showed that the AR concentration and affinity for steroid binding in the foreskin of boys (number not reported) with hypospadias was the same as that for the normal control boys, but just showed a different pattern of molecular sieve chromatography in the laboratory (Coulam et al., 1983). Gearhart et al., in 1988, examined 10 boys with isolated hypospadias and also found normal AR concentration and androgen binding properties (Gearhart et al, 1988). Terakawa, in 1990, showed normal binding capacity and dissociation constant in 16 boys with hypospadias (Terakawa et al., 1990). A study by Evans et al., in 1991, looked at androgen receptor binding in 19 boys with hypospadias and found normal androgen receptor concentration and steroid binding affinity (Evans et al., 1991). In 1992, Okamoto et al. reported decreased AR binding capacity in genital tissue fibroblasts cultured from the foreskin of a small group of 4 boys with hypospadias and an enlarged prostatic utricle (Okamoto, 1992). Most recently, in 1995, Bentvelsen et al. used immunoblotting and densitometry to analyze the AR in 15 patients with hypospadias, and all had normal AR analysis (Bentvelsen et al., 1995). Therefore, 5 of these 7 studies that have evaluated the AR quantity (concentration, or binding capacity) and quality (androgen binding affinity, or dissociation constant) in boys with hypospadias found no abnormalities in these parameters.

Other studies in the literature, involving DNA analysis to look at AR integrity, indicate that the AR gene is almost always normal in boys with isolated hypospadias. Analysis of 5 recent studies on androgen receptor mutations in boys with hypospadias over the last ten years reveals that only 5 out of 93 hypospadias patients studied (5%) had an androgen receptor mutation (Batch et al., 1993; Hiort et al., 1994; Allera et al., 1995; Sutherland et al., 1996; Muroya et al., 2001). Closer inspection of the 5 patients with AR mutations in these studies reveals that 4 patients had severe hypospadias (Batch et al., 1993; Hiort et al., 1994; Allera et al., 1995) and 1 patient had mild hypospadias ((Sutherland et al., 1996). Therefore, if severe hypospadias is excluded, only 1 of 93 patients (1%) with mild hypospadias in these 5 studies has been identified with an AR gene mutation.

Therefore, the collective data from these studies and prior reports in the literature indicate that AR mutations in cases of isolated hypospadias, occurs with an incidence of about 1%, and at most, including even severe cases, occurs with an incidence of about 5%. Androgen receptor mutations may be related to hypospadias, but this is a very rare cause of isolated hypospadias with an incidence that is much lower than that for testosterone biosynthesis defects (40-66%) (Aaronson, 1997) or 5α-reductase type 2 mutations (10-13%) (Silver and Russell, 1999; Zhou et al., 1999).

Table 2.   Androgen Receptor Evaluation in Boys with Isolated Hypospadias

Reference	Year	Type of AR analysis	No. of Patients / Controls	Results	Number of Patients with AR abnormalities
1. Svensson and Snochowski	1979	Androgen binding capacity	11 / 16	AR concentration in boys with hypospadias is lower than that in controls	4 subjects with lower total binding capacity (TBC) than controls
2. Coulam et al.	1983	Scatchard analysis Competitive binding assays Sedimentation analysis Isoelectric focusing Molecular sieve chromatography	Not reported	Normal AR concentration and affinity, abnormal molecular sieve chromatography	Not reported
3. Gearhart et al.	1988	Androgen binding capacity AR affinity	10 / 0	Normal AR concentration Normal AR affinity	0 / 10
4. Terakawa et al.	1990	Androgen binding capacity AR affinity	16 / 10	Normal AR binding Normal AR affinity	0 / 16
5. Evans et al.	1991	Androgen binding capacity AR affinity	19 / 0	Normal AR concentration Normal AR binding	0 / 19
6. Okamoto et al.	1992	Androgen binding capacity	4 / 4	Decreased AR concentration	Not reported
7. Batch et al.	1993	Androgen binding capacity DNA analysis	4 / 0 2 sets of twin brothers; 2 brothers with isolated perineal hypospadias, 2 brothers with perineal hypospadias and genital ambiguity (excluded)	2 sets of 2 identical AR mutations (1 set excluded)	2 / 2 DNA mutations
8. Hiort et al.	1994	DNA analysis	21 / 43	1 patient with severe hypospadias and bilateral undescended testes and a point mutation in exon 8	1 / 21 DNA mutations
9. Allera et al.	1995	DNA analysis	9 / 0 (severe hypospadias)	1 patient with severe hypospadias and a point mutation in exon 2	1 / 9 DNA mutations
10. Bentvelsen et al.	1995	Immunoblotting – AR concentration Densitometry – AR immunoreactivity	15 / 7	Normal AR concentration Normal AR immunoreactivity	0 / 15
11. Sutherland et al.	1995	DNA analysis	40 / 0	1 patient with mild hypospadias and a missense point mutation in exon 2	1 / 40 DNA mutations
12. Muroya et al.	2001	DNA analysis	21 / 0	Normal DNA analysis	0 / 21 DNA mutations
Totals					5 / 93 (5%) AR mutations 9 / 164 (5%) AR abnormals

6. Assisted Reproduction

Assisted reproduction involves laboratory technology to help an infertile couple achieve a pregnancy. One common form of assisted reproduction is "in vitro fertilization" (IVF). IVF protocols usually include progesterone administration after embryo transfer to support the pregnancy. Progesterone is an excellent substrate for 5α-reductase and can act as a competitive inhibitor of the T to DHT conversion reaction.

Maternal exposure to progesterone has been reported as a possible risk factor for the development of hypospadias, but reports in the literature have not always produced consistent data and conclusions. In 1979, Aarskog performed a retrospective review of human subjects exposed to progesterone during gestation and found an incidence of hypospadias of 8.5% (Aarskog, 1979). Conflicting reports include a prospective review by Mau, in 1981, which indicated that progestin exposure is not predictive of hypospadias (Mau, 1981) and an animal study by Briggs, in 1982, which showed that neither estrogen nor progesterone increased the rate of hypospadias in male offspring in a rat model (Briggs, 1982).

If progesterone exposure increases the risk of hypospadias, it might be expected that the use of birth control pills would be associated with an increased risk of hypospadias, since these agents contain progesterone. However, there is no evidence in the literature to suggest that the use of birth control pills is a risk factor for hypospadias, and some reports specifically disprove any association between oral contraceptives and hypospadias (Kallen et al., 1991).

However, a study by Macnab and Zouves, in 1991, suggested an association between IVF and hypospadias in a series of 53 babies born by IVF and gamete intra-fallopian transfer (GIFT), in which 2 of the 53 (3.8%) were born with hypospadias (Macnab and Zouves, 1991). This statistic represents a 10-fold increase in the incidence of hypospadias compared to the incidence in the general population. Later, Silver et al. published in 1999 a large, retrospective review involving male hypospadias patients born through IVF in the greater Baltimore area between 1988 and 1992. The data from this study indicates that the incidence of hypospadias in IVF patients is 1.46%, compared to a control group from the State of Maryland, in which the hypospadias incidence is 0.27%. Therefore, the results of this study indicate that there is a 5-fold increased risk of hypospadias in males born through IVF technology (Table 3) (Silver et al., 1999).

The only apparent difference between the IVF and control groups in this study was maternal progesterone exposure in the IVF group, to support the pregnancy, but which may also potentially interfere with the normal fetal endocrine milieu. Therefore, taking these two studies together, there seems to be a 5 to 10-fold increase in the risk of hypospadias for parents who use IVF or GIFT to conceive a child.

To examine this issue further, as a pediatric urology fellow at Johns Hopkins Hospital in 1996, I created an animal model to further test the human IVF data from Baltimore. Timed pregnant female rats were exposed to high-dose progesterone administration via subcutaneous delivery pellets; a control group received placebo. The progesterone pellet was implanted in the subcutaneous tissue at the nape of the neck on day 10 and left through the remainder of the pregnancy (the fetal rat urethra develops from day 10-21, the last half

Table 3.

<u>In Vitro Fertilization Is Associated With An Increased Risk Of Hypospadias</u>

(Data from Silver et al., 1999)

Incidence of Hypospadias from IVF Program in Baltimore Compared to Annual Incidence in Maryland, Years 1988-92

Years	IVF Program (GBMC) % (No./Total No.)	Maryland State % (No./Total No.)	p Value (years)
1988	1.12 (1/89)	0.26 (84/32,308)	0.58
1989	1.09 (1/92)	0.32 (106/33,125)	0.181 (1988-89)
1990	2.13 (2/94)	0.29 (105/35764)	<0.01 (1988-90)
1991	2.04 (2/98)	0.22 (80/35948)	<0.001 (1988-91)
1992	0.93 (1/108)	0.24 (86/35910)	<0.001 (1988-92)
1988-92	1.46 (7/481)	0.27 (461/173,055)	<0.001 (5 yr total)

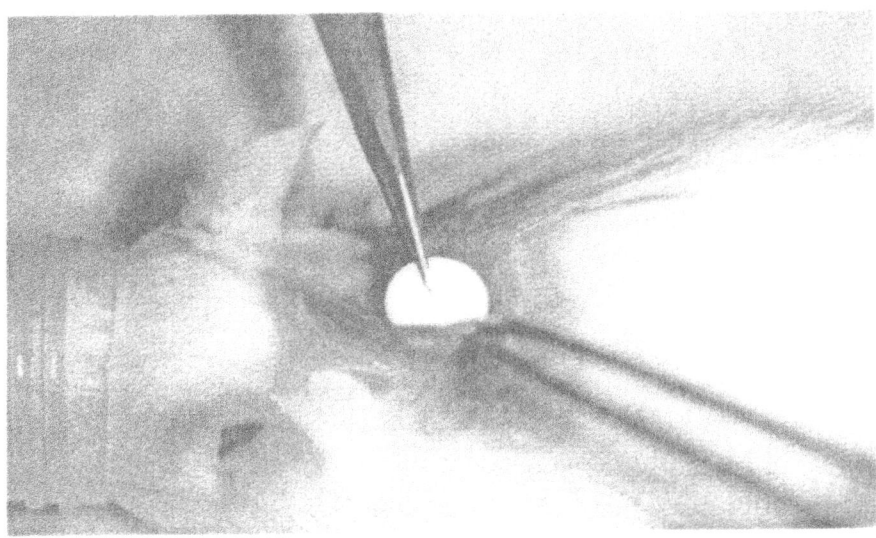

Figure 9a

a) A timed-pregnant female Sprague-Dawley rat dame was treated with a high-dose, zero-order delivery, subcutaneous progesterone pellet from days 10-21 of the gestation, during the interval when the male fetal rat genitalia develop. b) Photograph of a dissection of one of the male rat pups with female appearing external genitalia with hypospadias (octagonal outline), and male internal genital structures and gonads, including testes (arrows).

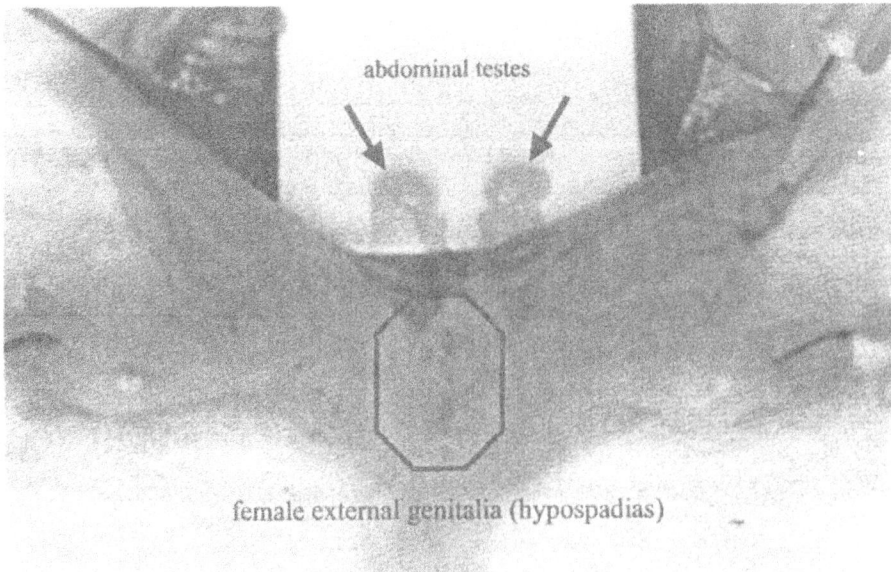

Figure 9b

of the gestation, in contrast to the first trimester of gestation in humans). The pups were delivered at 21 days by Caesarean section and the external and internal genitalia were examined. Of 82 male pups that developed in mothers exposed to prenatal progesterone during the period of urethral development, 3 out of 82 (4%) had hypospadias, while none of the control male pups (exposed to placebo) were born with hypospadias (Figure 9, a and b). Hypospadias does not naturally occur in male rat pups. The fact that progesterone administration seemed to cause hypospadias in some of the male pups was intriguing. However, the reason why only a small subset of rats was affected by maternal progesterone administration, when it would appear that all male rat pups should have been affected, remains a mystery at this time.

If maternal progesterone administration is not the cause of the increased risk of hypospadias in IVF, how can we explain the data suggesting an increased risk of hypospadias in IVF patients? The increased risk of hypospadias in boys born through assisted reproduction may also be related to underlying endocrine problems that contribute to infertility, or fetal endocrine abnormalities that are more common in infertile couples. In 1986, a case control study by Kallen et al. evaluated over 8,000 patients and concluded that fertility is inversely related to hypospadias (Kallen et al., 1986). If true, separating the risk factor of subfertility from other factors that may increase the risk of hypospadias in the IVF population may be difficult, but necessary, for future research endeavors.

7. Genetic Factors

Evidence for the Mendelian inheritance of uncomplicated hypospadias exists in the literature; some studies suggest autosomal recessive inheritance (Frydman et al., 1985), while others suggest an autosomal dominant inheritance pattern (Page, 1979). The familial tendency for hypospadias is well recognized, with an increased incidence of hypospadias in the family members of an affected boy. For example, for a family with a son with hypospadias, there is a 10% - 17% risk of hypospadias in a future son, who would be his brother (Table 4) (Bauer et al., 1979; Bauer et al., 1981; Editorial, 1972; Stoll et al., 1990). Since the risk in the general population is approximately 1/335 or about 0.32% (Belman, 2002; Bauer et al., 1979; Editorial, 1972), this range represents about a 45-fold increase in risk for a future male sibling to be born with hypospadias – a number that parents would probably be interested to know when planning to expand a family affected by hypospadias. However, since the recurrence risk in a single family appears to be no greater than 21%, the inheritance pattern appears heterogeneous, polygenetic, and multifactorial (Harris, 1993).

The genetic tendency for hypospadias is also supported by the historical reports that indicate that the incidence of hypospadias is higher in Caucasians than in African-Americans, and also most common in Jews and Italians (Welch, 1979). These racial and ethnic groups bear a common ancestry in Europe. Also, the fact that there is an 8.5% increase in the incidence of hypospadias in monozygotic twins suggests a strong genetic basis in the development of this condition (Roberts and Lloyd, 1973).

Recognition of the familial tendency for hypospadias is important because it may implicate or relate to hereditary genetic mutations in genes for proteins involved in critical androgen physiology and genital development as an etiology, as previously described. For example, if a 5α-reductase type 2 mutation is related to a phenotype of isolated hypospadias, then the expected frequency of transmission (assuming 100% penetrance) would be 25%, requiring the inheritance of a mutant 5α-reductase type 2 gene from either parent and the

Table 4

Genetics Aspects of Hypospadias

(Modified from Bauer et al., 1981)

Relationship to Hypospadias Patient	Incidence of Hypospadias
Second Family Member	21%
Brothers	14%
Fathers	7%
Male Cousins	2.5%
Uncles	1.8%
General population	0.32%

SRY gene from the father ($1/2 \times 1/2 = 1/4 = 25\%$). The fact that the familial tendency for hypospadias can be complicated by environmental factors should be recognized as a confounding variable in evaluating this issue, and is addressed below.

8. Family History

A positive family history is interesting and potentially useful to help identify specific genetic factors involved in the etiology of hypospadias. Although a positive family history may not be surprising in evaluating a specific patient, many – if not most - cases do not have a positive family history. As previous studies have shown, a family history of hypospadias may not be helpful to identify genetic and/or endocrine abnormalities that might be present in an individual with hypospadias (Silver and Russell, 1999).

Why is family history such a poor predictor of genetic abnormalities in   hypospadias patients? What could be the reasons for a potentially falsely negative family history? Reasons for this may include a missed diagnosis, poor documentation of medical findings, difficult access to medical records, family concealment of a potentially embarrassing situation, and early successful surgical repair. Additional factors include delicate family matters, such as the issue of true paternity, child adoption, and family separation (e.g., divorce) with divergent migration (Table 5).

Table 5

Causes of a Potentially Misleading Family History for Hypospadias

Falsely Negative Family History
- Missed diagnosis
- Poor documentation in medical records
- Difficult access to medical records
- Family concealment
- Early surgical repair
- Paternity, true source of paternal genes
- Adoption, incomplete medical records
- Family separation and divergent migration

Falsely Positive Family History
- Multifactorial causes of hypospadias
  - Chromosomal abnormalities
  - Genetic syndromes
  - Maternal drugs
  - Environmental exposures
  - Sporadic cases

Conversely, what could be the reasons for a falsely positive family history? Since the etiology of hypospadias is multifactorial, any one of a number of factors, not necessarily hereditary, could cause the development of hypospadias in two males in the same family. Therefore, e.g., maternal drugs, environmental factors, and sporadic cases could be misleading in the scientific evaluation for endogenous biological causes of hypospadias (Table 5).

Nevertheless, despite this discouraging information, obtaining a thorough and accurate family history remains important for further research endeavors. Given such information, it will be possible identify families with more than one member with hypospadias, and affecting more than one generation, to help identify specific real and/or potential endogenous factors that cause this condition.

9. Environmental Factors

In addition to intrinsic or endogenous endocrine derangements (fetal and/or maternal), hypospadias can also be caused by extrinsic or exogenous (maternal) factors that can indirectly cause a fetal endocrine derangement. In addition to IVF, extrinsic factors include maternal exposure to drugs or environmental agents that have estrogenic or anti-androgenic properties. A hot topic in pediatric andrology today is the potential impact of environmental pollutants on male genital anatomy and function. Such agents are receiving increasing attention as having an adverse effect on male genitourinary health, such as the development of hypospadias.

In a recent commentary on the incidence of hypospadias, Dolk refers to the concept of the "endocrine disruptor hypothesis" of environmental estrogens and anti-androgens (Dolk, 1998). Over the last few years, review articles (Kennedy and Snyder, 1999) and entire books (Colborn et al., 1997) are now focusing on the dangers of environmental pollutants on male sexual and genital health. The fact that environmental factors may contribute to the familial tendency seen for hypospadias is of obvious clinical significance.

Environmental factors implicated in the development of hypospadias include environmental estrogens such as pesticides (DDT), toxic substances (PCBs), and plant estrogens (soy). In addition, environmental anti-androgens may play a role, such as the fungicide Vinclozolin. Experimental evidence indicates that Vinclozolin is detrimental for sexual development of the male rat fetus, with subsequent hypospadias (Ostby et al., 1999; Gray et al., 1994). Another similar anti-fungal agent, Procymidone, has also been implicated as an androgen receptor antagonist that can have similar effects (Ostby et al., 1999). It is interesting to speculate that exposure to such agents, especially in rural areas, may contribute to an increased risk of hypospadias in boys with families who live in such areas, and some studies support that possibility (Angerpointer, 1984). Although there is some evidence that there may be an association between pesticides and hypospadias (Kristensen et al., 1997), at least one study shows that there is an increased risk of cryptorchidism, but no increased risk of hypospadias, in boys born to farmers or gardeners (Weidner, 1998).

As described by Skakkeback et al., experimental and epidemiological evidence is accumulating to suggest that environmental estrogens and anti-androgens may lead to the "testicular dysgenesis syndrome," with subsequent hypogonadism and male genital abnormalities, including testis cancer, infertility, cryptorchidism, and hypospadias (Skakkebaek, 2002; Skakkebaek et al., 2001). Investigative efforts have been taken into the laboratory, with the development of a human prostate cell line (PALM) containing a chemiluminescent reporter (luciferase) to demonstrate the effects of environmental estrogens and anti-androgens on human AR function (Sultan et al., 2001).

10. Miscellaneous Factors

In addition to the controversial implication of   progestin exposure (Aarskog, 1979; Calzolari, 1986), other factors implicated in the etiology of hypospadias include exposures to aspirin (Correy et al., 1991) and cocaine (Battin et al., 1995). Factors related to infertility have also been implicated, including extremes of age (older mothers, younger fathers) (McIntosh et al., 1995); early (Calzolari et al., 1986) or late (Polednak and Janerich, 1983) menarche; low parity (Kallen et al., 1991); subfertility (Czeizel and Toth, 1990) (especially in fathers); low weight of the placenta (Stoll et al., 1990);   threatened abortion (Angerpointer, 1984; Calzolari et al., 1986); and weak contractions or Cesarean section (Kallen, 1988).

Factors that indicate abnormal endocrine function in the parents, such as   subfertility or maternal problems in pregnancy, are perhaps not surprising as a risk factor for hypospadias. On first inspection, the idea that Cesarean section may be a risk factor for hypospadias lacks obvious scientific rationale. However, it may be statistically related to hypospadias because women who have endocrine abnormalities may fail to progress during labor and require a Cesarean delivery.

Exogenous factors that have not been implicated in the etiology of   hypospadias include tobacco, alcohol, birth control pills (Kallen et al., 1991; Polednak and Janerich, 1983; Kallen, 1988; Janerich et al., 1980), and spermicides (Polednak et al., 1982; Louik et al., 1987).

11. The Incidence of Hypospadias

A recent report by Paulozzi et al., in 1997, indicates that the incidence of   hypospadias is increasing, with a doubling of the incidence of hypospadias from 1970 to 1993 both in Atlanta and in the United States overall (Paulozzi et al., 1997). An increased rate of more severe cases of hypospadias was noted, contradicting the possibility of improved diagnosis or the detection of mild cases.   A recent report by Gallentine et al., in 2001, suggests that that there may be an increasing incidence of hypospadias in boys born into American military families, especially within minority groups (Gallentine et al., 2001).

Data from Europe, specifically Britain (Matlai and Beral, 1985) and Hungary (Czeizel, 1985), indicate similar trends in those countries. However, a recent report from Finland, in 2000, indicates that the prevalence of hypospadias in that country remained constant from 1970-1986, but appears to have been approximately three times higher than previously reported. Changes in the completeness of birth defect registration may account for a substantial proportion of the reported increases in the prevalence of hypospadias in Finland, and perhaps also elsewhere (Aho et al., 2000).

The small but   significant incidence of 5 α-reductase type 2 mutations in boys with hypospadias (10-13%) (Silver and Russell, 1999; Zhou et al., 1999), and the unexpectedly high incidence of compound heterozygotes with 5α-reductase deficiency (about 35%) (Wilson et al., 1993), suggests that the prevalence of mutant 5α-reductase type 2 genes is greater than that suspected by the relatively rare incidence of the 5α-reductase deficiency syndrome. Although the fertility of men with a history of hypospadias and a mutant 5α-reductase type 2 allele has not yet been studied, the fact that hypospadias surgery and assisted reproduction technology have improved significantly over the last 50 years suggests that an increase in the prevalence of 5α-reductase type 2 mutations may be developing in the general population due to improved techniques and technology in these fields. Therefore, it is interesting to speculate that the apparently increasing incidence of hypospadias may be related to the spread of genetically transmissible defects in androgen metabolism due to better medical care by both urologists and gynecologists, as well as the increasing use of environmental agents with detrimental effects on male genitourinary embryology. Further work will be necessary to determine if the apparently increasing incidence of hypospadias is real, here in the U.S. and elsewhere, and also to determine if better medical care and agricultural chemicals have caused that unintended outcome.

12. Summary

The multifactorial etiology of hypospadias is becoming more clearly defined with ongoing investigation. Endogenous endocrine abnormalities identified so far include testosterone biosynthesis defects, 5α-reductase type 2 mutations, and androgen receptor mutations (the rarest cause, even in cases of severe hypospadias). Other significant risk factors include IVF (because of progesterone administration or endocrine abnormalities associated with infertility) and environmental agents that can potentially cause testicular dysgenesis, disrupt the male androgen axis, and disturb normal male genital embryology (Table 6).

Table 6

<u>Risk Factors for Isolated Hypospadias</u>
Endocrine Abnormalities
      Hypogonadism
      5α-reductase mutations
      Androgen receptor mutations
Assisted Reproduction
      In Vitro Fertilization (IVF)
      Progesterone administration
      Infertility
Family History
      Genetics vs. environment
      (Nature vs. nurture)
Environmental Factors
      Environmental estrogens
            Pesticides (DDT)
            Toxic substances (PCBs)
            Plant estrogens (soy)
      Environmental anti-androgens
            Fungicides (Vinclozolin, Procymidone)
<u>Non-Risk Factors for Isolated Hypospadias</u>
      Tobacco
      Alcohol
      Birth Control Pills/Spermicides

It also seems that the incidence of hypospadias is increasing, both in the United States and in Europe – which may be due to better medical care for those with genital abnormalities and/or infertility problems, as well as environmental endocrine disruptors.

Hypospadias is a physical manifestation that may be a consequence of numerous physiological aberrations, and our ability to understand and to potentially prevent this congenital malformation will require a significant amount of additional work. Our challenge for the future remains to identify the various etiologies, provide prenatal counseling for affected families with a history of hypospadias, and minimize or eliminate exposure to environmental agents that may contribute to this problem. Perhaps one day we will be able to offer prenatal therapy to prevent hypospadias when the risk for this birth defect seems high.

How might this be possible? Consider the modern management of a family with a child born with the adrenogenital syndrome, another endocrine derangement that can cause abnormal genital development. In this situation, dexamethasone can be administered to the mother in subsequent pregnancies to prevent fetal virilization until the sex of the fetus can be determined or adrenal enzyme mutations can be excluded. Perhaps in the future a similar approach will be taken for those families with strong risk factors for hypospadias.

REFERENCES

Aaronson, I. A., Cakmak, M. A., Key, L. L., 1997, Defects of the testosterone biosynthetic pathway in boys with hypospadias. *Journal of Urology*, **157**: 1884-1888.

Aarskog, D., 1979, Maternal progestins as a possible cause of hypospadias. *New England Journal of Medicine*, **300**: 75-78.

Aho, M., Koivisto, A. M., Tammela, T. L. et al., 2000, Is the incidence of hypospadias increasing? Analysis of Finnish hospital discharge data 1970-1994 *Environmental Health Perspectives*, **108**: 463-465.

Allen, T. D., Griffin, J. E., 1984, Endocrine studies in patients with advanced hypospadias. *Journal of Urology*, **131**: 310-314.

Allera, A., Herbst, M. A., Griffin, J. E. et al., 1995, Mutations of the androgen receptor coding sequence are infrequent in patients with isolated hypospadias. Journal of Clinical Endocrinology & Metabolism, **80**: 2697-2699.

Angerpointner, T. A., 1984, Hypospadias--genetics, epidemiology and other possible aetiological influences. *Zeitschrift fur Kinderchirurgie*, **39**: 112-118.

Austin, P. F., Siow, Y., Fallat, M. E. et al., 2002, The relationship between mullerian inhibiting substance and androgens in boys with hypospadias. *Journal of Urology*, **168**: 1784-1788.

Baskin, L. S., 2000, Hypospadias and urethral development. *Journal of Urology*, **163**: 951-956.

Batch, J. A., Evans, B. A., Hughes, I. A. et al., 1993 Mutations of the androgen receptor gene identified in perineal hypospadias. *Journal of Medical Genetics*, **30**: 198-201.

Battin, M., Albersheim, S., Newman, D., 1995, Congenital genitourinary tract abnormalities following cocaine exposure in utero. *American Journal of Perinatology*, **12**: 425-428.

Bauer, S. B., Bull, M. J., Retik, A. B., 1979, Hypospadias: a familial study. *Journal of Urology*, **121**: 474-477.

Bauer, S. B., Retik, A. B., Colodny, A. H., 1981, Genetic aspects of hypospadias. *Urologic Clinics of North America*, **8**: 559-564.

Belman, A. B., 2002, Hypospadias and Chordee. In: Clinical Pediatric Urology, 4th ed. Edited by A. B. Belman, L. R. King, S. A. Kramer. London: Martin Dunitz Ltd., pp. 1061-1092.

Bentvelsen, F. M., Brinkmann, A. O., van der Linden, J. E. et al., 1995, Decreased immunoreactive androgen receptor levels are not the cause of isolated hypospadias. *British Journal of Urology*, **76**: 384-388.

Berman, D. M., Tian, H., Russell, D. W., 1995, Expression and regulation of steroid 5 alpha-reductase in the urogenital tract of the fetal rat. *Molecular Endocrinology*, **9**: 1561-1570.

Boehmer, A. L., Nijman, R. J., Lammers, B. A. et al., 2001, Etiological studies of severe or familial hypospadias. *Journal of Urology*, **165**: 1246-1254.

Briggs, M. H., 1982, Hypospadias, androgen biosynthesis, and synthetic progestogens during pregnancy. *International Journal of Fertility*, **27**: 70-72.

Calzolari, E., Contiero, M. R., Roncarati, E. et al., 1986, Aetiological factors in hypospadias. *Journal of Medical Genetics*, **23**: 333-337.

Colborn, T., Dumanoski, D., Myers, J. P., 1997, Our Stolen Future: Are We Threatening Our Fertility, Intelligence, and Survival? - A Scientific Detective Story. New York: Penguin, p. 336.

Correy, J. F., Newman, N. M., Collins, J. A. et al., 1991, Use of prescription drugs in the first trimester and congenital malformations. *Australian & New Zealand Journal of Obstetrics & Gynaecology*, **31**: 340-344.

Coulam, C. B., Razel, A. J., Kelalis, P. P. et al., 1983, Androgen receptor in human foreskin. II. Characterization of the receptor from hypospadiac tissue. *American Journal of Obstetrics & Gynecology*, **147**: 513-520.

Crocoll, A., Zhu, C. C., Cato, A. C. et al., 1998, Expression of androgen receptor mRNA during mouse embryogenesis. *Mechanisms of Development*, **72**: 175-178.

Czeizel, A., 1985, Increasing trends in congenital malformations of male external genitalia. *Lancet*, **1**: 462-463.

Czeizel, A., Toth, J., 1990, Correlation between the birth prevalence of isolated hypospadias and parental subfertility. *Teratology*, **41**: 167-172.

Dolk, H., 1998, Rise in prevalence of hypospadias. *Lancet*, **351**: 770.

Editorial: Genetics of hypospadias. *British Medical Journal*, **4**: 189-190, 1972.

Evans, B. A., Williams, D. M., Hughes, I. A., 1991, Normal postnatal androgen production and action in isolated micropenis and isolated hypospadias. *Archives of Disease in Childhood*, **66**: 1033-1036.

Feyaerts, A., Forest, M. G., Morel, Y. et al., 2002, Endocrine screening in 32 consecutive patients with hypospadias. *Journal of Urology*, **168**: 720-725.

Frydman, M., Greiber, C., Cohen, H. A., 1985, Uncomplicated familial hypospadias: evidence for autosomal recessive inheritance. *American Journal of Medical Genetics*, **21**: 51-60.

Gallentine, M. L., Morey, A. F., Thompson, I. M., Jr., 2001, Hypospadias: a contemporary epidemiologic assessment. *Urology*, **57**: 788-790.

Gearhart, J. P., Linhard, H. R., Berkovitz, G. D. et al., 1988, Androgen receptor levels and 5 alpha-reductase activities in preputial skin and chordee tissue of boys with isolated hypospadias. *Journal of Urology*, **140**: 1243-1246.

Gearhart, J. P., Donohoue, P. A., Brown, T. R. et al., 1990, Endocrine evaluation of adults with mild hypospadias. *Journal of Urology*, **144**: 274-277.

Gray, L. E., Jr., Ostby, J. S., Kelce, W. R., 1994, Developmental effects of an environmental antiandrogen: the fungicide vinclozolin alters sex differentiation of the male rat. *Toxicology & Applied Pharmacology*, **129**: 46-52.

Harris, E. L., Beaty, T. H., 1993, Segregation analysis of hypospadias: a reanalysis of published pedigree data. *American Journal of Medical Genetics*, **45**: 420-425.

Hiort, O., Klauber, G., Cendron, M. et al., 1994, Molecular characterization of the androgen receptor gene in boys with hypospadias. *European Journal of Pediatrics*, **153**: 317-321.

Janerich, D. T., Piper, J. M., Glebatis, D. M., 1980Oral contraceptives and birth defects. *American Journal of Epidemiology*, **112**: 73-79.

Kallen, B., Bertollini, R., Castilla, E. et al., 1986, A joint international study on the epidemiology of hypospadias. *Acta Paediatrica Scandinavica Supplement*, **324**: 1-52.

Kallen, B., 1988, Case control study of hypospadias, based on registry information. *Teratology*, **38**: 45-50.

Kallen, B., Castilla, E. E., Kringelbach, M. et al., 1991, Parental fertility and infant hypospadias: an international case-control study. *Teratology*, **44**: 629-634.

Kallen, B., Mastroiacovo, P., Lancaster, P. A. et al., 1991, Oral contraceptives in the etiology of isolated hypospadias. *Contraception*, **44**: 173-182.

Kalloo, N. B., Gearhart, J. P., Barrack, E. R., 1993, Sexually dimorphic expression of estrogen receptors, but not of androgen receptors in human fetal external genitalia. *Journal of Clinical Endocrinology & Metabolism*, **77**: 692-698.

Kennedy, W. A., Snyder, H. M., 1999, Paediatric andrology: the impact of environmental pollutants. *BJU International*, **83**: 195-200.

Kim, K. S., Liu, W., Cunha, G. R. et al., 2002, Expression of the androgen receptor and 5 alpha-reductase type 2 in the developing human fetal penis and urethra. *Cell & Tissue Research*, **307**: 145-153.

Kristensen, P., Irgens, L. M., Andersen, A. et al., 1997, Birth defects among offspring of Norwegian farmers, 1967-1991. *Epidemiology*, **8**: 537-544.

Levine, A. C., Wang, J. P., Ren, M. et al., 1996, Immunohistochemical localization of steroid 5 alpha-reductase 2 in the human male fetal reproductive tract and adult prostate. *Journal of Clinical Endocrinology & Metabolism*, **81**: 384-389.

Louik, C., Mitchell, A. A., Werler, M. M. et al., 1987, Maternal exposure to spermicides in relation to certain birth defects. *New England Journal of Medicine*, **317**: 474-478.

Macnab, A. J., Zouves, C., 1991, Hypospadias after assisted reproduction incorporating in vitro fertilization and gamete intrafallopian transfer. *Fertility & Sterility*, **56**: 918-922.

Matlai, P., Beral, V., 1985, Trends in congenital malformations of external genitalia. *Lancet*, **i**: 108.

Mau, G., 1981, Progestins during pregnancy and hypospadias. *Teratology*, **24**: 285-287.

McIntosh, G. C., Olshan, A. F., Baird, P. A., 1995, Paternal age and the risk of birth defects in offspring. *Epidemiology*, **6**: 282-288.

Muroya, K., Sasagawa, I., Suzuki, Y. et al., 2001, Hypospadias and the androgen receptor gene: mutation screening and CAG repeat length analysis. *Molecular Human Reproduction*, **7**: 409-413.

Nonomura, K., Fujieda, K., Sakakibara, N. et al., 1984, Pituitary and gonadal function in prepubertal boys with hypospadias. *Journal of Urology*, **132**: 595-598.

Okamoto, E., 1992, A study of quantitative and qualitative abnormality of androgen receptor in patients with hypospadias associated with enlarged prostatic utricle. *Japanese Journal of Urology*, **83**: 1593-1599.

Ostby, J., Monosson, E., Kelce, W. R. et al., 1999, Environmental antiandrogens: low doses of the fungicide vinclozolin alter sexual differentiation of the male rat. *Toxicology & Industrial Health*, **15**: 48-64.

Ostby, J., Kelce, W. R., Lambright, C. et al., 1999, The fungicide procymidone alters sexual differentiation in the male rat by acting as an androgen-receptor antagonist in vivo and in vitro. *Toxicology & Industrial Health*, **15**: 80-93.

Page, L. A., 1979, Inheritance of uncomplicated hypospadias. *Pediatrics*, **63**: 788-790.

Paulozzi, L. J., Erickson, J. D., Jackson, R. J., 1997, Hypospadias trends in two US surveillance systems. *Pediatrics*, **100**: 831-834.

Pelletier, G., Labrie, C., Labrie, F., 2000, Localization of oestrogen receptor alpha, oestrogen receptor beta and androgen receptors in the rat reproductive organs. *Journal of Endocrinology*, **165**: 359-370.

Polednak, A. P., Janerich, D. T., Glebatis, D. M., 1982, Birth weight and birth defects in relation to maternal spermicide use. *Teratology*, **26**: 27-38.

Polednak, A. P., Janerich, D. T., 1983, Maternal characteristics and hypospadias: a case-control study. *Teratology*, **28**: 67-73.

Raboch, J., Pondelickova, J., Starka, L., 1976, Plasma testosterone values in hypospadiacs. *Andrologia*, **8**: 255-258.

Roberts, C. J., Lloyd, S., 1973, Observations on the epidemiology of simple hypospadias. *British Medical Journal*, **1**: 768-770.

Shima, H., Ikoma, F., Yabumoto, H. et al., 1986, Gonadotropin and testosterone response in prepubertal boys with hypospadias. *Journal of Urology*, **135**: 539-542.

Shima, H., Yabumoto, H., Okamoto, E. et al., 1992a, Testicular function in patients with hypospadias associated with enlarged prostatic utricle. *British Journal of Urology*, **69**: 192-195.

Shima, H., Okamoto, E., Ikoma, F., 1992b, Pituitary and gonadal functions in patients with chordee without hypospadias. *International Urology & Nephrology*, **24**: 69-73.

Silver, R. I., Russell, D. W., 1999, 5alpha-reductase type 2 mutations are present in some boys with isolated hypospadias. *Journal of Urology*, **162**: 1142-1145.

Silver, R. I., Rodriguez, R., Chang, T. S. et al., 1999 In vitro fertilization is associated with an increased risk of hypospadias. *Journal of Urology*, **161**: 1954-1957.

Skakkebaek, N. E., Rajpert-De Meyts, E., Main, K. M., 2001, Testicular dysgenesis syndrome: an increasingly common developmental disorder with environmental aspects. *Human Reproduction*, **16**: 972-978.

Skakkebaek, N. E., 2002, Endocrine disrupters and testicular dysgenesis syndrome. *Hormone Research*, **57**: 43.

Stoll, C., Alembik, Y., Roth, M. P. et al., 1990, Genetic and environmental factors in hypospadias. *Journal of Medical Genetics*, **27**: 559-563.

Sultan, C., Balaguer, P., Terouanne, B. et al., 2001, Environmental xenoestrogens, antiandrogens and disorders of male sexual differentiation. *Molecular & Cellular Endocrinology*, **178**: 99-105.

Sutherland, R. W., Wiener, J. S., Hicks, J. P. et al., 1996, Androgen receptor gene mutations are rarely associated with isolated penile hypospadias. *Journal of Urology*, **156**: 828-831.

Svensson, J., Snochowski, M., 1979, Androgen receptor levels in preputial skin from boys with hypospadias. *Journal of Clinical Endocrinology & Metabolism*, **49**: 340-345.

Teixeira, J., Fynn-Thompson, E., Payne, A. H. et al., 1999, Mullerian-inhibiting substance regulates androgen synthesis at the transcriptional level. *Endocrinology*, **140**: 4732-4738.

Terakawa, T., Shima, H., Yabumoto, H. et al., 1990, Androgen receptor levels in patients with isolated hypospadias. *Acta Endocrinologica*, **123**: 24-29.

Thigpen, A. E., Silver, R. I., Guileyardo, J. M. et al., 1993, Tissue distribution and ontogeny of steroid 5 alpha-reductase isozyme expression. *Journal of Clinical Investigation*, **92**: 903-910.

Tian, H., Russell, D. W., 1997, Expression and regulation of steroid 5 alpha-reductase in the genital tubercle of the fetal rat. *Developmental Dynamics*, **209**: 117-126.

Walsh, P. C., Curry, N., Mills, R. C. et al., 1976, Plasma androgen responce to hCG stimulation in prepubertal boys with hypospadias and cryptorchidism. *Journal of Clinical Endocrinology & Metabolism*, **42**: 52-59.

Weidner, I. S., Moller, H., Jensen, T. K. et al., 1998, Cryptorchidism and hypospadias in sons of gardeners and farmers. *Environmental Health Perspectives*, **106**: 793-796.

Welch, K., 1979, Hypospadias. In: Pediatric Surgery. Edited by M. Ravitch, K. Welch, C. Benson et al. Chicago: Year Book Medical Publishers, pp. 1353-1376.

Wilson, J. D., Griffin, J. E., Russell, D. W., 1993, Steroid 5 alpha-reductase 2 deficiency. *Endocrine Reviews*, **14**: 577-593.

Zhou, L., Mei, H., Liu, T. et al., 1999, Identification of mutations of SRD5A2 gene and SRY gene in patients with hypospadias. *Chung-Hua i Hsueh i Chuan Hsueh Tsa Chih*, **16**: 311-314.

GENETIC AND CLINICAL STUDIES ON HYPOSPADIAS

Agneta Nordenskjöld*

1. INTRODUCTION

Hypospadias, one of the most common malformations in man, is characterized by an abnormal position of the urethral meatus and different degrees of ventral curvature of the penis. Hypospadias arises due to a defect in the urethral development during fetal week 8 to 16 causing the meatus to be located on the underside of the penis. Earlier disturbance of the urethral development will result in a more severe hypospadias (figure 1a and b). In severe cases the penis is curved, short and with the urethral opening in the scrotum or perineum. The most severe cases are boys born with ambiguous genitalia. The registered incidence in Sweden according to the Swedish Malformation Registry is 1:500 boys, but the malformation is probably underreported and the estimated incidence is likely to be about 1:300 (Källén and Winberg 1982).

Figure 1a

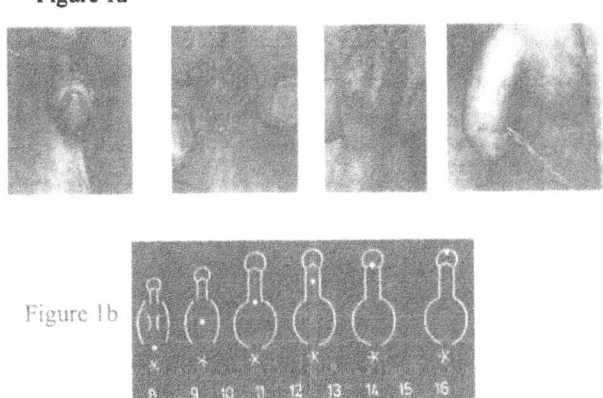

Figure 1a. Different severity of hypospadias as a continuum (perineal, scrotal, penile and glandular hypospadias).
Figure 1b. The male urethral development during fetal weeks 8-16. The urethral folds are successively fused in the midline and the meatus is located on the glans after 16 weeks. The last event during fetal organogenesis is the fusion of the prepuce over glans.

* Department of Molecular Medicine, Karolinska Institutet, Stockholm, Sweden, Ph# +46-8-5177 6408, Fax# +46-8-5177 3620, e:mail, agneta.nordenskjold@cmm.ki.se

Hypospadias and Genital Development, edited by
L. Baskin, Kluwer Academic/Plenum Publishers, 2004

2. HYPOSPADIAS AND GENETICS

There are several factors speaking in favor of genetic involvement in the pathogenesis of hypospadias. A familial aggregation is found in between 4 and 25% of hypospadiac cases (Sørensen 1953, Sweet et al., 1974, Czeizel et al., 1979, Bauer et al., 1981, Monteleone et al., 1981) and a few families show an autosomal dominant inheritance pattern for hypospadias (Lowry and Kliman 1976, Page 1979). In ethnic groups with a high degree of consanguinity a recessive mode of inheritance can be observed (Frydman et al., 1985, Tsur et al., 1987). Furthermore, hypospadias is part of more than one hundred genetic syndromes (McKusick 2002). Several single gene traits can cause this malformation, including the partial androgen insensitivity syndrome, testosterone 5-α reductase deficiency and Drash syndrome (Wilson et al., 1993, Quigley et al., 1995, Pelletier et al., 1991). However the majority of cases are sporadic and several lines of evidence indicate a multifactorial etiology with contribution of genetic and environmental factors. The recurrence risk for a son to a father with hypospadias is 5% and the recurrence risk for the second brother in a family with one affected boy is 10% (Chen and Wolley 1971, Bauer et al., 1981, Källén 1986, Stoll 1990). If two first-degree relatives have hypospadias (i.e. father and one son) the recurrence risk for the next boy is as high as 25%.

The degree of genetic contribution can be evaluated by calculating the "relative risk" (Risch 1990). The recurrence risk for a brother to a hypospadiac boy is about 10% as compared to incidence in the general population of about 0.3. The relative risk is then $10/0.3 = 30$, which speaks in favor of a strong genetic influence. The heritabilty of hypospadias has been estimated to be 0.7, which means that 70% of the phenotypic variance is due to the additive effect of many genes (Haines and Pericak-Vance 1998, Chen and Wolley 1971, Czeizel et al., 1979, Monteleone et al 1981, Calzolari 1986, Stoll 1990). Taken together, hypospadias is considered a complex disorder, characterized by a familial aggregation but no simple Mendelian inheritance pattern. Such disorders are in most cases caused by an interaction between genes and environmental factors.

3. Dissecting genetics in complex disorders

Although it is well established that genetic factors are major contributors to the etiology of hypospadias, the genes involved are still unknown. Different strategies can be used to find such genes. One method is to screen for mutations in candidate genes. Candidate genes may have a specific expression pattern in the affected tissue during the critical embryonic stage. In hypospadias, this approach has been attempted for several genes known to be of importance in sexual differentiation. One example is the androgen receptor gene (AR). Several hundred different mutations in the AR gene have been demonstrated in partial androgen insufficiency but in only one case of isolated hypospadias (Allera et al., 1995, Hiort et al., 1994, Nordenskjöld et al., 1999). Another candidate gene is the testosterone 5-α reductase gene. This enzyme converts testosterone to dihydrotestosterone, a key hormone for the development of the male external genitalia. The recessively inherited 5-α reductase deficiency results in a more severe phenotype than isolated hypospadias (Nordenskjöld et al 1999). Recently, a few heterozygous carriers with hypospadias were reported (Silver and Russel 1999).

A second approach to search for disease genes is linkage analysis in large families with hypospadias transmitted as a dominant trait. "Linkage" refers to the fact that genetic markers in the same chromosomal region as a disease gene are linked, i.e. are inherited together. Finding a linked marker points at the location of the disease gene and gives an opportunity to characterize the molecular background for the particular malformation in that particular family. When the disease gene is identified the gene can be studied in order to determine if the specific mechanism is shared. However, the drawback is that hypospadias families suitable for linkage analysis are rare and therefore not easy to ascertain.

An alternative to locate disease genes is to use sib pair analysis or affected relative pair analysis first proposed by Penrose (1935). This method is based on the assumption that affected sibs share the chromosomal segment involved in the disease more often than expected. It has been successful in mapping genes involved in other complex disorders like diabetes, mb Crohn and asthma (Horikawa et al., 2000, Hugot et al., 2001, Ogura et al., 2001, Van Eerdewegh et al., 2002). In sib pair analysis, a genome-wide linkage screening with polymorphic markers is performed. The method is based on allele sharing, which means that deviations from the expected Mendelian distribution of alleles are calculated. In Figure 2 the basic idea of sib pair analysis is shown. This pedigree of a family with two affected sibs shows how four alleles of one polymorphic marker on any chromosome from the parents can be inherited among their children. According to the Mendelian law, 25% the affected sibs share two alleles by chance, 25% they do not share any allele and half of the cases the share one allele by chance. The sib pair analysis will then statistically show chromosomal regions where affected relative pairs are observed to share alleles more often than expected from this expected distribution (1:2:1). The informativity of the analysis is improved by including genotyping of the parents since you then can differ between identical by descent and identical by state. Genotyping has to be performed for several hundred markers evenly distributed over the whole genome and in many families with more than one affected member.

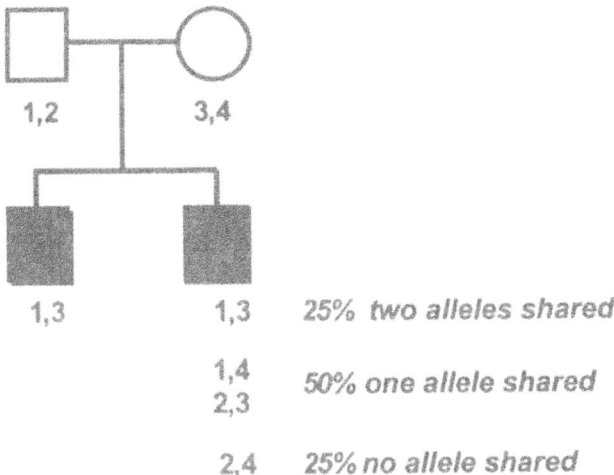

Figure 2. The pedigree with a family with two affected brothers illustrates the Mendelian inheritance of different alleles of an informative marker.

The chance of success with sib pair analysis depends on several factors. Generally around 350 polymorphic markers with about 10cM marker distance has to be analyzed. The number of sib pairs is also crucial. It is impossible to know in advance how many affected pairs that are needed since the underlying genetic model is unknown. In addition the number of susceptibility loci and their relative risk is not known. Fewer than 40 sip pairs are unlikely to give significant results. Usually several hundred sibs are collected for analysis. To some extent a small sample size can be compensated with a denser marker map. Other factors that can reduce the chance for success in the analysis are reduced penetrance, phenocopies and genetic heterogeneity.

The affected relative pair analysis gives results depicted in Figure 3. The chromosome, represented by its polymorphic markers is placed along the X-axis. The result is presented as a likelihood of a candidate gene located in the respective chromosomal region on the Y-axis. The likelihood of a candidate gene located in every region can in every specific study be calculated and expressed in terms of significance (Lander and Kruglyak 1995).

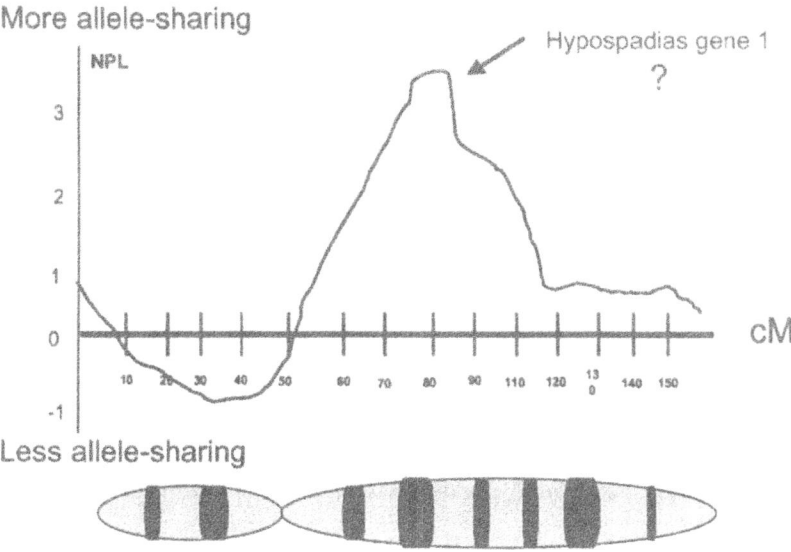

Figure 3: The figure illustrates the principle of sib pair analysis with the markers representing the chromosome on the x-axis and the relative allele-sharing on the y-axis. NPL- non parametric lod score.

4. SEARCHING FOR MECHANISMS CAUSING HYPOSPADIAS

In order to identify mechanisms causing hypospadias we have studied both genetic and non-genetic factors. The basis for this whole project is a large nation-wide material consisting of 2503 cases of hypospadias in Sweden (Fredell et al., 2002). This represents half of the registered number of hypospadias in Sweden. These patients are ascertained from departments of Pediatric Surgery in Stockholm, Uppsala, Gothenburg and Lund and from departments of Plastic Surgery and Urology in Örebro and Umeå. Questionnaires were sent to these families asking for additional family members with hypospadias. We have also obtained information from Statistics Sweden, on birth weight, ethnic origin, number of brothers and their birth weight. We specially collected both concordant and discordant twins. As many as possible of the hypospadias boys were phenotyped from hospital records. The Ethics Committee at all participating hospitals has approved the study.

4.1 Recurrence risk

The final reply rate of the questionnaire was 88%. Initially 202 families responded a positive family history. After telephone interviewing the fraction of cases with at least one affected relative was 7%. The families were predominantly small with two affected (112 families) but 24 families had 3 affected and in 8 families between 4 and 11 cases had hypospadias (Fredell et al., 2002). In this material the recurrence risk for a brother was 6%. The phenotypes were registered in 27% of the whole material and 6 % had a severe form of hypospadias (perineal, scrotal and penoscrotal forms), 39% were penile, 53% glandular and 2% had cleaved prepuce (a mild variant of hypospadias). Familial clustering was more common in milder forms, which is probably due to a moderately reduced fertility in the more severe forms. The focus was on non-syndromic hypospadias cases but some associated malformations were found to be over-represented in this material. Sixteen boys also had anal atresia, limb or cardiac malformations or cleft lip palate.

4.2 Birth weight

We compared the birth weight between the boy with hypospadias and their respective brother, which gives a highly significant difference. The mean birth weight for affected boys is 3.322 kg as compared with the brothers of 3.485 kg ($p = 5 \times 10^{-13}$). We also compared birth weight in four different gestational groups and the results were highly significant in the groups older than 30 gestational weeks.

4.3 Ethnicity

Currently, 19% of the Swedish population has one or both parents with a non-Swedish origin according to Statistics, Sweden. In this material we observed 22% with a non-Swedish parent. Hypospadias appear to be over-represented in boys with a Middle Eastern descent. In this material 6% were from the Middle East and in that group 22% of the cases were familial.

4.4 Twin studies

In order to outline non-genetic factors we studied monozygotic twins discordant for hypospadias. A total of 40 male-male twin pairs were ascertained as compared with the expected number 9 (1/80 pregnancies). The twins were monozygotic in two thirds of the cases and dizygotic in one third. Usually the number of monozygotic and dizygotic twins is equal, indicating a skewed relationship with an overrepresentation of monozygotic twins in this material. The concordance rate in monozygotic twins was 27% and in dizygotic twins 9%, which support the notion that genetic factors are important. During the initial sampling we noticed that among discordant twins, it was the smallest that had hypospadias. In the initial sampling we obtained 28 twin pairs of which 18 were monozygotic and discordant. Zygosity was confirmed by DNA fingerprinting in 15 twin pairs and we relied on placenta examination or earlier performed fingerprinting in 3 cases. In 16 of the 18 pairs it was the smallest boy that had hypospadias, in one pair they were of the same size and in one pair the boy with higher birth weight was affected. On average the birth weight of the smallest boy was 78% of the larger brother corresponding to a difference of 498 grams (p<0.01). Since these boys have identical genes we concluded that an environmental factor associated with birth weight is important in the pathogenesis of hypospadias. The factor may be a humeral factor like human chorionic gonadotropin (hCG) that diffuses to the two fetuses in different amounts due to different size or function of the placenta.

4.5 Family studies

In order to locate and identify genes predisposing to hypospadias we performed family studies of two kinds. Larger families with an autosomal dominant mode of inheritance were analyzed by linkage analysis while smaller families were studied by affected relative pair analysis. The pedigree of the largest autosomal dominant family with hypospadias is shown in Figure 4.

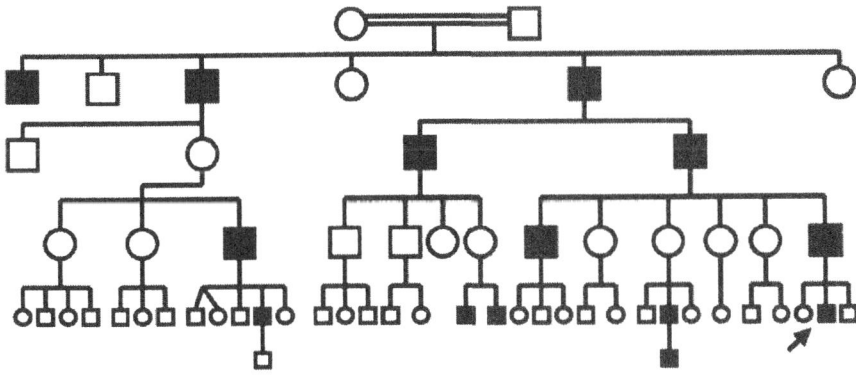

Figure 4: A large family with autosomal dominant inherited hypospadias in five generations.

We have performed a genome wide linkage study on 29 available persons altogether, among which 15 were affected. We recently observed linkage to markers on chromosome 7 and have identified the specific mutation in this family (Frisén et al, submitted). One additional family contains eight boys with hypospadias (Figure 5). The linkage analysis is not yet completed in this family.

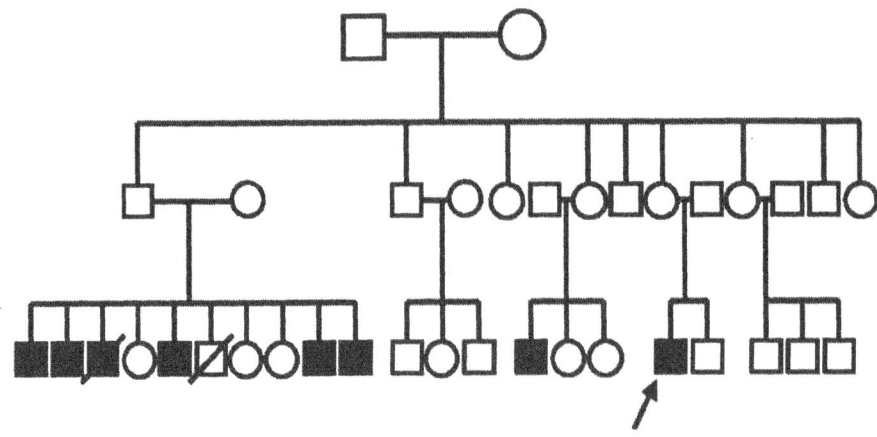

Figure 5. The pedigree of a family with 8 boys with hypospadias.

4.6 Sib pair analysis

One hundred and fifty of the original 200 families approved to participate by donating blood for DNA studies. Some of the families remained incomplete and we finally obtained blood from 69 families (83 affected relative pairs) for sib pair analysis. Isolation of DNA was performed according to standard protocols. Altogether, we have genotyped 309 individuals of which 153 were affected. We used 360 markers spread over the whole genome with 9.5 cM between. Marker heterozygosity was about 0,78 (Weber set). Fluorescent-labeled markers in different colors permit simultaneously allele scoring by gel electrophoresis with the ABI system. Data were interpreted by software program GENOTYPER and statistical analysis was performed with the GENEHUNTER software.

The preliminary result after analyzing half of the material is shown for some chromosomes (Figure 6).

The statistical data does not support hypospadias genes on any of the chromosomes shown. On some chromosomes the markers are too distant, or there were technical problems with the PCR for some markers and thus sufficient informativity was not obtained. The result from three chromosomes is however more promising (Figure 7).

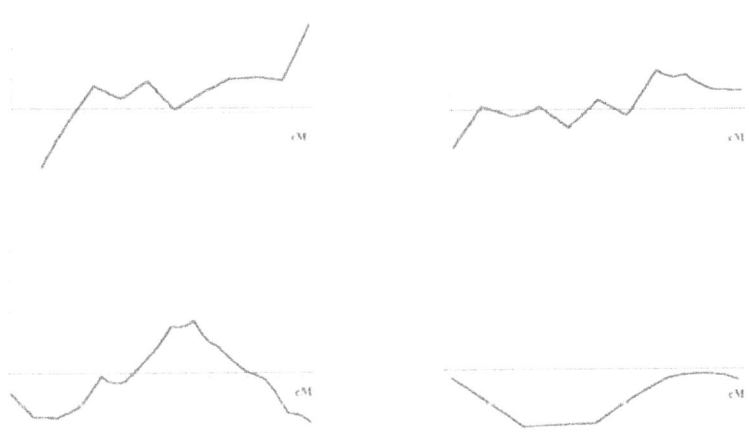

Figure 6. Sib pair analysis on some chromosomes with "negative" result. Horizontal lines indicates p>0.05.

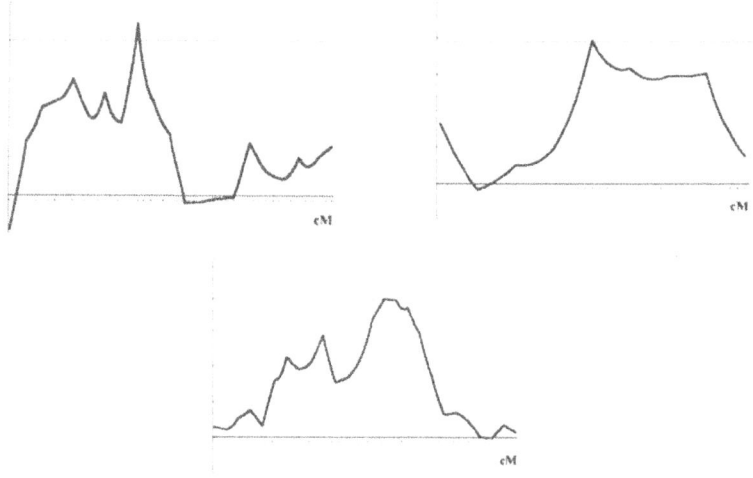

Figure 7: On these three chromosomes the sib pair analysis gives chromosomal regions that could harbor hypospadias genes. Horizontal lines indicates p>0.05.

The statistical evidence is unfortunately so far to weak to draw conclusions after analysis of only half of the material. In addition, every single "peak" still contains many genes that are candidate genes only due to their chromosomal location. In general only a few of the genes have been sufficiently studied for function or expression pattern in order to make them good candidate genes. Unfortunately, the vast majority of genes are not previously studied sufficiently enough in order to justify more extensive studies of mutations in patients.

4.7 Future plans

The sib pair analysis will be completed shortly. Since the "candidate regions" still are extensive these need to be studied with more densely located markers. In order to obtain a higher power in the sib pair analysis we will also aim at including additional families. In the future two main strategies can be used, first obviously mutational screening of candidate genes and secondly further characterization of novel genes in the chromosomal candidate regions. Hopefully these studies will enlighten the etiology of hypospadias.

REFERENCES

Allera, A., Herbs,t M.A., Griffin, J.E., Wilson, J.D., Schweikert, H.U., McPhaul, M.J., 1995, Mutations of the androgen receptor coding sequence are infrequent in patients with isolated hypospadias, *J Clin Endocrinol Metab*. **80**: 2697.

Bauer, S.B., Retik, A.B., Colodny, A.H., 1981, Genetic aspects of hypospadias., *Urol Clin North Am*. **8**: 559.

Calzolari ,E., Contiero, M.R., Roncarati, E., Mattiuz, P.L., Volpato S., 1986, Aetiological factors in hypospadias, *J Med Genet*. **23**: 333.

Chen, Y.C., Woolley, P.V., Jr., 1971, Genetic studies on hypospadias in males, *J Med Genet*. **8**: 153.

Czeizel, A., Toth, J., Erodi, E., 1979, Aetiological studies of hypospadias in Hungary, *Hum Hered*. **29**: 166.

Fredell L., Lichtenstein,P., Pedersen N.L.,, Svensson J., Nordenskjöld A., 1998, Hypospadias is related to birth weight in discordant monozygotic twins, *J Urol*. **160**: 2197.

Fredell, L., Kockum, I., Hansson, E., Holmner, S., Lundquist, L., Lackgren, G., Pedersen, J., Stenberg, A., Westbacke, G., Nordenskjold, A., 2002, Heredity of hypospadias and the significance of low birth weight, *J Urol* **167**: 1423.

Frydman, M., Greiber, C., Cohen, H.A., 1985, Uncomplicated familial hypospadias: evidence for autosomal recessive inheritance, *Am J Med Genet*. **21**: 51.

Haines, J.L., Pericak-Vance, M.A., 1998, *Approaches to Gene Mapping in Complex Human Diseases*, Wiley-Liss, New York.

Hiort, O., Klauber, G., Cendron, M., Sinnecker, G.H., Keim, L., Schwinger, E., Wolfe, H.J., Yandell, D.W., 1994, Molecular characterization of the androgen receptor gene in boys with hypospadias, *Eur J Pediatr*. **153**: 317.

Horikawa, Y., Oda, N., Cox, N.J., Li ,X., Orho-Melander, M., Hara, M., Hinokio, Y., Lindner, T.H., Mashima, H., Schwarz, P.E., del Bosque-Plata, L., Oda, Y., Yoshiuchi, I., Colilla, S., Polonsky, K.S., Wei, S., Concannon, P., Iwasaki, N., Schulze, J., Baier, L.J., Bogardus, C., Groop, L., Boerwinkle, E., Hanis, C.L., Bell, G.I., 2000, Genetic variation in the gene encoding calpain-10 is associated with type 2 diabetes mellitus, *Nat Genet*. **26**: 163.

Hugot, J.P., Chamaillard, M., Zouali, H., Lesage, S., Cezard, J.P., Belaiche, J., Almer, S., Tysk, C., O'Morain, C.A., Gassull, M., Binder ,V., Finkel ,Y., Cortot ,A., Modigliani, R., Laurent-Puig, P., Gower-Rousseau, C., Macry, J., Colombel, J.F., Sahbatou, M., Thomas, G., 2001, Association of NOD2 leucine-rich repeat variants with susceptibility to Crohn's disease, *Nature* **411**: 599.

Källén ,B., Winberg, J., 1982, An epidemiological study of hypospadias in Sweden, *Acta Paediatr Scand Suppl*. **293**: 1.

Källén, B., Bertollini, R., Castilla, E., Czeizel , A., Knudsen, L.B., Martinez-Frias, M.L., Mastroiacovo, P., Mutchinick, O., 1986, A joint international study on the epidemiology of hypospadias, *Acta Paediatr Scand Suppl* **324**: 1.

Lander, E., Kruglyak , L., 1995, Genetic dissection of complex traits: guidelines for interpreting and reporting linkage results, *Nat Genet*. **11**: 241.

Lowry, R.B., Kliman, M.R., 1976, Hypospadias in successive generations - possible dominant gene inheritance, *Clin Genet*. **9**: 285.

McKusick V.A., 2002, OMIM. http://www.ncbi.nlm.nih.gov/omim.

Monteleone Neto, R., Castilla, E.E., Paz, J.E., 1981, Hypospadias: an epidemiological study in Latin America, *Am J Med Genet*. **10**: 5.

Nordenskjöld, A., Friedman, E., Tapper-Persson, M., Söderhäll , C., Leviav, A., Svensson, J., Anvret, M., 1999, Screening for mutations in candidate genes for hypospadias, *Urol Res*. **27**: 49.

Ogura, Y., Bonen, D.K., Inohara, N., Nicolae, D.L., Chen, F.F., Ramos, R., Britton, H., Moran, T., Karaliuskas, R., Duerr, R.H., Achkar, J.P., Brant, S.R., Bayless, T.M., Kirschner, B.S., Hanauer, S.B., Nunez ,G., Cho, J.H., 2001, A frameshift mutation in NOD2 associated with susceptibility to Crohn's disease, *Nature* **411**: 603.

Page, L.A., 1979, Inheritance of uncomplicated hypospadias, *Pediatrics.* **63**: 788.

Pelletier, J., Bruening, W., Kashtan, C.E., Mauer, S.M., Manivel, J.C., Striegel, J.E., Houghton, D.C., Junien, C., Habib, R., Fouser, L., Fine, R.N., Silverman, B.L., Haber, D.A., Housman, D., 1991, Germline mutations in the Wilms' tumor suppressor gene are associated with abnormal urogenital development in Denys-Drash syndrome, *Cell* **67**: 437.

Penrose, L.S., 1935, The detection of autosomal linkage in data which consist of pairs of brothers and sisters of unspecified parentage, *Annual Eugenics* **6**: 133.

Quigley, C.A., De Bellis, A., Marschke, K.B., el-Awady M.K., Wilson E.M., French F.S., 1995, Androgen receptor defects: historical, clinical, and molecular perspectives, *Endocr Rev,* **16**: 271.

Risch, N., 1990, Linkage strategies for genetically complex traits. I. Multilocus models, *Am J Hum Genet.* **46**: 222.

Silver, R.I., Russell, D.W., 1999, 5alpha-reductase type 2 mutations are present in some boys with isolated hypospadias, *J Urol.* **162**: 1142.

Stoll, C., Alembik ,Y., Roth, M.P., Dott, B., 1990, Genetic and environmental factors in hypospadias, *J Med Genet.* **27**: 559.

Sweet, R.A., Schrott ,H.G., Kurland,R., Culp, O.S., 1974, Study of the incidence of hypospadias in Rochester, Minnesota, 1940-1970, and a case-control comparison of possible etiologic factors, *Mayo Clin Poc.* **49**: 52.

Sørensen, H.R., 1953, Hypospadias with special reference to etiology. Thesis, Munksgaard, Copenhagen

Tsur, M., Linder, N., Cappis, S., 1987, Hypospadias in a consanguineous family, *Am J Med Genet.* **27**: 487.

Van Eerdewegh, P., Little, R.D., Dupuis, J., Del Mastro, R.G., Falls, K., Simon, J., Torrey, D., Pandit ,S., McKenny ,J., Braunschweiger ,K., Walsh, A., Liu, Z., Hayward, B., Folz, C., Manning S.P., Bawa A., Saracino L., Thackston M., Benchekroun Y., Capparell N., Wang, M., Adair, R., Feng, Y., Dubois, J., FitzGerald, M.G., Huang, H., Gibson, R., Allen, K.M., Pedan, A., Danzig, M.R., Umland, S.P., Egan, R.W., Cuss, F.M., Rorke, S., Clough, J.B., Holloway, J.W., Holgate, S.T., Keith, T.P., 2002, Association of the ADAM33 gene with asthma and bronchial hyperresponsiveness. *Nature* **418**: 426.

Wilson, J.D., Griffin, J.E., Russell, D.W., 1993, Steroid 5 alpha-reductase 2 deficiency, *Endocr Rev.* **14**: 577.

Mechanism of
Genital Development

DEVELOPMENT OF THE PENILE URETHRA

Gerald R. Cunha*, Laurence Baskin

1. INTRODUCTION

In humans hypospadias consists of an abnormal ventral   urethral meatus, incomplete formation of the prepuce and penile curvature (Baskin et al., 1998).   The cause of hypospadias is thought to be failure of formation or fusion of the urethral folds bounding the urethral groove.   While failure of urethral fold fusion may be the ultimate cause of hypospadias, there are many other possible causes of this malformation.   In this review, the development of the penile urethra will be described for both the human and the mouse. The purpose of this paper is to review the developmental anatomy of the external genitalia to provide a framework for subsequent chapters that will focus on mechanisms of penile development. The second goal of this review is to discuss the development of the external genitalia of the human and the mouse so as to evaluate the mouse as a relevant animal model for development of the external genitalia.

2. Development of the External Genitalia

During gastrulation the bi-layered embryo acquires a third layer through the migration of mesoderm cells through the primitive streak.   The consequence of gastrulation is the formation of the tri-laminar embryo consisting of ectoderm, endoderm and an intervening layer of mesoderm.   However, there are two areas in the embryo that remain bi-layered: the oropharyngeal membrane cranially and the cloacal membrane caudally (Carlson, 1994).   Both of these membranes consist of ectoderm in intimate contact with the underlying endoderm without an intervening layer of mesoderm.   The oropharyngeal membrane is the future site of the mouth, whereas the cloacal membrane is the future site of structures in the perineum such as urogenital ostium and the anus.

During the 4th week of development in the human the flattened tri-laminar embryo is converted through a series of complicated morphogenetic movements into a tubular embryo containing a primitive tubular gastrointestinal system subdivided into three zones: foregut, midgut and hindgut.   The hindgut terminates caudally in an expansion

---
* Department of Anatomy, University of California, San Francisco, CA 94143-0452, Fax = 415-502-2270, e-mail = grcunha@itsa.ucsf.edu

Hypospadias and Genital Development, edited by
L. Baskin, Kluwer Academic/Plenum Publishers, 2004

called the cloaca having a ventral diverticulum, the allantois extending up the anterior body wall to terminate blindly in the umbilical cord (Figure 1).

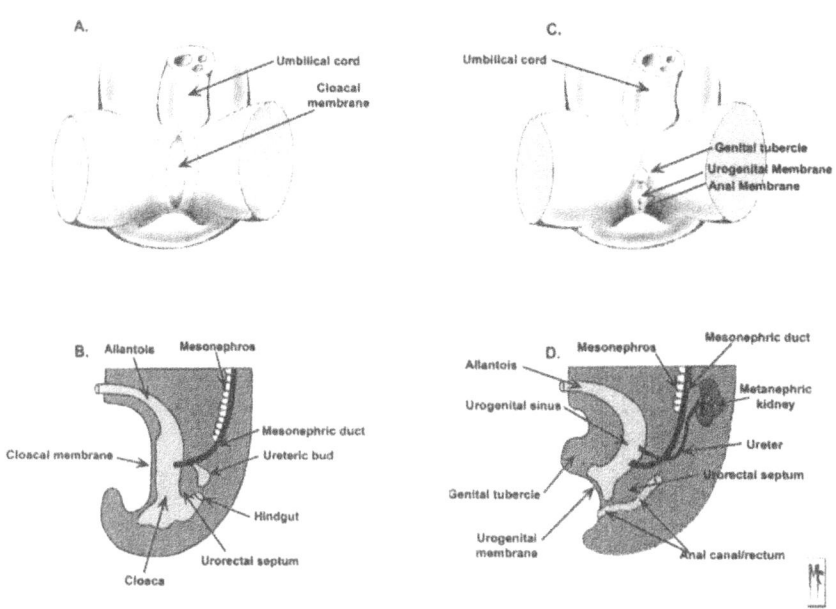

Figure 1. Development of the external genitalia. (A) Surface view of the perineum, anterior body wall and cloacal membrane of a 4-week human fetus. Note that the cloacal membrane extends from the perineum up the anterior body wall to the base of the umbilicus. (B) Mid-sagittal view through the pelvis of a 4-week human fetus showing the cloaca, the allantois and the urorectal septum. The cloaca is the blind caudal terminus of the hindgut and has a ventral diverticulum, the allantois, extending up the anterior body wall and terminating in the umbilical cord. The urorectal septum is growing caudally towards cloacal membrane to subdivide the cloaca. (C) Surface view of the perineum and anterior body wall of a 7-week human fetus. The cloacal membrane has "retracted" into the perineum and has been divided into the urogenital membrane and anal membrane. Note the genital tubercle immediately cranial to the urogenital membrane. (D) Mid-sagittal view through the pelvis of a 7-week human fetus showing division of the cloaca into the urogenital sinus and rectum and anal canal.

The cloacal membrane separates the cavity of the cloaca from the amniotic cavity. Initially the cloacal membrane is an extensive elongated midline structure whose cranial extremity approaches the root of the umbilical cord (Moore and Persaud, 1993). The caudal extremity of the cloacal membrane is located in the future perineum and is the site of the anus. During development, the bi-layered cloacal membrane "retracts" into the perineum as a result of cranial and medial migration of mesodermal cells into the anterior body wall between the ectoderm and endoderm of the cloacal membrane. This mesenchymal migration brings about closure of the inferior part of the anterior abdominal wall and causes the cloacal membrane to "retract" caudally into the perineal region (Brookes and Zietman, 1998; Moore and Persaud, 1993). These migrating mesodermal cells give rise to the musculature of the medial portion of the anterior abdominal wall, the anterior wall of the bladder, the pubic symphysis, and to the rudiments of the external genitalia. Failure of migration of these mesodermal cells into the midline results in exstrophy of the bladder and associated malformations: defect in the anterior abdominal wall, absence of the anterior wall of the bladder, defect in the pubic symphysis and epispadias, a sever malformation of the external genitalia (Langer, 1993; Mollohan, 1999; van der Werff et al., 2000; Vermeij-Keers et al., 1996).

Migrating mesenchymal cells spread around the cloacal membrane and pile up to form three swellings. One swelling, the genital tubercle, appears in the midline between the cloacal membrane and the umbilicus. Bi-lateral genital swellings appear on each side of the cloacal membrane (Brookes and Zietman, 1998; Moore and Persaud, 1993; Snell, 1975) (Figure 2). These external features are related internally to the caudal terminus of the gastrointestinal tract, the cloaca and its diverticulum the allantois (Figure 1). The cloaca becomes divided coronally by the urorectal septum into the urogenital sinus ventrally and the rectum and anal canal dorsally (Marshall, 1978). Accordingly, the cloacal membrane becomes subdivided into the urogenital membrane and the anal membrane (Figures 1 and 2).

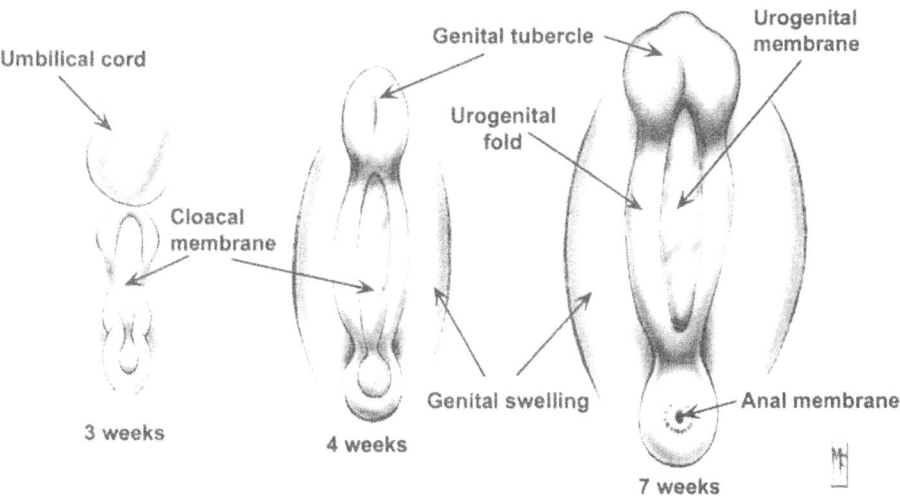

Figure 2. Detail of the cloacal membrane, its division and the development of the genital tubercle and genital swellings. At three weeks the cloacal membrane extends to the base of the umbilicus. Mesenchymal cells are migrating toward the midline to elicit "retraction" of the cloacal membrane. At four weeks, division of the cloacal membrane into the urogenital and anal membranes is nearly complete, and the genital tubercle and genital swellings are recognizable. At seven weeks the cloacal membrane is divided by the urorectal septum into the urogenital and anal membranes. The genital tubercle and genital swelling have formed, and the urogenital membrane is demarcated laterally by the urogenital folds.

When this occurs the urogenital membrane is bounded cranially by the genital tubercle and laterally by the urogenital folds and the genital swellings (Figure 2). Subsequently, the urogenital membrane (and anal membrane) breaks down to establish continuity between internal structures (urogenital sinus and anus) and the amnionic cavity.

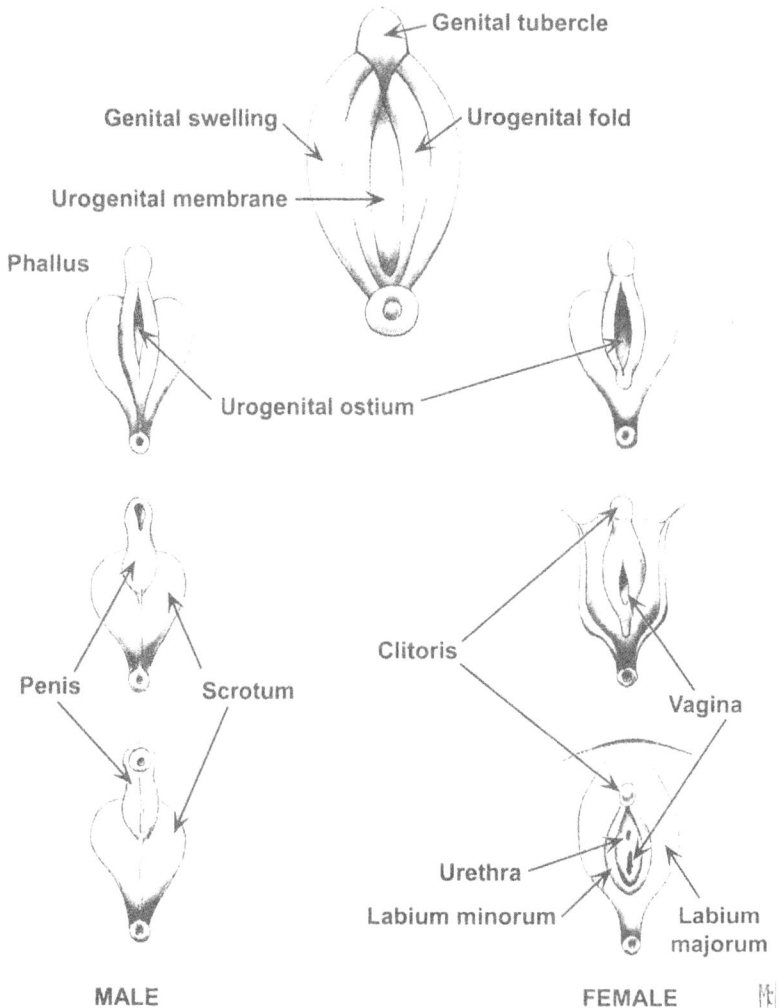

Figure 3. Sex differentiation of the external genitalia. The ambisexual stage (top center) consists of the genital tubercle, the urogenital folds bounding the urogenital ostium and the more laterally placed genital swellings. During male development (left), the genital tubercle elongates to form the penis, while the genital swellings migrate caudally and fuse in the midline to form the scrotum. As the phallus elongates the urethral groove forms on the ventral aspect of the genital tubercle. The urethral groove extends distally as a result of canalization of the solid urethral plate. Finally, the urethral folds bounding the urethral groove grow to the midline and fuse with each other thus forming a tubular urethra. During female development (right), only minor changes occur from the original ambisexual stage. The genital tubercle undergoes minimal growth to form the clitoris. The urogenital folds and genital swellings remain unfused to form the labia minora and majora, respectively. The urogenital folds (labial minora) define the vestibule of the vagina into which open the urethra and vagina.

Before 12 weeks of gestation in human fetuses the external genitalia of both male and female fetuses are identical and consist of the genital tubercle, the urogenital folds bounding the urogenital ostium and the more laterally situated genital swellings (Figures 2 and 3).

These undifferentiated structures in both male and female embryos constitute the ambisexual stage of sex differentiation of the external genitalia. The genital tubercle is the primordium of the penis in males and the clitoris in females. In females the urogenital folds and the genital swellings remain separate and form the labia minora and labia majora, respectively, while the genital tubercle forms the clitoris. In males the genital swellings migrate caudally and fuse in the midline to form the scrotum (Figure 3). As the genital tubercle elongates in males, a groove appears on its ventral aspect called the urethral groove (Figures 4 and 5).

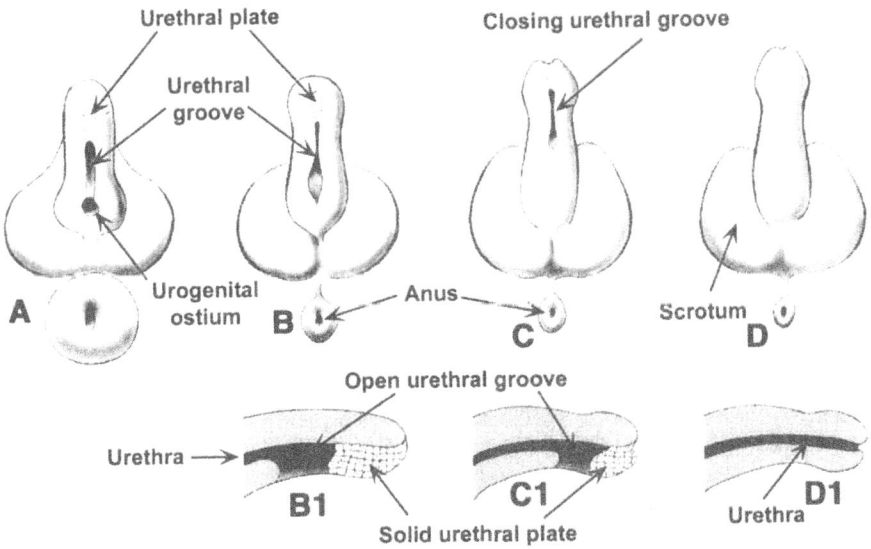

Figure 4. Detail of the urethral groove, urethral folds and urethral plate. In (A) at the beginning of the ambisexual stage the urethral groove bounded by the urethral folds extends about half way distally along the ventral aspect of the elongating genital tubercle. At the distal aspect of the urethral groove is the solid urethral plate (dotted lines) that extends to the glans of the developing penis. The urethral groove extends distally to the glans by canalization of the urethral plate (B-C). The urethral folds grow to the midline where they fuse to extend the penile urethra distally to the glans (B-D). B1, C1, D1 are mid-sagittal sections showing canalization of the urethral plate and formation of the penile urethra.

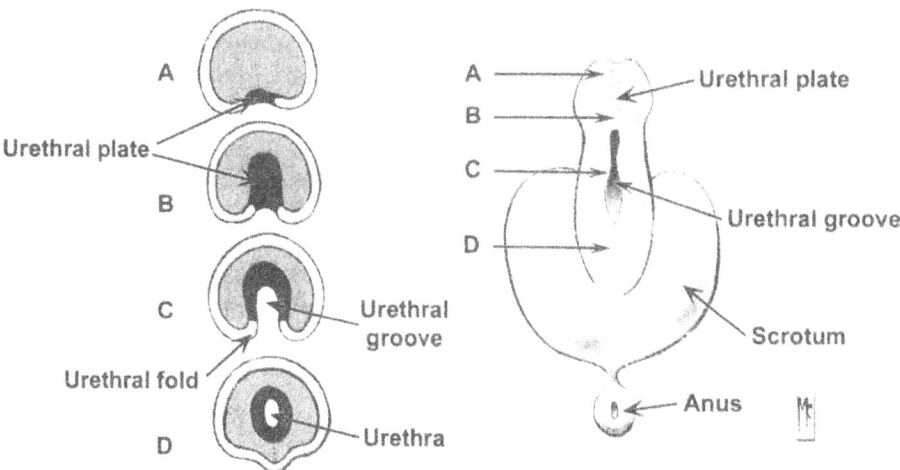

Figure 5. Detail in transverse section of the urethral groove, urethral folds and urethral plate. Transverse sections (A-D) are shown at the levels depicted (arrows). The solid urethral plate is present in the distal part of the phallus (A-B). Canalization of the solid urethral plate (C) forms the urethral groove bounded by urethral folds, which grow to the midline (arrows) and fuse to form the tubular urethra (D). After fusion of the urethral folds, the resultant epithelial seam is removed (D).

The urethral groove is defined laterally by urethral folds, which are continuations of the urogenital folds surrounding the urogenital ostium (Figures 4, 5). Initially, the urethral groove and folds extend only part of the way distally along the shaft of the elongating phallus (Carlson, 1994). The distal portion of the urethral groove terminates in a solid epithelial plate called the urethral plate that extends to the glans penis (Fig. 4 and 5) (Hamilton et al., 1972). The solid urethral plate canalizes and thus extends the urethral groove distally to the glans (Figures 4 and 5). As the phallus elongates, the urethral folds grow towards each other in the midline and fuse, thus converting the urethral groove into a tubular penile urethra (Figures 4 and 5). This fusion process begins proximally in the perineal region and extends distally towards the glans penis to form the penile urethra. Thus, from proximal to distal there is the progressive formation of a urethral groove by canalization of the urethral plate, followed by fusion of the urethral folds in the midline. In so doing the urethral groove is converted into a tubular penile urethra (Figure 5).

The urethral groove, which is continuous with the   urogenital folds and   urogenital ostium, is thought to be lined by endoderm even though this conclusion is based mostly upon histological observations (Kurzrock et al., 1999a). Likewise, the solid urethral plate, the distal precursor of the urethral groove, is believed to be derived from endoderm. The implication of this idea is that almost all of the urethra may be endodermal in origin. The urethra is lined by urothelium. Urothelium is a unique epithelium found only in the ureters, bladder and urethra. In the ureter, urothelium is derived from mesodermal epithelium arising from the Wolffian duct. In the bladder and pelvic portions of the urethra the urothelium is derived from endoderm (Marshall, 1978). Since epithelium of the urethral folds is precisely at the interface of ectoderm and endoderm, it is conceivable that the urethral groove and thus the penile urethra could be derived from either ectoderm or endoderm. However, the consensus is that most (if not all) of the penile urethra is endodermal in origin, with the possible exception of the distal urethra within the glans whose epithelium may be ectodermal in origin (Figure 6).

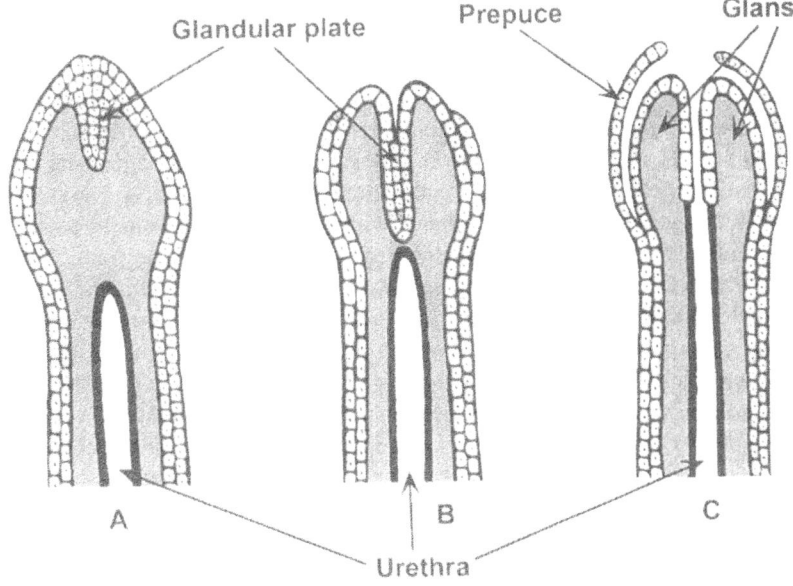

Figure 6. An alternate theory of urethral development in the glans. (A) The penile urethra forms as a result of fusion of the urethral folds throughout the shaft of the penis, but penile urethral formation stops in the glans. The last segment of urethra develops from a solid ingrowth of ectoderm (A-B), which after making contact with the penile urethra (B), canalizes to complete the urethra (C).

In regard to the terminal portion of the penile urethra within the glans, there are two theories concerning the development of this distal most portion of the urethra (Kurzrock et al., 1999a). One theory (Glenister, 1954; Hart, 1908; Jones, 1910) states that the distal portion of the urethra is formed by a solid in-growth of ectoderm (the glandular plate) into the glans (Figure 6), which makes contact with the endodermally derived tubular urethra. After canalization of the solid glandular plate the continuity of the urethra to the exterior is established. An alternate theory (Felix, 1912; Herzog, 1904; Johnson, 1920; Kurzrock et al., 1999a) suggests that the solid endodermal urethral plate extends distally to the tip of the glans (Figures 4 and 5). Canalization of the urethral plate extends the urethral groove distally into the glans where subsequent fusion of the urethral folds extends the tubular urethra all the way distally to the tip of the glans (Figures 4 and 5). Hypospadias results from failure of formation or fusion of the urethral folds. The site of failure of urethral fold fusion will dictate the position of the abnormal opening of the urethra from perineal positions proximally to more distal openings along the shaft of the penis.

The elongating phallus (penis) is covered externally by ectoderm that gives rise to penile skin, while most of the substance of the penis is derived from mesodermal cells forming the large corporal bodies, connective tissue and dermis. During the course of development, the mesoderm of the penis segregates into these derivatives. Corporal tissue is first recognized as distinct dense mesenchymal condensations within the shaft of the developing penis. The outer aspect of the corporal bodies differentiates into thick investing connective tissue capsules called the tunica albuginea. Surrounding the tubular urethra the mesenchyme differentiates into smooth muscle contributing to the mucosa and sub-mucosa of the urethra (Gilpin and Gosling, 1983), which in humans is in turn surrounded by erectile tissue of the corpus spongiosum. In some species penile mesenchyme also forms an os penis, composed of both bone and cartilage (Bolton et al., 1996; Izumi et al., 2000; Rasmussen et al., 1986). Little is known concerning the cellular and molecular mechanisms of differentiation of penile mesenchyme into its various derivatives. However, penile development involves an outgrowth somatic tissue from the body surface, and thus has been likened to development of the limb. Since limb development and patterning involves interactions between epithelium and mesenchyme, investigations of penile development have utilized similar approaches. While these studies of penile development are limited in scope, it is evident that differentiation of penile mesenchyme into its derivatives is dependent upon epithelial-mesenchymal interactions (Kurzrock et al., 1999b; Murakami and Mizuno, 1986) (Masters et al, 2003).

Fusion of the urethral folds is the key step in the formation of the penile urethra. A prerequisite of urethral fold fusion is the formation of the urethral groove bounded by the urethral folds. Clearly, if the urethral groove and the urethral folds do not form properly, the possibility of urethral fold fusion will be impaired. Assuming normal development of the urethral folds, the process of development of the penile urethra involves the movement or growth of the urethral folds towards the midline so that they make contact. When the urethral folds contact each other in the midline, the opposing epithelial surfaces must adhere to one another resulting in the formation of a midline epithelial seam. Finally, the midline epithelial seam must be removed so that mesenchymal continuity is established across the midline. This process is analogous to development of the secondary palate. Thus, the voluminous literature on palatal development may provide a relevant conceptual and experimental paradigm for investigating penile development and understanding the etiology of hypospadias.

3.   Endocrine Aspects of External Genitalia Development

A feature that distinguishes development of the external genitalia from development of the limb is the sexual dimorphism inherent in the external genitalia. The molecular basis of this sexual dimorphism has been extensively studied and is based upon the presence or absence of signaling via the androgen receptor. The fetal testes produce testosterone, which is the primary serum androgen in male fetuses. Cells within the fetal external genitalia express the enzyme 5 alpha-reductase which converts testosterone to dihydrotestosterone, a particularly potent androgen (Kim et al., 2002; Levine et al., 1996; Tian and Russell, 1997). Androgen receptors are present in cells of the developing external genitalia and are prominently expressed in genital tubercle mesenchymal cells (Kim et al., 2002; Murakami, 1987a). The action of dihydrotestosterone acting via androgen receptors masculinizes the developing external genitalia.

The penis has a particular external and internal form, which is radically different from the clitoris. The developmental fate (penis versus clitoris) of the genital tubercle is determined by androgen action. Penile development requires (a) the production of testosterone by the fetal testes, (b) the conversion of testosterone to dihydrotestosterone by 5 alpha-reductase within the genital tubercle and (c) signaling through the androgen receptor following binding of dihydrotestosterone to the androgen receptor (Renfree et al., 2002a; Wilson et al., 1993; Wilson, 1992; Wilson et al., 1995a). If either of the above events (a-c) does not occur, penile development and growth will be reduced to a greater or lesser degree, in which case the external genitalia will be feminized or ambiguous. Individuals having defects in the ability to produce androgenic steroids will also be feminized. Finally, females with congenital adrenal hyperplasia producing abnormally high levels of androgens during gestation will have varying degrees of masculinization of the external genitalia (New, 1990; Villee and Crigler, 1976).

The key role of androgen in sexually dimorphic development of the external genitalia has been corroborated through experimental studies. In utero exposure of laboratory rodents to anti-androgenic drugs inhibits the binding of testosterone and DHT to the androgen receptor, and thus reduces size of the genital tubercle, reduces anogenital distance, and inhibits development of pigment cells in the scrotum (Gray et al., 1994; Toppari and Skakkebaek, 1998; Wilson, 1992; Wilson, 1983). Likewise, in utero exposure of rats to 5 alpha-reductase inhibitors reduces anogenital distance and inhibits distal development of the urethra resulting in hypospadias (Anderson and Clark, 1990; Clark et al., 1993; Clark et al., 1990). The importance of androgen receptor signaling in genital tubercle development has also been verified by genetic analysis. Spontaneous mutations in the gene encoding the androgen receptor in mice and humans (testicular feminization locus, Tfm) cause complete feminization of the external genitalia (Feldman, 1992; Murakami, 1987b; Wilson, 1992). XY humans with 5 alpha-reductase deficiency have ambiguous genitalia which at birth are more female-like than male-like and are usually hypospadic (Imperato-McGinley, 1984; Wilson et al., 1993). Thus, masculine development of the genital tubercle requires (a) the production of testosterone by the testes, (b) the conversion of testosterone to DHT by 5 alpha-reductase within the genital tubercle and (c) signaling through the androgen receptor (Renfree et al., 2002b; Wilson et al., 1993; Wilson, 1992; Wilson et al., 1995b).

The actions of androgens are elicited by a variety of down-stream factors whose production, receptor binding and action must be intact in order to carry out androgen action. Downstream mediators of the androgen receptor action include growth factors, adhesion molecules, transcription factors and downstream mediators of these secondary factors. Other papers in this book will focus on these downstream mediators of androgen action in development of the external genitalia.

4. Development of the mouse external genitalia

Inferences on developmental mechanisms have been drawn from examination of anatomical and histological aspects of developing human external genitalia. However, given the scarcity of developing human tissues and the ethical and administrative impediments associated with the study of human fetal external genitalia, progress in understanding the cellular and molecular mechanisms of development the external genitalia and hypospadias has been meager. It is for this reason that relevant animal models of external genitalia development are essential. In this regard, the mouse requires special attention because of the wealth of transgenic and mutant strains available. The key question is whether penile development in the mouse is comparable to that in the human.

At first sight penile development in the mouse appears to be very different than that in the human. Whereas the developing human penis has an extensive and broadly open urethral groove, the comparable structure is almost indiscernible in the mouse fetus (Figure 7, comparison of mouse and human).

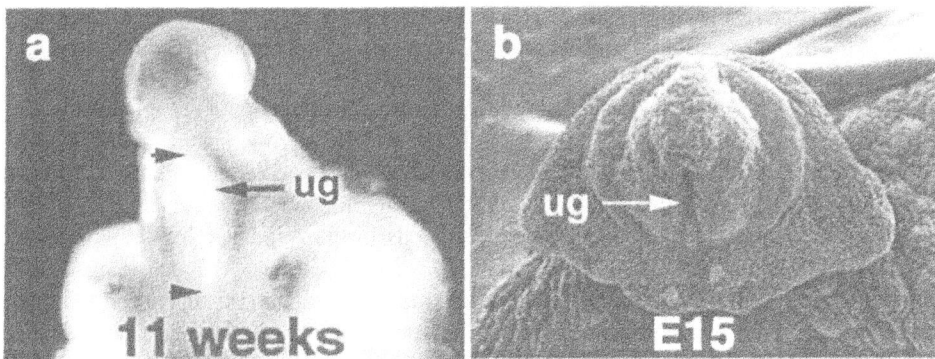

Figure 7. The urethral groove (ug) on the ventral aspect of the embryonic genital tubercle is prominent in the human (a) at 11 weeks of gestation, and discernable in the mouse genital tubercle by scanning electron microscopy (b) at 15 days of gestation.

However, close examination of the developing external genitalia in the mouse reveals that species differences in development are mostly differences in scale. The mouse has a urethral groove on the ventral aspect of the developing penis, which is difficult to discern even with the aid of a dissecting microscope. However, the murine urethral groove can be revealed by scanning electron microscopy or in serial sections (Figures 7-9). The opening of the urethral groove is the terminus of a patent channel from the bladder down the pelvic urethra to the amnionic cavity. This continuity has been definitively revealed by injection of the embryonic bladder with dye, which exists freely from the murine urethral groove (Baskin and Cunha, unpublished). As in the human, distal to the open urethral groove is a solid cord of epithelial cells in the midline that constitute the urethral plate (Figure 8a).

Examination of penile development in the mouse during the last third of gestation indicates that initially the opening of the murine urethral groove is located proximally, and with time this opening extends distally. In the mouse genital tubercle the solid urethral plate distal to the urethral groove undergoes canalization to extend the urethral groove distally (Figures 8b-d). This distal extension of the urethral groove is balanced by the proximal to distal fusion of the urethral folds, which lays down a tubular penile urethra in its wake (Figure 9b-e).

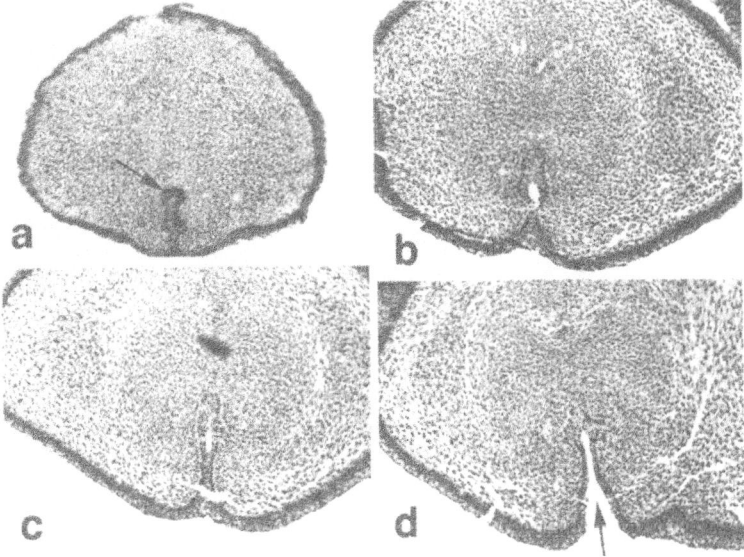

Figure 8. Transverse sections through the embryonic mouse genital tubercle at 15 days of gestation. Sections (a-d) are from distal to proximal. In (a) note the solid urethral plate in the glans. In more proximal zones (b-c) the urethral plate is canalizing, and in (d) the urethral groove is open and bounded by the urethral folds.

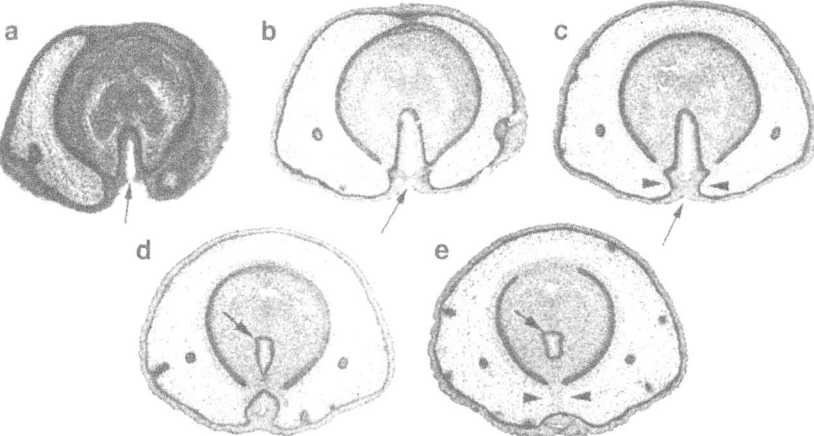

Figure 9. Transverse sections through the embryonic mouse genital tubercle at 17 days of gestation. Sections (a-e) are from distal to proximal. In (a) note the open urethral groove. In more proximal zones (b-c) the urethral folds are fusing, and in (d-e) the epithelial seam is being removed, thus establishing mesenchyme continuity across the midline and leaving the definitive tubular penile urethra (e).

This means that the urethral groove is in a dynamic state of tissue remodeling. Distally, canalization of the solid urethral plate occurs to form the urethral groove. Proximally the urethral folds fuse to form the urethra, and the midline epithelial seam is then removed.

The most informative method of studying the development of the mouse penis has been the examination of serial sections. Because the process represents a wave of developmental events that progress from proximal to distal, it is possible to see the entire developmental process unfolding in a single specimen. For example, at 15 days of gestation, serial sections through the distal portion of the genital tubercle demonstrate the process of formation of the urethral groove as a result of canalization of the solid urethral plate. Transverse sections in figures 8a-d progress from distal to proximal. In Fig 8a the solid urethral plate is demonstrated. Figures 8 b-d move proximally down the shaft of the developing penis and reveal the process of canalization and urethral groove formation and finally an open urethral groove bounded by the newly formed urethral folds (Fig 8d). Likewise, examination of serial sections along the proximal aspect of the penis demonstrates fusion of the urethral folds and the formation of the tubular urethra. In Figure 9 sections (a) – (e) progress from distal to proximal. In Figure 9a the urethral folds are separate and define the open urethral groove. Moving proximally, Figures 9b-c demonstrate fusion of the urethral folds in the midline. Figure 9d reveals removal of the midline epithelial seam, and Figure 9e demonstrates the tubular urethra. Examination of an even closer set of serial sections indicates that closure of the urethral groove and formation of the tubular urethra involves two epithelial fusion events in the mouse (Figure 10).

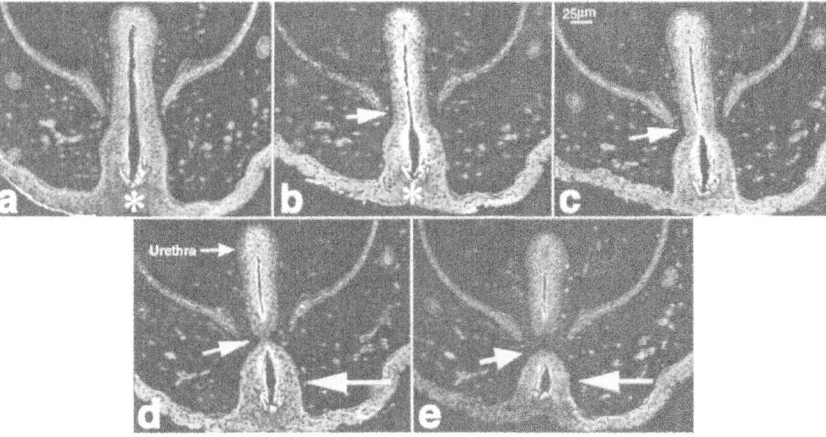

Figure 10. Transverse sections through the embryonic mouse genital tubercle at 17 days of gestation showing two epithelial fusion events. Sections (a-e) are from distal to proximal. In (a and b) the urethral folds have fused (*) to establish the epidermal surface. This generates an extensive internal epithelial lined chamber, which is subsequently subdivided by a second epithelial fusion (small arrows, c-d). Finally in the region of the second fusion event the resultant epithelial seam becomes attenuated (small arrow, d) and is entirely removed (e), thus establishing mesenchymal continuity across the midline. After segregation of the urethra, the remaining epithelial seam (large arrows, d-e) is incorporated into the epidermis.

The initial fusion of the urethral folds occurs superficially to define the epidermal surface of the developing penis (Figures 10a-b).

Subsequently, a second epithelial fusion (Figures 10c-d) further divides the internalized epithelial chamber to form the definitive urethra (Figure 10e). Formation of the definitive urethra also requires the removal of midline epithelial seam (Figures 9c-d and 10d-e) and thus requires dynamic tissue remodeling. In this regard, apoptotic activity has been revealed in mesenchyme cells precisely in the midline just proximal to the epithelial seam (Figure 11).

In summary, while there are subtle differences in the development of the external genitalia of the human and mouse, the basic processes of formation of the urethral groove and urethral folds occurs via canalization of the solid urethral plate. In both species the tubular penile urethra develops as a results of midline fusion of the urethral folds. Likewise, in both species the midline epithelial seam is removed resulting in the formation of an internal tube (the urethra), an intervening mesenchyme, and surface epithelium which ultimately forms the epidermis of the penis. Between the tubular urethra and the investing layer of skin is tissue of mesenchymal origin continuous across the midline which will give rise to the corporal bodies, the dermis and the fibro-muscular wall of the urethra and erectile tissue of the corpus spongiosum (if present). Based upon careful examination of penile development in the mouse, it is evident that the mouse provides a relevant animal model of external genitalia development. For this reason, the examination of mutant mice defective in specific molecular pathways should provide a fertile area of future research into the etiology and causes of hypospadias.

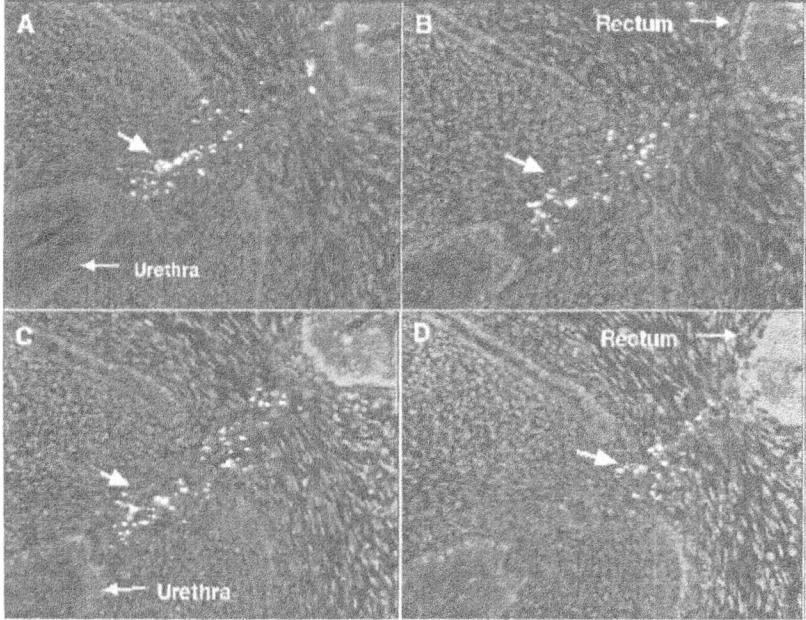

Figure 11. Transverse sections through the embryonic mouse genital tubercle at 17 days of gestation showing apoptotic activity (white cells, arrows). These serial sections are at the level depicted in Figure 10e, just proximal to the region of epithelial seam removal. This is an area of dynamic tissue remodeling, and apoptotic activity is observed precisely in the midline in mesenchymal cells.

Acknowledgments

We would like to thank Michael McLaughlin for preparation of the illustrations. This research was supported by the following NIH grants: CA89520, CA91967, DK058105 and DK57246.

REFERENCES

Anderson, C. A., and Clark, R. L., 1990, External genitalia of the rat: normal development and the histogenesis of 5 alpha-reductase inhibitor-induced abnormalities. *Teratology* **42**: 483-496.

Baskin, L. S., Erol, A., Li, Y. W., and Cunha, G. R., 1998, Anatomical studies of hypospadias. *J Urol* **160**: 1108-1115.

Bolton, L. A., Camby, D., and Boomker, J., 1996, Aberrant migration of Ancylostoma caninum to the os penis of a dog. *J S Afr Vet Assoc* **67**: 161-2.

Brookes, M., and Zietman, A., 1998, "Clinical Embryology." CRC Press, New York.

Carlson, B. M., 1994, Human Embryology and Developmental Biology. Mosby, St. Louis.

Clark, R. L., Anderson, C. A., Prahalada, S., Robertson, R. T., Lochry, E. A., Leonard, Y. M., Stevens, J. L., and Hoberman, A. M., 1993, Critical developmental periods for effects on male rat genitalia induced by finasteride, a 5 alpha-reductase inhibitor. *Toxicol Appl Pharmacol* **119**: 34-40.

Clark, R. L., Antonello, J. M., Grossman, S. J., Wise, L. D., Anderson, C., Bagdon, W. J., Prahalada, S., MacDonald, J. S., and Robertson, R. T., 1990, External genitalia abnormalities in male rats exposed in utero to finasteride, a 5 alpha-reductase inhibitor. *Teratology* **42**: 91-100.

Feldman, S. R., 1992, Androgen insensitivity syndrome (testicular feminization): a model for understanding steroid hormone receptors. *J Am Acad Dermatol* **27**: 615-619.

Felix, W., 1912, The development of the urogenital organs. *In* "Manual of Human Embryology" (R. Kiebel and F. P. Mall, Eds.), pp. 869-972. Lippincott, Philadelphia.

Gilpin, S. A., and Gosling, J. A., 1983, Smooth muscle in the wall of the developing human urinary bladder and urethra. *J Anat* **137 (Pt 3)**: 503-12.

Glenister, T., 1954, The origin and fate of the urethral plate. *J. Anat.* **88**: 413-424.

Gray, L. E., Jr., Ostby, J. S., and Kelce, W. R., 1994, Developmental effects of an environmental antiandrogen: the fungicide vinclozolin alters sex differentiation of the male rat. *Toxicol Appl Pharmacol* **129**: 46-52.

Hamilton, W. J., Boyd, J. D., and Mossman, H. W., 1972, Human Embryology. Williams & Wilkins, Baltimore.

Hart, D., 1908, On the role of the developing epidernis in forming sheaths and lumina to organs, illustrated specially in the development of the prepuce and urethra. *J. Anat. Lond.* **42**: 50-56.

Herzog, F., 1904, Beitrage zur Entwicklungsgeschichte und Histologie der mannlichen Harnrohre. *Arch. fur mikr. Anat. und Entw* **63**: 710-747.

Imperato-McGinley, J., 1984, 5a Reductase deficiency in man. *Prog. Cancer Res. and Therap.* **31**: 491-496.

Izumi, K., Yamaoka, I., and Murakami, R., 2000, Ultrastructure of the developing fibrocartilage of the os penis of rat. *J Morphol* **243**: 187-91.

Johnson, F. P., 1920, The later development of the urethra in the male. *J. Urol.* **4**: 447-501.

Jones, F., 1910, The development and malformations of the glans and prepuce. *Brit. Med. J.* **1**: 137-138.

Kim, K. S., Liu, W., Cunha, G. R., Russell, D. W., Huang, H., Shapiro, E., and Baskin, L. S., 2002, Expression of the androgen receptor and 5 alpha-reductase type 2 in the developing human fetal penis and urethra. *Cell Tissue Res* **307**: 145-53.

Kurzrock, E. A., Baskin, L. S., and Cunha, G. R., 1999a, Ontogeny of the male urethra: theory of endodermal differentiation. *Differentiation* **64**: 115-122.

Kurzrock, E. A., Baskin, L. S., Li, Y., and Cunha, G. R., 1999b, Epithelial-mesenchymal interactions in development of the mouse fetal genital tubercle. *Cells Tissues Organs* **164**: 125-130.

Langer, J. C., 1993, Fetal abdominal wall defects. *Semin Pediatr Surg* **2**: 121-8.

Levine, A. C., Wang, J. P., Ren, M., Eliashvili, E., Russell, D. W., and Kirschenbaum, A., 1996, Immunohistochemical localization of steroid 5 alpha-reductase 2 in the human male fetal reproductive tract and adult prostate. *J Clin Endocrinol Metab* **81**: 384-9.

Marshall, F. F., 1978, Embryology of the lower genitourinary tract. *Urol Clin North Am* **5**: 3-15.

Mollohan, J., 1999, Exstrophy of the bladder. *Neonatal Netw* **18**: 17-26.

Moore, K. L., and Persaud, T. V. N., 1993, The Developing Human. W.B. Saunders Company, Philadelphia.

Murakami, R., 1987a, Autoradiographic studies of the localisation of androgen-binding cells in the genital tubercles of fetal rats. *J Anat* **151**: 209-19.

Murakami, R., 1987b, A histological study of the development of the penis of wild-type and androgen-insensitive mice. *J Anat* **153**: 223-31.

Murakami, R., and Mizuno, T., 1986, Proximal-distal sequence of development of the skeletal tissues in the penis of rat and the inductive effect of epithelium. *J Embryol Exp Morphol* **92**: 133-143.

New, M. I., 1990, Prenatal diagnosis and treatment of adrenogenital syndrome (steroid 21-hydroxylase deficiency). *Dev Pharmacol Ther* **15**: 200-10.

Rasmussen, K. K., Vilmann, H., and Juhl, M., 1986, Os penis of the rat. V. The distal cartilage process. *Acta Anat* **125**: 208-212.

Renfree, M. B., Wilson, J. D., and Shaw, G., 2002a, The hormonal control of sexual development. *Novartis Found Symp* **244**: 136-52; discussion 152-6, 203-6, 253-7.

Renfree, M. B., Wilson, J. D., and Shaw, G., 2002b, The hormonal control of sexual development. *Novartis Found Symp* **244**: 136-152.

Snell, R. S., 1975, Clinical Embryology for Medical Students. Little, Brown and Company, Boston.

Tian, H., and Russell, D. W., 1997, Expression and regulation of steroid 5 alpha-reductase in the genital tubercle of the fetal rat. *Dev Dyn* **209**: 117-126.

Toppari, J., and Skakkebaek, N. E., 1998, Sexual differentiation and environmental endocrine disrupters. *Baillieres Clin Endocrinol Metab* **12**: 143-156.

van der Werff, J. F., Nievelstein, R. A., Brands, E., Luijsterburg, A. J., and Vermeij-Keers, C., 2000, Normal development of the male anterior urethra. *Teratology* **61**: 172-83.

Vermeij-Keers, C., Hartwig, N. G., and van der Werff, J. F., 1996, Embryonic development of the ventral body wall and its congenital malformations. *Semin Pediatr Surg* **5**: 82-9.

Villee, D. B., and Crigler, J. F., Jr., 1976, The adrenogenital syndrome. *Clin Perinatol* **3**: 211-20.

Wilson, J., Griffin, J., and Russell, D. W., 1993, Steroid 5a-reductase deficiency. *Endocrine Rev.* **14**: 577-593.

Wilson, J. D., 1992, Syndrome of androgen resistance. *Biol. Reprod.* **46**: 168-173.

Wilson, J. D., George, F. W., and Renfree, M. B., 1995a, The endocrine role in mammalian sexual differentiation. *Recent Prog Horm Res* **50**: 349-64.

Wilson, J. D., George, F. W., and Renfree, M. B., 1995b, The endocrine role in mammalian sexual differentiation. *Recent Prog Horm Res* **50**: 349-364.

Wilson, M. J., 1983, Inhibition of development of both androgen-dependent and androgen-independent pigment cells in scrotal skin dermis of the rat by antiandrogen treatment during fetal growth. *Endocrinology* **112**: 321-325.

ANATOMICAL STUDIES OF THE MOUSE GENITAL TUBERCLE

Laurence S. Baskin*, Wenhui Liu, Jacob Bastacky, and Selcuk Yucel

1. Abstract

1.1. Background

To study the etiology of hypospadias, we propose the use of a mouse model, the embryonic mouse genital tubercle. In this study, we define the development of the mouse genital tubercle with special emphasis on urethral formation demonstrating anatomical similarities to human development.

1.2 .Materials and Methods

Serial sections of genital tubercles from embryonic male and female mice ages 14 to 21 days gestation from timed pregnant animals, newborn and adult mice were immunohistochemical stained with antibodies to E-cadherin, cytokeratins 7, 10, and 14. Patency of the urethral was assessed by india ink injection via the bladder. Urethral lumen morphology was determined by the creation of plastic resin cast. Surface morphology of the genital tubercle was defined by scanning electron microscopy.

1.3. Results

India Ink injection into the bladder showed that the urethral lumen was patent from 14 days gestation. Plastic resin casts revealed that the male urethra

---

*Laurence S. Baskin, M.D., Department of Urology, University of California, San Francisco, CA, 94143-0738, (415) 476-1611, (415) 476-8849(FAX), lbaskin@urol.ucsf.edu

Hypospadias and Genital Development, edited by
L. Baskin, Kluwer Academic/Plenum Publishers, 2004

was characterized by a S shaped curve, the presence of the bulbar urethral gland and a longer length than age matched females. The ontogeny of the genital tubercle development revealed two epithelial edges that subsequently touched and fused into the completed urethra. During development cytokeratin immunohistochemical staining demonstrated that the epithelial cells of the urethral lumen are of bladder origin and the surface cells of skin origin.

1.4. Conclusion

The functional and developmental anatomy of the mouse genital tubercle provides a useful model to study normal and abnormal human urethral development.

2. INTRODUCTION

The mouse genital tubercle is a useful model for understanding the process of normal urethral development (Baskin et al., 2001). During development the urethral forms via fusion of the urethral folds to form a midline epithelial seam which subsequently remodels into the tubular urethra (Baskin et al., 2001). Post-natally, mouse genital development diverges from the human in that the male, mouse genital tubercle differentiates into cartilage and bone under the influence of androgens and epithelial mesenchymal interactions (Kurzrock et al., 1999). During the embryonic period, however, both the human and the mouse urethra develop by formation of a urethral seam. The mechanism of urethral seam formation and subsequent remodeling into the tubular urethra is not completely defined. Previous work supports the hypothesis that urethral seam remodeling occurs via cellular migration and not by epithelial mesenchymal transformation or epithelial apoptosis (Baskin et al., 2001).

The mechanism of urethra formation is relevant to the study of congenital abnormalities of abnormal urethral development such as hypospadias (Baskin, 2000). Hypospadias is one of the most common birth defects in human affecting between 0.008 – 0.0004 of newborn males (Baskin and Colborn et al., 2001). Hypospadias is a spectrum of genital abnormalities from mild to extremely severe, where the gender of the patient is in question (Figure 1). Anatomically, hypospadias is defined as a defect in the development of the urethral and surrounding muscular structures or supporting urethral spongiosum. Associated defects include an arrest in normal ventral closure of the prepuce resulting in an excess of dorsal skin characterized by the dorsal hooded foreskin of hypospadias. Hypospadias is also associated with penile curvature typically in the ventral direction resulting in the appearance of a smaller than normal phallus that is buried within the perineal skin. Recent reports suggest a possible doubling in the incidence of hypospadias in Western countries over the last 20 years (Paulozzi et al, 1997; Paulozzi, 1999). A leading hypothesis is that the increase in hypospadias is due to an increase in environmental exposure to natural and synthetic endocrine disruptors (Baskin and Colborn et al., 2001).

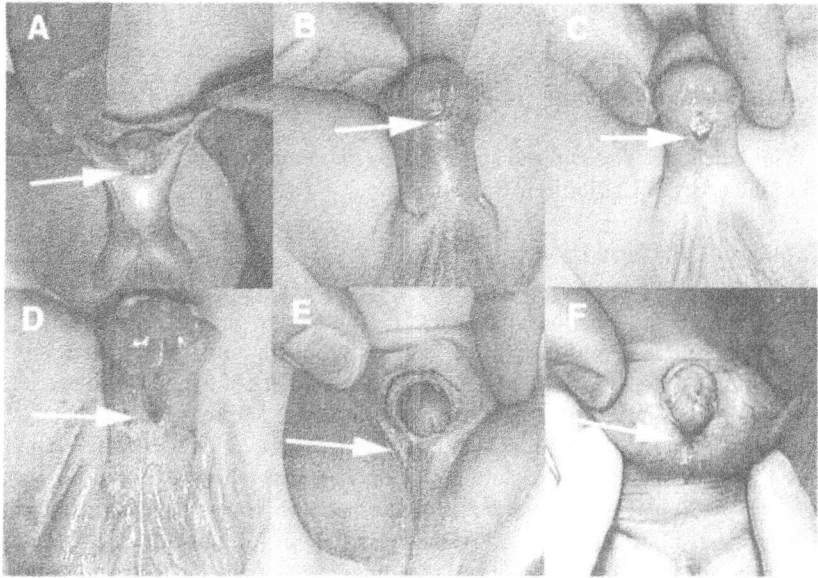

Figure 1.
The spectrum of hypospadias. A. Anterior, where the urethral meatus is on the inferior surface of the glans penis. B. Coronal, where the meatus is in the balanopenile furrow. C. Distal, on the distal third of the penile shaft. D. Peno-scrotal, at the base of the shaft in front of the scrotum. E. Scrotal, on the scrotum or between the genital swellings. F. Perineal where the meatus is behind the scrotum or genital swellings. The more severe forms of hypospadias are often associated with penile curvature.

To study the etiology of hypospadias, we have proposed that the embryonic mouse genital tubercle has anatomical similarities to human development. For example, both the mouse and the human have a ventral urethra relative to the dorsal corporeal bodies. In both the human and the mouse, the urethra forms in a proximal to distal fashion with closure of a urethra seam. In critical evaluation of the mouse as an experimental model for urethral development, we have noted a paucity of functional and morphologic anatomical detail (Baskin et al., 2001). Historically, it has been difficult to determine whether urethral abnormalities exist in the genital tubercle of normal, mutant or mice exposed to teratogenic compounds. This is because the tubular urethra is hidden under the skin thus requiring histologic analysis to determine abnormalites in the urethral opening (Baskin et al., 2001). In reported animal models of hypospadias, the definition of "hypospadias" is typically based on an increase in ano-genital distant from gross descriptions without histologic analysis of the urethra and urethral seam abnormalities. Ano-genital distance per se is a poor measure of hypospadias, which is a urethral defect. In humans, abnormalities in anogenital distance are not associated with hypospadias but occur in anorectal malformations such as imperforate anus and urogenital sinus anomalies.

The goal of this study is to define the functional anatomy of the mouse genital tubercle as a model for normal urethral development. Normal mouse urethral development is critical for comparison to animals with specific genetic alterations and in animal models of hypospadias induced by exogenous chemicals.

2.1 Materials and Methods

The scientific protocol was approved by the committee on animal research at the University of California at San Francisco.

2.2. India Ink Urethral Patency

The patency and functional location of the embryonic urethral meatus on the genital tubercle was checked by injecting India ink into the bladder and capturing the appearance of the dye on the ventral aspect of the genital tubercle. The experiment was repeated three times in both male and female specimens at gestational ages 16 days, through the newborn period. Under the dissecting microscope the pelvis of the embryonic mouse was opened to expose the bladder. A 30 gauge needle attached to a tuberculin syringe filled with India ink was injected into the bladder. At the exact time of injection a video link via NIH image software was used to capture a movie at 24 frames per second of the functional location of the urethral meatus opening in both male and female specimens. Select frames pre and post urethral dye transit were captured from the movie to document the functional location of the urethral meatus on the ventral aspect of the genital tubercle.

Patency of the urethral lumen was also checked in the genital tubercle embryos by passing a human hair from the exposed bladder out through the urethral lumen. The bladder was exposed and the hair gently passed into the open bladder neck using a fine forceps.

2.3 Urethral Lumen Resin Cast

The three dimensional morphology of the urethral lumen and associated structures (prostate and bulbar urethral gland) was assessed by creating plastic resin cast of the urethral lumen. At least three specimens from time pregnant male and female fetuses at gestational days 14 through the newborn period and adult mice were analyzed. Batson's No.17 Plastic Replica and Corrosion Kit (Polysciences, Inc. Warrington, PA) was prepared immediately after exposing the bladder of the fetuses under the dissecting microscope. The ratio of Base Solution A: Catalyst B: and Promoter was 8.3:1.6:1. After adding the catalyst the liquid plastic was injected into the bladder via a 30 gauge needle or microglass needle connected by a Silastic tubing (Dow-Corning Co. Midland, MI) attached to a tuberculin syringe. The liquid plastic injection was terminated on visualization of the plastic resin material at the urethral opening. The specimens were air dried at room temperature for 2 hours, immersed in 10N NaCl overnight followed by 10M HCl for 24 hours and rinsed in 70% ethanol. The stringent chemical processing facilitated tissue destruction around the hardened replica of the urethra without damaging the resistance casts. Using the dissecting microscope and micro forceps any remaining tissue was removed.

The bladder aspect of the hardened urethral replicas were superglued to toothpicks to facilitate handling for digital photography with a macro lens of a Nikon 900 Digital camera.

The casts were quantitated by capturing a lateral photographic image. Using NIH imaging software, measurements of the cast were taken to quantify the size of the cast using multiple variables corresponding to total pixel count, pixel count per mm, area, length, perimeter, convex perimeter and equivalent diameter. This analysis allowed the different cast to by distinguished by a reproducible and quantitative numerical valve.

2.4 Scanning Electron Microscopy

Scanning electron microscopy (SEM) was performed on male embryonic genital tubercle specimens embryonic ages E14 through the newborn period. Specimens were fixed in 2.3% glutaraldehyde with 0.05 M Sodium Cacodylate (EMS Brand) pH 7.4 at 4° C for 24 hrs followed by repetitive rinse buffer overnight at 4°C (0.05 M Na Cacodylate pH 7.4 with 8.125% sucrose). The specimens were then fixed with 1% OsO_4 in 0.05 M Na Cacodylate, pH 7.4 with 7.28% sucrose for 2 hours followed by 3 washing for 15 minutes each in rinse buffer. The specimens were dehydrated in graded ethanol's (10,25,35,50,70,80,90,100%), followed by critical point drying with CO_2. After mounting on SEM stubs and sputter coating with gold the specimens were analyzed and photographed using a scanning electron microscope.

2.5 Immunohistochemistry

Genital tubercles from embryonic male and female mice ages 14 to 21   days gestation from timed pregnant animals, newborn and adult mice were fixed in formalin, and processed into paraffin. Fetuses from at least three different liters at each time point were analyzed. Sex was determined by gonadal histology and the present or absence of internal female organs. Six µm transverse sections were cut using a microtome. Serial sections were immunostained with antibodies to E-cadherin, cytokeratins 7, 10, and 14, and with Hematoxylin and Eosin (Baskin et al., 1997). The avidin-biotin-peroxidase procedure was employed using Vectastatin ABC kits (Vector Laboratories, Burlingame, CA) with cobalt intensification. Primary antibodies to E-cadherin, and cytokeratins 7, 10, and 14, were obtained from Sigma (St Louis, MO). Biotinylated anti-rabbit and anti-mouse IgG were obtained from Amersham International (Arlington Heights, IL). Peroxidase linked avidin/biotin complex reagents were obtained from Vector Laboratories (Burlingame, CA). Negative controls utilized IgG of the same species in place of and at the same dilution as the primary antibody. The purified mouse IgGs were obtained from Zymed Corp. (So. San Francisco, CA). Slides were imaged using a Leaf Systems (Southborough, MA) Lumina scanner system attached to a Zeiss microscope. Images were collected on an Apple Power Macintosh G-4 computer using Adobe PhotoShop software.

2.6. Results

2.1.1. India Ink Injections

India Ink injection into the bladder showed that the   urethral lumen was patent from the earliest gestational day analyzed, E-16. At the E16 time period the ink was noted to exit on the proximal aspect of the ventral surface of both the male and female genital tubercle (Figure 2).   The size of the ink droplets were larger than the size of the genital tubercle thereby making it impossible to define the exact opening of the urethral meatus.   The specimens older than E-16 exhibited the same findings.

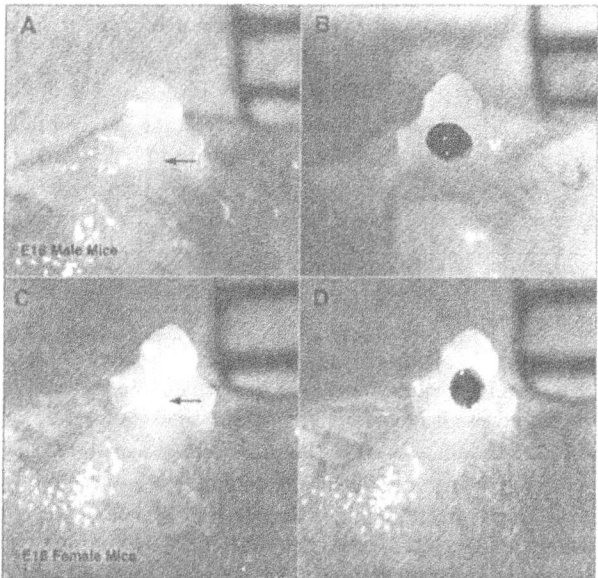

Figure 2.
India Ink injection experiments to verify the patency of the male and female genital tubercle at 16 days gestation. The genital tubercles appear similar in the male (A & B) and female (C & D). Note the urethral groove in both the male and female (arrow). The functional urethral opening appears to be in the same place at this stage of development.

A human hair confirmed patency of the urethral lumen from gestational age 15 days until the newborn period in both male and female genital tubercle specimens (Figure 3). Resistance within the lumen of the urethra was not encountered consistent with the India Ink studies.

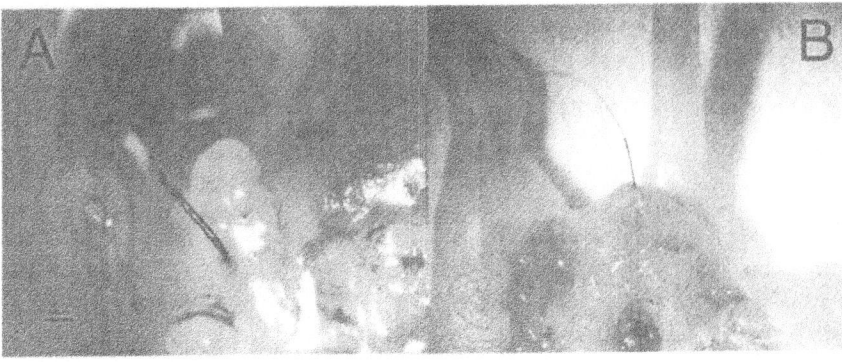

Figure 3.
Patency of the male genital tubercle was confirmed by the ability to pass a human hair through the open bladder neck out through the urethral meatus.   A. 15 days gestation   B. Newborn mouse genital tubercle (Note the hair within the exposed bladder).

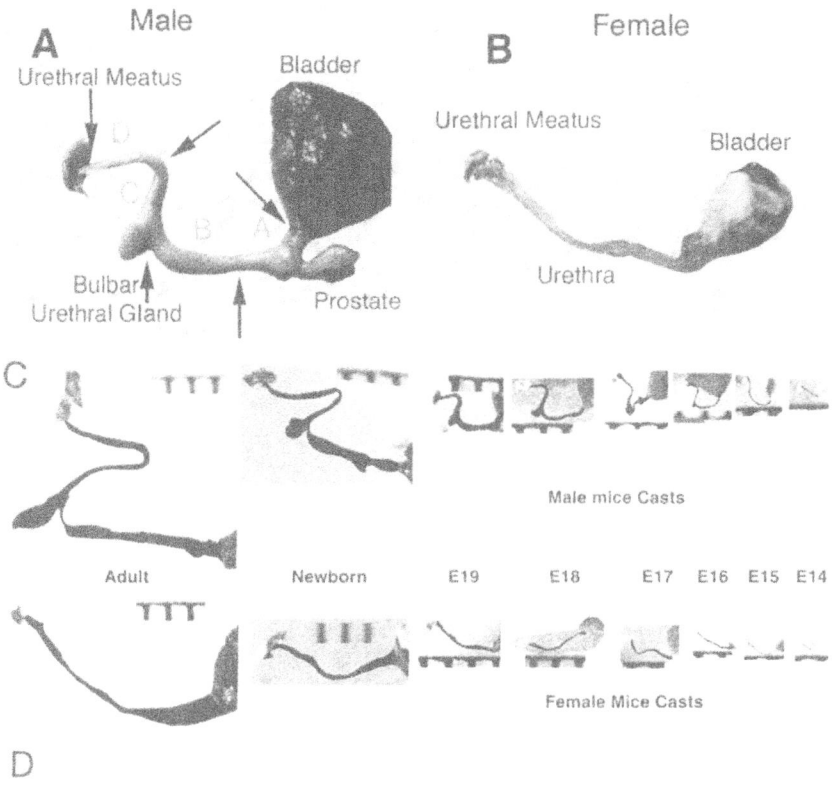

Figure 4.

Plastic resin cast of the male (A) and female (B) internal urethral and bladder anatomy. A. In the male mouse (19 days gestation) note the tortuous course of the urethra compared to the female (B) and the location of key anatomical structures such as the urethral meatus and bulbar urethra. The urethra can be divided into distinct anatomical zones (A to D). A=bladder neck to prostatic rudiment, B=prostatic rudiment to bulbo-urethral gland , C= BUG to point of angulation, D=penile urethra. C. The ontogeny of the male and female urethra from gestational day 14 to an adult. In the adult GT and late in gestation (E17-newborn) the differences between the urethra are obvious. Earlier in gestation (E14 and E15), however, differences between the male and female more subtle. D. lateral views for NIH imaging software, used to generate reproducible values (Table 1) corresponding to pixel count, pixel count per mm, area, length, perimeter, convex perimeter and equivalent diameter.

Table 1. Quantitative measurements of plastic resin casts of the urethra of adult, newborn and embryonic male and female mouse at different gestational stages.

	Pixels	Pixels Counting #	Urethra length	Area	Perimeter	Convex Perim	Equiv. Diam.
	Countin g #	294 pixels per mm	(mm)	(mm^2)	(mm)	(mm)	(mm)
Adult Male	4599	15.64	15.64	7.50	68.89	23.87	3.09
Newborn Male	2734	9.30	9.30	3.94	30.24	15.36	2.24
E19 Male	1314	4.47	4.47	1.13	10.64	7.11	1.20
E18 Male	1037	3.53	3.53	0.68	9.32	6.21	0.93
E17 Male	801	2.72	2.72	0.45	7.61	4.74	0.75
E16 Male	715	2.43	2.43	0.18	5.74	3.96	0.48
E15 Male	528	1.80	1.80	0.08	4.20	3.43	0.32
E14 Male	230	0.78	0.78	0.03	2.00	1.82	0.21
Adult Female	2603	8.85	8.85	4.54	.45.27	20.41	2.40
Newborn Female	1574	5.35	5.35	1.56	16.81	11.67	1.41
E19 Female	813	2.77	2.77	0.38	6.43	5.82	0.69
E18 Female	713	2.43	2.43	0.25	5.91	5.27	0.56
E17 Female	601	2.04	2.04	0.19	5.21	4.19	0.49
E16 Female	459	1.56	1.56	0.11	3.89	3.58	0.37
E15 Female	371	1.26	1.26	0.03	2.95	2.60	0.20
E14 Female	209	0.71	0.71	0.02	1.43	1.28	0.14
	1mm=294 Pixels						

2.1.2 Plastic Resin Cast

Plastic Resin cast of the urethra revealed characteristic differences between the male and the female urethral lumen (Figure 4A). The male urethra was characterized by acute curvatures, the presence of the bulbar urethral gland and a longer length than the female specimens. Differences between the male and female were more subtle at the E14 stage becoming more noticeable as a function of age. At E14, the bulbar urethral gland was not present in the male, becoming visible at the E15 stage. The acute angulation of the male urethra was first noted at E16. Computer measurements of lateral views of the casts revealed quantitative changes in the males and females as a function of age and sex (Figure 4B). Non-linear relationships existed in every category (Table 1).
Scanning electron microscopic analysis at 15 days gestation showed the developing genital tubercle in the male (Figure 5). The two epithelial surfaces of urethral seam were well visualized. Note the exact location of the urethral lumen is hidden inside the seam area.

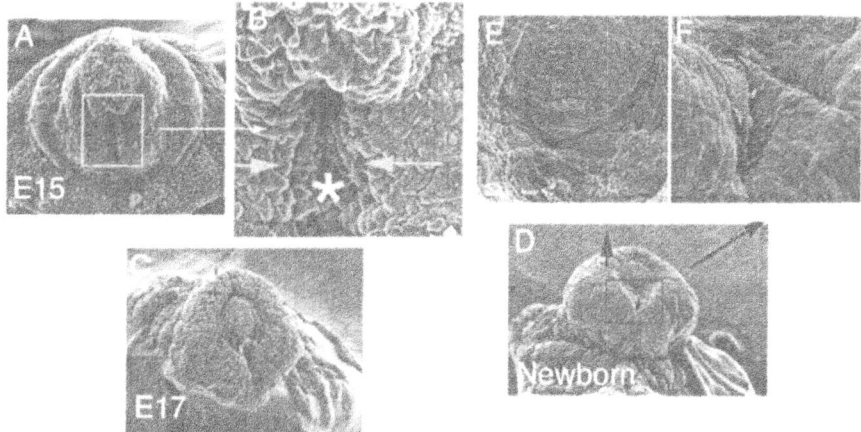

Figure 5.
Scanning electron microscopy of the male mouse genital tubercle. A & B 15 days gestation. C. 17 days gestation D-F. newborn. Note the progressive closure of the urethral seam (*) by fusion of the epithelial edges (arrows).

2.1.3 Immunohistochemistry

At 14 days gestation immunohistochemical staining did not reveal any difference between the male and female GT (Figure 6). Note the differential expression of cytokeratin 7 and 10 in the bladder, proximal and distal urethra. On the distal genital tubercle (GT) the seam is closed consistent with patency studies (Figure 2 and 3). At 16 days gestation the epithelial fusion progresses within the male GT (Figure 7). Note the epithelial cell fusion with subsequent migration of the discarded epithelial cells to the ventral epithelial surface of the GT. Figure 8 and 9 shows the immunostaining of the male and female newborn GT, respectively. Note the progression of the urethra development. In the male,

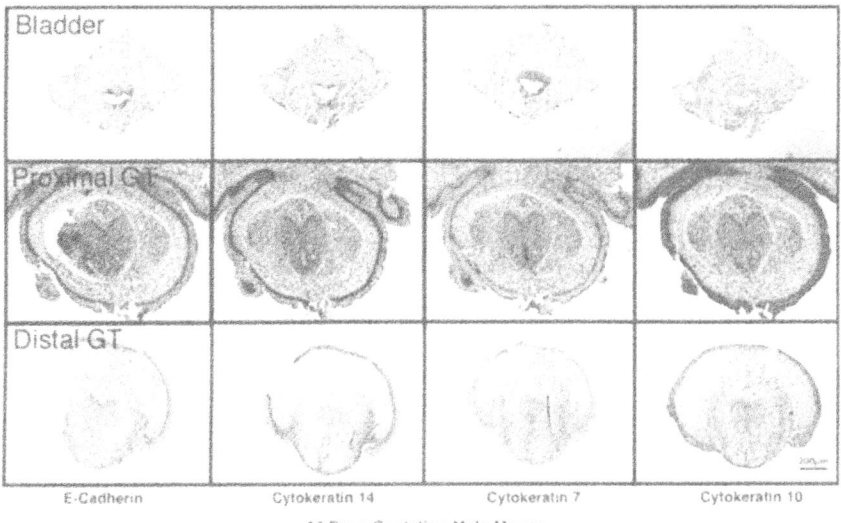

Figure 6.
Immunohistochemical staining of the male genital tubercle at 14 days gestation with E-cadherin, cytokeratin, 14, 7 and 10 , respectively. The top row is the bladder, middle row the proximal genital tubercle and bottom row the distal tubercle. Note that the cytokeratin 7 localizes to the bladder epithelium and urethra in contrast, to cytokeratin 10 which localizes to the skin.

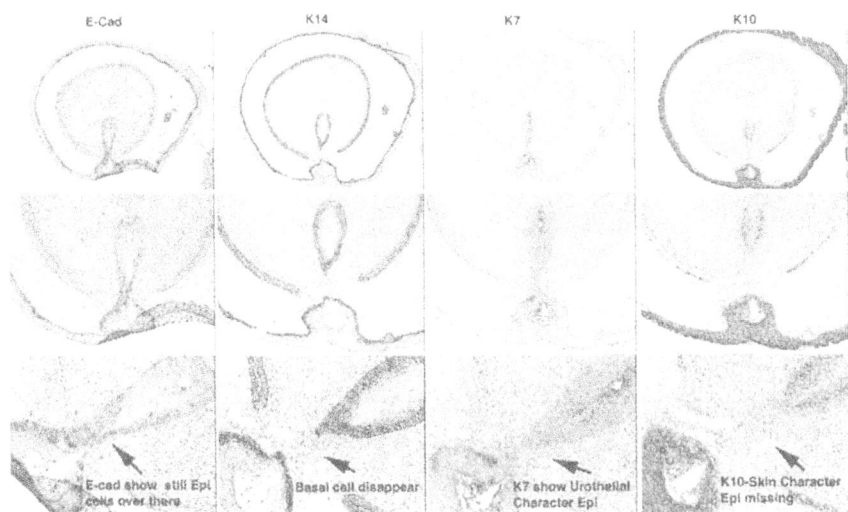

Figure 7.
Immunohistochemical staining with E-cadherin, cytokeratin, 14, 7 and 10, respectively of the male genital tubercle at 16 days gestation at the site of the urethra formation. Successive rows from top to bottom are of greater magnification. Note that the cytokeratin 7 localizes to the bladder epithelium and urethra in contrast, to cytokeratin 10 which localizes to the skin.

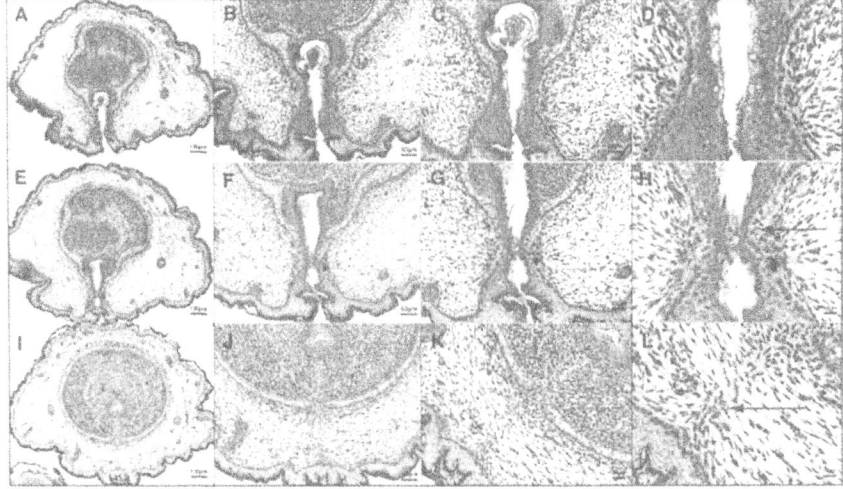

Figure 8.
Normal Newborn Male Mouse Genital Tubercle. Serial section histologic analysis. Note the closed urethra (A-C) and the remodeling urethral seam (D black arrow). The distal open urethra is shown in figures E-L. Note in figures G and H (high power) the skin appears to touch, yet two epithelial surfaces are clearly visualized. Compare the normal male newborn to the normal female (Figure 9).

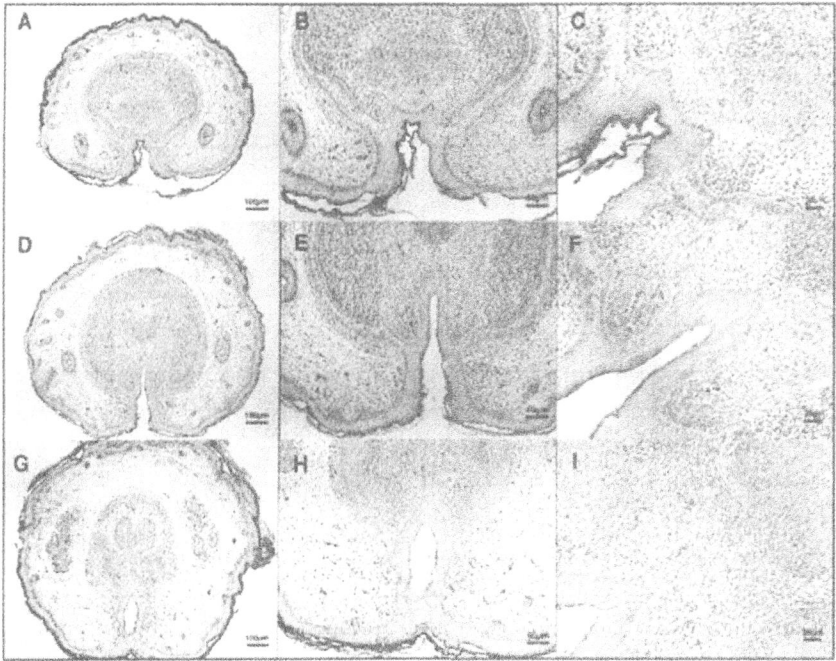

Figure 9.
Normal Newborn Female Mouse Genital Tubercle. Note the open urethra (A-F). The epithelial edges fuse proximally although they do not remodel into a mesenchymal seam (G-I). (C, F, I high power) Compare to normal male (Figure 8).

with the urethral opening is not yet at the tip of the GT. Proximally the epithelial seam has been reabsorbed into the ventral skin or formed into the completed urethra. In contrast, the female urethral remains proximal without epithelial fusion within the GT. Table 2 summarizes the ontogeny of the immunostaining for E-cadherin, cytokeratin 7, 10 and 14. Note the reciprocal expression for cytokeratin 7 and 10.

Table 2.
The differential expression of Cytokeratin 7, 10 and 14, and E- Cadherin of the urinary system epithelium of mouse as a function of gestational age.

Mouse Gestations Days		14	15	16	17	18	19	20	Newborn
Antibodies	Tissues								
Cytokeritin 7	Distal Urethra	-	-	-	-	±	±	+	+
	Proximal Urethra	+	+	+	+	+	+	+	+
	Bladder Epithelium	+	+	+	+	+	+	+	+
Cytokeritin 10	Distal Urethra	+	+	+	+	+	±	±	-
	Proximal Urethra	-	-	-	-	-	-	-	-
	Bladder Epithelium	-	-	-	-	-	-	-	-
Cytokeritin 14	Distal Urethra	+	+	+	+	+	+	+	+
	Proximal Urethra	+	+	+	+	+	+	+	+
	Bladder Epithelium	+	+	+	+	+	+	+	+
E-Cadherin	Distal Urethra	+	+	+	+	+	+	+	+
	Proximal Urethra	+	+	+	+	+	+	+	+
	Bladder Epithelium	+	+	+	+	+	+	+	+

Distal Urethra = Urethral
Plate Area
Proximal Urethra = Urethra
Closed Area
Bladder Epithelium = From Same
Embryonic Body

3. Discussion

The worldwide frequency of developmental abnormalities of the urethra such as hypospadias is increasing over the last three decades without any obvious reason. In the United States, data from two birth defects surveillance systems has also shown an unexplained doubling in the incidence of hypospadias (Paulozzi and Erickson et al., 1997; Paulozzi, 1999). Recently, environmental contamination has been proposed as one of the leading causes (Baskin and Colborn et al., 2001). Because of the lack of precise knowledge on normal human urethral development, the molecular events and their morphological results leading to hypospadias still remains unknown. An ideal animal model for urethral development, analogous to human urethral development would be useful.

The mouse is one of the most commonly utilized research animals yet its application to normal and abnormal urethral development has been ignored because of the divergent postnatal development with formation of the os penis. The urethral morphology in the mouse is also difficult to visualize grossly secondary to the developing prepuce. Our data supports the concept that development of the embryonic mouse genital tubercle and urethra is analogous in many ways to the human and is a useful model for urethral development abnormalities such as hypospadias.

A detailed macroscopic and microscopic evaluation of the developing fetal mouse genital tubercle has demonstrated many similarities to the morphogenesis of the human phallus and urethra (Figure 8). The structure of mouse and human corpora is very similar with sinusoidal anatomy surrounding by a tunica composed of connective tissue. The foreskin develops in a comparable fashion to the human with advancement over the glans and proximal extension of preputial lamellae, which delaminate postnatally to form the preputial glanular space. Moreover, in both the human and the mouse, the urethra forms in a proximal to distal fashion with closure of the two epithelial surfaces of the urethral folds to form the urethral seam.

In humans, by the end of first month of gestation, the hindgut and future urogenital system reach the ventral surface of the embryo as the cloacal membrane. The cloacal membrane is divided by the urorectal septum into a posterior or anal half and an anterior half, called the urogenital membrane. Three protuberances appear around the urogenital membrane. The most cephalad protuberance is called the genital tubercle. The other two protuberances are described as the genital swellings, which flank the urogenital membrane at each side. Up to this point the male and female genitalia are indistinguishable. Masculinization of the external genitalia occurs under the influence of testosterone in response to a surge of luteinizing hormone from the pituitary gland. An initial sign of masculinization is an increase in the distance between anus and genital structures, followed by elongation of the phallus, formation of the penile urethra from the urethral groove and development of the prepuce (Jirasek et al., 1968; Hinman, 1993). In mouse, this stage of embryological development is the same as humans with the outgrowth and swelling of the genital tubercle. As seen in Figure 4C, the cast studies of the male and female urethra at 14 days gestation are similar. In the mouse, the early masculinization of the urethra is characterized by the appearance of the bulbourethral glands at the E15 stage (Figure 4C). Plastic resin cast studies reveal the discordant development of the urethral structure in both sexes at E16, with the male undergoing acute angulations through its course from the bladder neck to the external urethral meatus (Fig 4A, B, and C).

As in humans, gestational age and sex are the main variables that define the gross morphology of the mouse genital tubercle. The use of plastic resin cast technique for internal urethral morphology provided accurate quantitative measurements to compare normal male and female urethral development (Fig 4D and Table 1). Defining normative data is critical to distinguish mouse models of abnormal urethral development such as exogenous endocrine disruptor exposure or genetically altered mice.

Human external genitalia development is characterized by the indifferent stage at 8 weeks of gestation. The urethral groove on the ventral surface of the phallus lies between the paired urethral folds. The penile urethra forms as a result of fusion of the medial edges of the endodermal urethral folds. The ectodermal edges of the urethral groove fuse to form the median raphae. By 12 weeks the coronal sulcus separates the glans from the penile shaft. The urethral folds have completely fused in the midline on the ventrum of the penile shaft. During week 16 of gestation the glandular urethra appears. At about 8 weeks of gestation, low preputial folds appear on each side of the penile shaft, which join dorsally to form a flat ridge at the proximal coronal edge. The ridge does not completely encircle the glans because it is blocked at the ventrum by incomplete development of glandular urethra. Thus, the preputial fold is transported distally by active growth of the mesenchyme between it and glanular lamella. The process continues until the preputial fold or foreskin covers the whole glans. Fusion is usually present at birth but subsequent desquamation of the epithelial fusion allows the prepuce to retract. If the genital folds fail to fuse the preputial tissues do not form ventrally. Consequently in hypospadias preputial tissue is absent on the ventrum and it is excessive dorsally (Jirasek et al., 1968; Hinman, 1993).

The continuation of urethral plate and the fusion of genital folds around the plate to the tip of phallus has been recently described by immunohistochemical studies in the human fetus (Kurzrock and Baskin et al., 1999). The characteristic cytokeratin expression of the epithelial lining of distal urethra has been shown to be endodermal in origin in contrast to the classically described ectodermal embryological origin of anterior urethra (Paulozzi and Erickson et al., 1997). These findings lead to the concept of the endodermal differentiation theory in which the phenotype of the epithelial cell is determined by the underlying mesenchymal signaling (Kurzrock and Baskin et al., 1999). Kurzrock et al. demonstrated that urothelium that is endodermal origin can differentiate into stratified squamous epithelium (Kurzrock and Baskin et al., 1999). In the development of mouse urethra, a similar pattern of cytokeratin expression is observed. As depicted in Figures 6 and 7, the epidermis at the tip of the mouse phallus revealed ectodermal cytokeratin 10 expression, whereas, the anterior urethra showed cytokeratin 7 expression, consistent with an endodermal origin. This data supports the theory that the distal urethra in the mouse is an extension of the endodermal lined urogenital sinus as in human.

Glenister is cited as the first author to describe that the urethral plate grows to reach the distal tip of the phallus in both sexes meeting the ingrowth of ectoderm. Glenister also noted that the urethral plate extended to the tip of the genital tubercle from its inception, and that it developed from a thickening in the anterior wall of the anterior cloaca and later from the fused converging walls of the anterior portion of the cloaca. Because he observed stratified squamous epithelium in the glandular urethra, he deduced that glandular urethra is derived from ectoderm rather than endoderm. Hence, he concluded a statement contradictory to his observations (Glenister, 1954). The endodermal differentiation theory is consistent with the perplexing behavior of urethral plate which extends from urogenital sinus which is endodermal in origin (Kurzrock and Baskin et al., 1999). The change of epithelial structure and cytokeratin expression differences in the anterior urethra is, also observed in mouse model (Table 2). The molecular basis of this epithelial-mesenchymal interaction in the anterior urethra remains to be defined.

As in human, in the mouse we observed the fusion of genital folds to form the entire urethra including into the glans. In Figure 5, scanning electron microscopy images reveal the progressive genital fold fusion to the tip of the phallus as the mouse embryo develops. The entire mouse urethra develops from the urethral plate, an extension of the urogenital sinus, extended to tip of the phallus, and maintains patency and continuity throughout urethral development. According to the old ectodermal invagination theory, a solid ectodermal ingrowth of epidermis, plugs the glandular urethra and later canalizes to form the patent anterior urethra (Glenister, 1954). However, human fetal studies fail to show any solid plug at the external urethral meatus in any of the gestational stages of human (Kurzrock and Baskin et al., 1999). Additionally, the uninterrupted endodermal origin of the urethral plate is also supported by Kluth's observation that a tiny canal can often be found along the entire plate which connects the genito-urinary sinus to the tip of the human phallus (Kluth et al., 1988). We did not see any evidence of ectodermal ingrowth and/or a plug of epithelial cells within the urethral lumen. Our Indian ink studies revealed a patent, developing mouse urethra at all stages (Figure 2). Also, we were able to pass a human hair through the urethra from the external meatus of the urethra into the bladder in the mouse embryo (Figure 3). These physical finding support the concept that the lumen of the mouse urethra remains patent as in human.

The mouse genital tubercle is also a useful model to study the formation of the urethral seam. Figure 8 shows the mouse epithelial seam with two epithelial surfaces fusing and remodeling into a mature urethra. Histologically, this resembles the folding of the urethra in the human with fusion and subsequent seam remodeling (Baskin et al., 1997; Baskin et al., 1998). We observed that the embryonic mouse goes through a natural state of hypospadias with the urethral opening on the ventral site of the genital tubercle as seen in human. With time, the urethral opening migrates to a terminal position, as in man. The main difference between the male and female genital tubercle is the lack of fusion of the urethral folds in the female (Figure 9). The histological analysis of the developing mouse urethra documents the importance of the epithelial seam remodeling during morphogenesis (Figure 7 and 8). In the mouse, gender seems to be the main factor deciding the length of urethral fusion, possibly via sex specific hormonal cascades.

There are differences between the development of the human and mouse urethra. In humans, the edges of the urethral folds fuse to form the completed urethra whereas, in the mouse, the true seam forms half way down the closed groove (Figure 7 and 8). This seam is subsequently remodeled with a forward progression of the urethral opening. The epithelial cells that are ventral to the seam are excluded from the true urethra. The excluded epithelial cells are seen as a tail of remodeling cells (Figure 8).

Another example of tissue morphogenesis characterized by epithelial seam formation is the palate formation (Griffith and Hay, 1992). Epithelial-mesenchymal interaction has been shown to play a major role in this fusion process of palatal plates, however, similar findings such as a simultaneous loss of cytokeratin expression and gain of either smooth muscle alpha - actin or vimentin has not been confirmed for urethral plate fusion (Baskin et al., 2001). Rather, epithelial cell migration into the true urethra and the subsequently reabsorbed epithelial tail appears to be the dominant process (Figure 7). The disappearance of the epithelial seam occurs concurrently with a highly focalized wave of apoptotic activity suggestive of a band of epithelial apoptosis. However, the apoptosis was found to arise in the mesenchymal cells that had replaced the migrating epithelial seam cells (Baskin et al., 2001). Further studies are necessary to delineate the molecular mechanism of urethral seam formation.

4. Conclusions

To study the pathogenesis of urethral developmental abnormalities we have shown that the mouse genital tubercle is a useful animal model. To understand abnormal development, we have defined the anatomy of the normal mouse genital tubercle and urethra. Defining normal anatomy is critical to understanding abnormal development such as malformations in the urethral seam or hypospadias.

Supported by a grant from the National Institute of Health #DK058105.

REFERENCES

Baskin, L. S., Y. T. Lee, et al., 1997, Neuroanatomical ontogeny of the human fetal penis, *Br J Urol*.79(4): 628-40.

Baskin, L. S., A. Erol, Li Y.W., Cunha G.R., 1998, Anatomical studies of hypospadias, *J Urol*. 160(3 Pt 2): 1108-15; discussion 1137.

Baskin, L. S., 2000, Hypospadias and urethral development, *J Urol*. 163: 951 – 956.

Baskin, L. S., Erol, A., Jegatheesan, P., Li, Y., Liu, W., Cunha, G.R., 2001, Urethral seam formation and hypospadias, *Cell Tissue Res*. 305: 379-387.

Baskin, L. S., Colborn, T., et al., 2001, Hypospadias and endocrine disruption: is there a connection? *Environmental Health Perspectives*. 109: 1175-1183.

Glenister, T. W., 1954, The origin and fate of the urethral plate in man, *J Anat*. 88: 413-25.

Griffith, C. M., Hay, E. D., 1992, Epithelial-mesenchymal transformation during palatal fusion: carboxyfluoroscein traces cells at light and electron microscopic levels, *Development*. 116: 1087-1099.

Hinman, F. J., 1993, Penis and male urethra, *UroSurgical Anatomy*. Philadelphia: WB Saunders Chapt. 16, p. 418.

Jirasek, J. E., Raboch, J. and Uher, J., 1968, The relationship between the development of gonads and external genitals in human fetuses, *Am J Obstet Gynecol.* **101:** 830.

Kluth, D., Lambrecht, W., Reich, P. , 1988, Pathogenesis of hypospadias – more questions than answers, *J Pediatric Surg.* **23:** 1095 – 1101.

Kurzrock, E., Baskin, L., et al., 1999, Ontogeny of the Male Urethra: Theory of endodermal differentiation, *Differentation.* **64:** 115-122.

Kurzock, E., Baskin, L. S., Li, Y., Cunha, G. R., 1999, Epithelial–mesenchymal interactions in development of mouse genital tubercle, *Cell Tissues and Organs.* **164:** 1015-1020.

Paulozzi, L., Erickson, D., et al., 1997, Hypospadias Trends in Two US Surveillance Systems, *Pediatrics.* **100(5):** 831-834.

Paulozzi, L. J., 1999, International trends in rates of hypospadias and cryptorchidism, *Environ Health Perspect.* **107(4):** 297-302.

ANATOMICAL STUDIES OF THE FIBROBLAST GROWTH FACTOR-10 MUTANT, SONIC HEDGE HOG MUTANT AND ANDROGEN RECEPTOR MUTANT MOUSE GENITAL TUBERCLE

Selcuk Yucel, Wenhui Liu, Dwight Cordero, Anne Donjacour, Gerald Cunha and Laurence S. Baskin*

1. Abstract

1.1. Objectives

Congenital genital abnormalities have a diverse spectrum from hypospadias to cloacal anomalies. The molecular events in the normal and abnormal development of the genital tubercle (GT) are still obscure. Genetically engineered mice with specific gene deletions that affect genital anatomy are a useful tool to better understand the etiology of genital abnormalities. In this study, we compared the genital tubercle anatomy of the androgen receptor (AR) deficient, fibroblastic growth factor (FGF)–10 deficient and Sonic HedgeHog (Shh) deficient mutant male mice to that of the wild type male and female mouse.

1.2. Materials and Methods

The lower pelvis of the androgen receptor deficient, FGF-10 deficient, Shh deficient mutant male and wild type male and female mouse at different gestational days (E13 - 21) and post natal ages (1day - 1 week) were studied. GTs were imaged, serially sectioned and stained immunohistochemically with antibodies raised against E-Cadherin, Cytokeratin 7, 10 and 14. Serial sections of the GTs were selected and three-dimensional computerized images were created to better elucidate the anatomy.

1.3. Results

AR deficient mutant male mouse revealed a distinctive GT anatomy, different from both sexes. The corporal bodies and glans remained hypoplastic

* Laurence S. Baskin, M.D., Department of Urology, University of California, San Francisco, CA, 94143-0738, (415) 476-1612,(415) 476-8849 (FAX), lbaskinl@urol.ucsf.edu

whereas the urethral spongiosa was more developed than the wild type female counterpart. This finding is consistent with the AR mutant mouse being a unique morphologic phenotype distinct from the normal male and female. FGF-10 deficient mutant male mouse revealed normal corporal bodies with failure of the urethral plate to fuse ventrally consistent with hypospadias. The Shh deficient mutant mouse demonstrated complete agenesis of GT outgrowth and a persistent cloaca.

1.4. Conclusion

Animal models bred by gene knockout technology or natural occurring mutants contribute to the basic understanding of normal and abnormal GT development. The anatomy of the these three mutant mice confirms the importance of the androgen receptor , FGF-10 and Shh in genital development.

2. INTRODUCTION

Hypospadias is one of the most common birth defects involving the external genitalia in humans with a rate of 1 in 250 births. In recent surveys conducted in the United States and Europe, the incidence of hypospadias is reported doubled without addressing any obvious reason (Paulozzi, 1997; Paulozzi, 1999). Although environmental contamination has been claimed as a possible explanation, the exact mechanism of events leading to hypospadias is not known (Baskin, 2001). The molecular events leading to other more extensive congenital genital abnormalities such as persistent common cloaca are also unknown.

The early stage of genital tubercle (GT) development in both sexes is morphologically indistinguishable with the potential to differentiate into either male or female genitalia. Since disruption of androgen signaling can frequently result in feminization of the genitalia which frequently includes hypospadias, androgens are known to play a major role in male type GT development (Kim et al., 2001; Anderson and Clark, 1009; Baskin et al., 1997; Kurzrock et al., 2000; Brinkmann, 2001). Similarly, in the normal mouse model, development of the os penis, which consists of a proximal segment of hyaline cartilage and membrane bone is also dependent on androgen function (Murakami, 1987). However, outgrowth and patterning of the genitalia starts to be executed before the onset of androgen synthesis. Human genital development begins almost 2 weeks before the testosterone production in fetal testes (Kalloo et al., 1993). Hence, early patterning and preprogramming of the GT is regulated independent of androgens. The genetic mechanisms regulating the early development of the GT are still obscure.

The development of vertebrate limb bud has many similarities with the development of GT. Early development of both structures is initiated by budding from the lateral plate mesoderm. Sustained outgrowth of the budding in limbs and GT are controlled by overlying epithelium as shown in the study by Murakami and Mizuno (Murakami and Mizuno, 1986). Even though tissue recombination studies using cross matching of epithelium and mesenchyma of limbs and GT indicate that the epithelial – mesenchymal interaction is crucial in the GT morphogenesis, the responsible mechanisms are still unknown. Additionally, similarities between GT and limb development extend to the molecular level. The posterior Hox genes are involved in patterning structures at the terminus of the main body axis specifically the genitalia and hindgut, and at the terminus of the limbs that are digits (Dolle et al., 1993; Kondo et al., 1997; Mortlock and Innis, 1997; Warot et al., 1997; Zakany et al., 1997). Hoxa13 or Hoxd13 defective homozygous mutants show agenesis of the GT and digits, and heterozygous mutants reveal patterning defects in limbs and genitalia (Dolle et al., 1993; Warot et al., 1997; Zakany et al., 1997). In humans, HOXA 13 gene mutations has been found in phenotypes observed in Hand-Foot-Genital Syndrome, which affects development of the distal limbs and urogenital system (Goodman et al., 2000). Recently, conserved functions of signaling molecules have been reported in limb and genital development. For example, Wnt5a expression is preserved in a distal to proximal fashion in both the limb bud and GT development (Yamaguchi et al., 1999).

Fibroblastic growth factors (FGF) are important signaling molecules in embryogenesis and morphogenesis of many tissues (Cohn et al., 1995; Wall and Hogan, 1995; Yamasaki et al., 1996; Bellusci et al., 1997; Martin, 1998; Ornitz, 2000). Large FGF family signaling elements have recently been found to command the growth and differentiation of the GT. It has been suggested that FGF-10 functions in limb, brain and lung development and also wound healing (Yamasaki et al., 1996; Ohuchi et al., 1997). Parallel to the defects in limb development, FGF-10 knockout mutant animal studies reveal developmental GT abnormalities (Ohuchi et al., 1997).

Sonic hedgehog is another critical gene, which is necessary for normal limb, lung, gut and GT development (Riddle et al., 1993; Ramalho-Santos et al., 2000; Roberts et al., 1995; Motoyoma et al., 1998; Bellusci et al., 1997; Dolle et al., 1991; Podlasek et al, 1999). During mouse GT development, Shh is initially expressed in the urogenital sinus and the distal tip of the urethral epithelium (Haraguchi et al., 2001). Gut and GT share some similar tissue patterning processes such as endodermal tubular structures. Moreover, embryonic gut development involves both regionalization along the anteroposterior (AP) axis, as well as radial patterning of the gut tube to achieve development of epithelium, connective tissue, muscle layers and glands. Shh has been demonstrated as the crucial element of this growth and differentiation in gut (Ramalho-Santos et al., 2000; Riberts et ak,m 1995). Similarly, overexpression of Shh in lung epithelium has been reported to cause epithelial and mesenchymal proliferation (Bellusci et al., 1997). Haraguchi et al. has pointed out the role of Shh expression in the initiation of GT outgrowth and the morphological differentiation of GT (Haraguchi et al., 2001).

A precise comparative functional anatomy of the GTs of genetic mutant and natural occurring mouse models has not been performed. The goal of this study is to define the functional anatomy of the FGF-10, Shh and androgen receptor deficient mouse GTs as a model for abnormal genitalia and urethral development.

2.1. Materials and Methods:

This investigation was approved by the Committee on Animal Research at our institution.

2.2. Androgen Receptor Deficient Mutant Mouse

The mice carrying the AR deficient mutation (Tfm) and Tabby (Ta) as a coat colour marker were obtained from as active colony in the Cunha laboratory (Department of Anatomy, University of California San Francisco, CA) . Female mice, heterozygous for the Tfm and the Ta ($X^{Tfm, +}$ / $X^{+,Ta}$), were mated with male mice ($X^{+,Ta}$ / Y) (Murakami,

1987). Gestational day of the pups was designated as 0 day of gestation on the day of vaginal plug in the mother mouse. Age of the newborns was designated as 0 days old on the day of birth. Whole pelvis of the mice specimens in both sexes were fixed in formalin and embedded in paraffin. Sex differentiation of mouse fetuses and AR deficient mutant mice were done by direct examination of testes under the dissecting microscope. The GT specimens were serially sectioned at 4 microns. The whole fetal specimens containing the GT was preserved with 2 to 4 histological sections per glass slide. Each 3rd section was stained with hematoxylin and eosin, and Masson's trichrome at each group.

2.3. FGF-10 Gene Deficient Mutant Mouse

Multiple GTs at 16, 18 and 20 days of gestation and newborn FGF-10 deficient mutant and wild type mouse were microdissected, imaged, fixed in formalin and processed into paraffin. FGF-10 deficient mice were identified as described previously (Sekine et al., 1999). Gestational day, age and sex were designated as stated above. The GT specimens were serially sectioned at 4 microns. The whole fetal specimens containing the GT was preserved with 2 to 4 histological sections per glass slide. Each third section was stained with hematoxylin and eosin and immunohistochemically using antibodies to cytokeratin 7, 10, 14 and to E-cadherins as described below. 3D reconstruction of the developing GT focusing on seam formation and remodeling was performed as stated below.

2.4. Sonic Hedge Hog Gene Deficient Mutant Mouse

Multiple embryonic mice at thirteen days gestation null for Shh gene were analyzed. Shh deficient mutant mouse with a targeted deletion of exon 2 of the gene were used (Chiang et al., 1996). The pelvic specimens were fixed in formalin, embedded in paraffin, and serially sectioned at 4 microns. The whole fetal specimen containing the pelvis and the GT was preserved with 2 to 4 histological sections per glass slide. Each third section was stained with hematoxylin and eosin. Immunohistochemical staining was performed on select sections for the epithelial proteins Cytokeratin 7 and 10. 3D reconstruction of serial sections stained with cytokeratin markers allowed in depth analysis of the whole pelvis and perineal anatomy of the Shh knock-out as described below. Results were compared to age matched wild type animals.

2.5. Immunohistochemistry

GTs from embryonic male and female mice ages 13 to 21 day gestation, newborn and adult mice were fixed in formalin, and processed into paraffin. Fetuses from at least three different liters at each time point were analyzed. Sex was determined by gonadal histology and the presence or absence of internal female organs. 4 μm transverse sections were cut using a microtome (RM 2135 Leica, Bannackburne, IL). Serial sections were immunostained with antibodies to E-cadherin, cytokeratins 7, 10, and 14, and with hematoxylin and eosin (Baskin et al., 1998). The avidin-biotin-peroxidase procedure was employed using Vectastatin ABC kits (Vector laboratories, Burlingame, CA) with cobalt intensification. Primary antibodies to E-cadherin, and cytokeratins 7, 10, and 14 were obtained from Sigma (St Louis, MO). Biotinylated anti-rabbit and anti-mouse IgG were obtained from Amersham International (Arlington Heights, IL). Peroxidase linked avidin/biotin complex reagents were obtained from Vector Laboratories (Burlingame, CA). Negative controls utilized IgG of the same species in place of and at the same dilution as the primary antibody. The purified mouse IgGs were obtained from Zymed Corp. (So. San Francisco, CA). Slides were imaged using a Leaf Systems (Southborough, MA) Lumina scanner system attached to a Zeiss microscope (Thornwood, NY). Images were collected on an Apple Power Macintosh G-4 computer (Sunnyvale, CA) using Adobe Photoshop version 7.0 software (Adobe Systems, Mountain View, CA).

2.6. Three-Dimensional Computerized Image Reconstruction

Structural anatomical computer reconstruction images were created with a digital camera (Nikon, Melville, NewYork), SURFdriver 3.5 software (SURFdriver, University of Hawaii and University of Alberta) and a Apple Power Macintosh G4 (Sunnyvale, California). Every third to fifth section was digitized. The corporal bodies, urethra, rectum, bone,and skin were manually outlined and checked against the original histologic sections. Three-dimensional (3D) analysis of the size and localization corporal bodies, glans, urethra, urethral spongiosum, bone, hindgut and skin was performed in the x and y axes as an animated motion picture and views of interest were captured as static images (Yucel and Baskin, in press).

2.7. Results:

2.1.1. Androgen Receptor (AR) Deficient Mutant Model

The gross morphology of the fetal GT's in wild type male, female and AR deficient mutants at 19 day of gestation is shown serially (Figure 1).

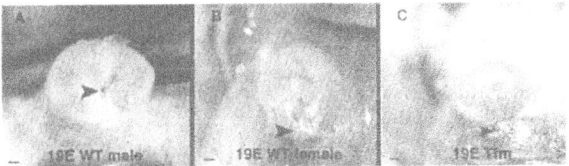

Figure 1
Genital tubercles at 19 days gestation of (A) wild type male, (B) wild type female, (C) AR deficient mutant (Tfm) mouse. Arrowheads shows the urethral openings. Bar represents 200 μm.

In direct examination, AR deficient mutant and wild type female GT appeared to be similar. Male and female type GTs were distinct with different urethral meatus shape and localization. Tfm GT was noted to be smaller compared to wild type male and female (Figure 2).

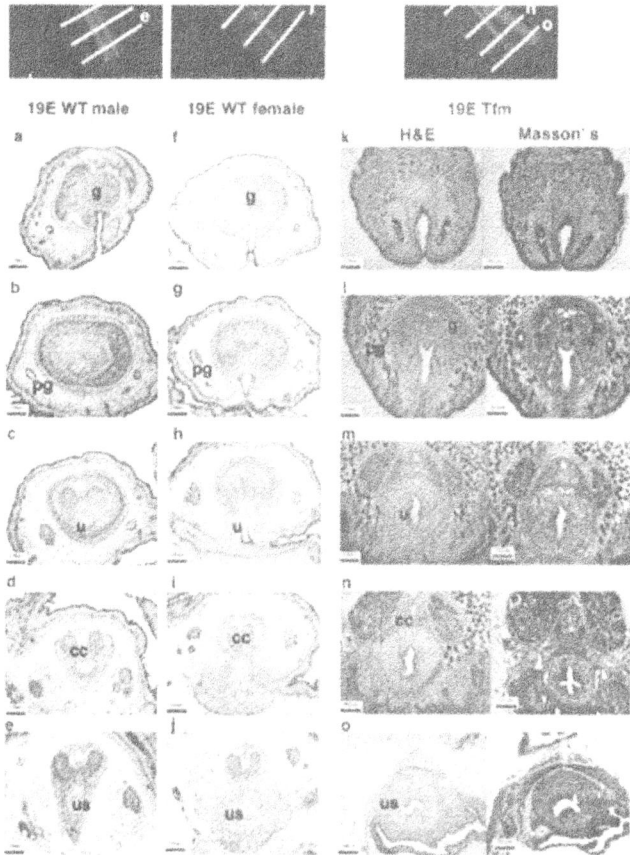

Figure 2
Representative sections of 3D reconstructed model of fetal mouse genital tubercles are seen. Left, middle and right columns show wild type male, wild type female and AR deficient mutant (Tfm) genital tubercles, respectively. The bar represents 100 μm.

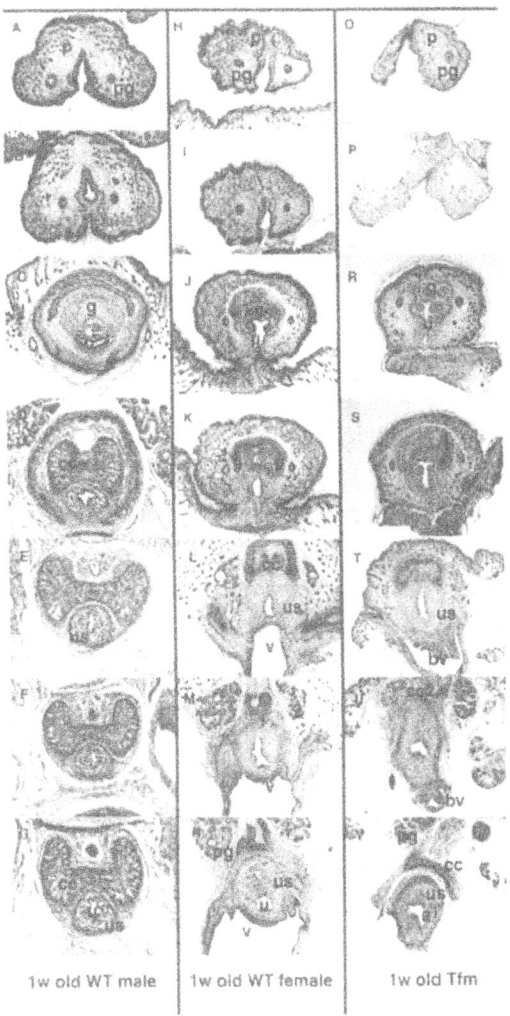

1w old WT male | 1w old WT female | 1w old Tfm

Figure 3
Adult mouse genital tubercle, oriented from proximal to distal. From A to G, wild type male genital tubercle
sections are seen with well demarcated corpus cavernosum and glans, highly vascular corpus spongiosum
covering mainly the ventral side of the urethra. From H to N, wild type female genital tubercle sections are
observed to have well demarcated but thinner corpus cavernosum and glans, corpus spongiosum rich in fibers
covering mainly the dorsal side of urethra. From O to V, AR deficient mutant (Tfm) genital tubercle sections
showing poorly developed corpus cavernosum and glans, and a broader corpus spongiosum rich in fibers
covering both sides of urethra.

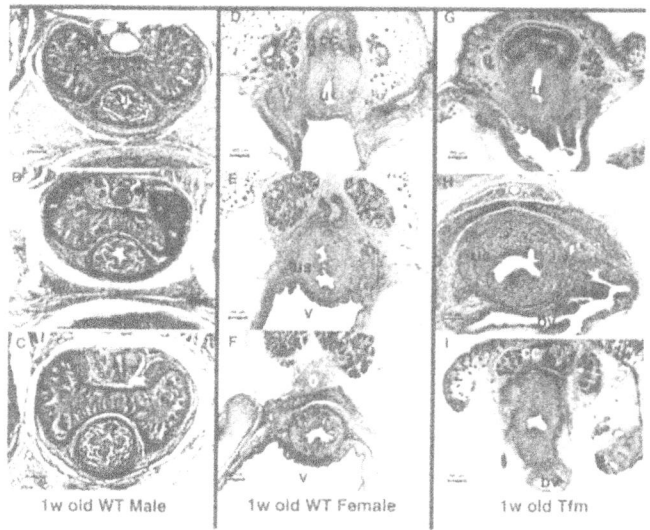

Figure 4
Three representative sections from three different phenotypes stained with Masson's trichrome are depicted. From A to C, wild type male sections are shown to have a highly vascular corpus spongiosum covering mainly the ventral side of the urethra. From D to F, corpus spongiosum is rich in fibers, mainly vascular dorsally and covering mainly the dorsal side of urethra in wild type female genital tubercle. From G to I, AR deficient mutant (Tfm) mouse sections reveal a unique structure of the corpus spongiosum covering the entire urethra compared to the male and female wild type.

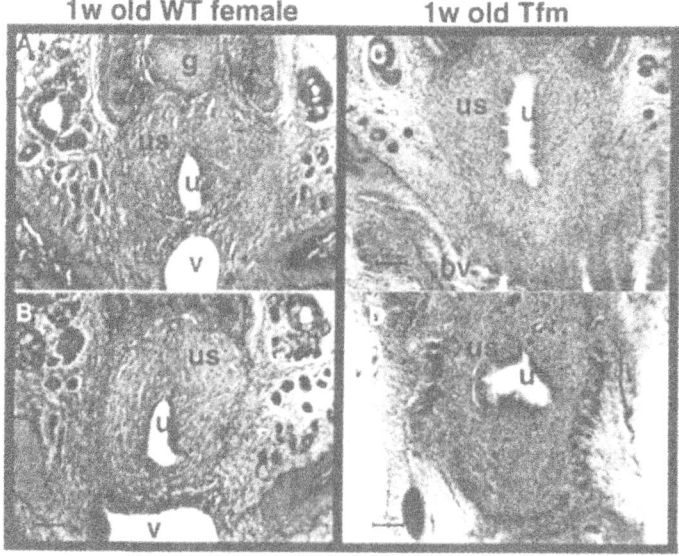

Figure 5
40X magnification of different sections from the wild type female (A and B) and AR deficient mutant (Tfm) (B and D) mouse depicting differences in the corpus spongiosum.   Sections A, C and B, D are at the same level of the genital tubercles. Note structural differences in the spongiosal tissue thickness and pattern of vascularity.

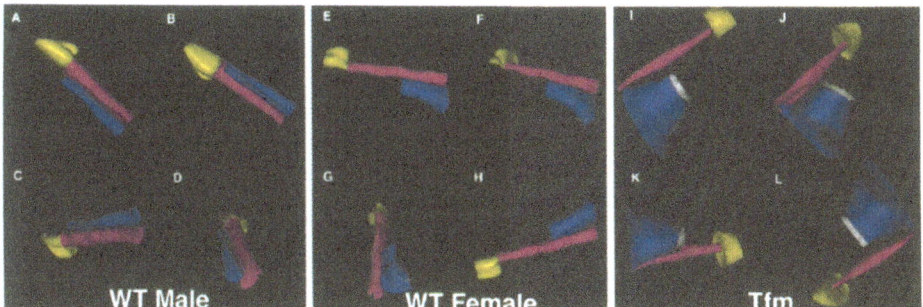

Figure 6

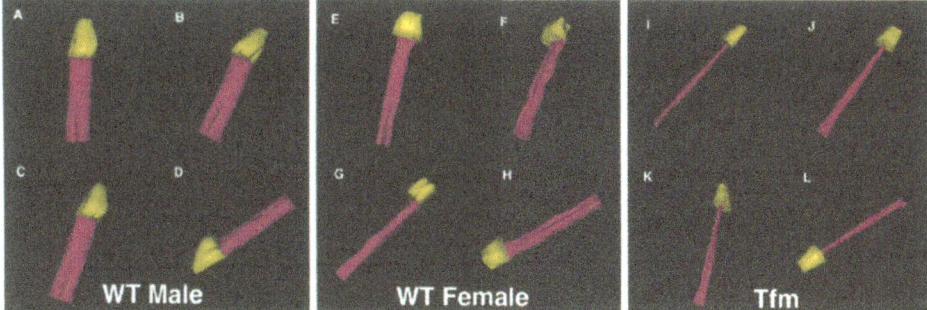

Figure 7

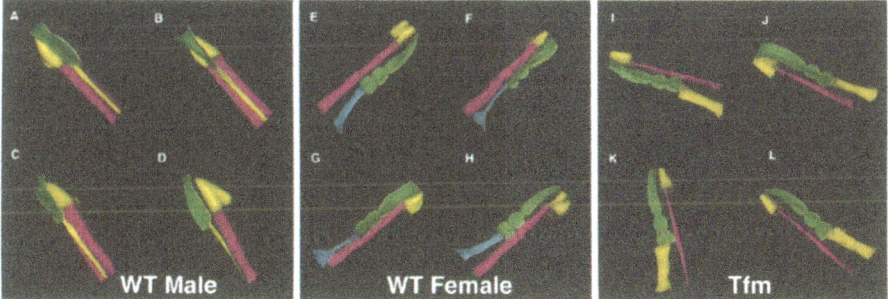

Figure 8

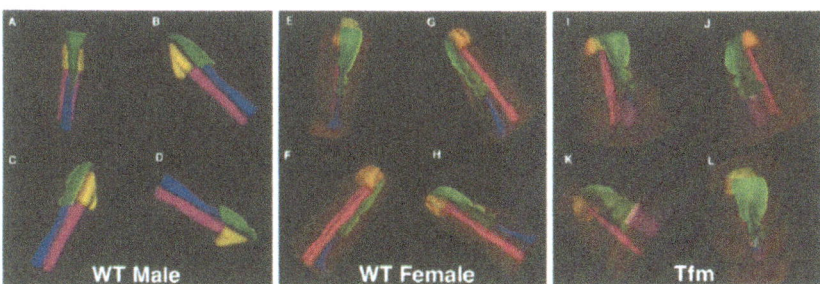

Figure 9

Figure 6
Selected static images of animated genital tubercle 3D reconstruction in three different phenotypes (A to D, wild type male; E to H, wild type female; I to L, Tfm) are seen. Only glans and corpus cavernosum are depicted. The significant discrepancy between the sizes of corporal bodies and glans are shown. AR deficient mutant (Tfm) mice is observed to have the smallest corpora cavernosa and glans in size among the three different phenotypes.

Figure 7
Selected static images of genital tubercle 3D reconstructions in three different phenotypes (A to D, wild type male; E to H, wild type female; I to L, AR deficient mutant (Tfm)). Only the corpus cavernosum, glans, corpus spongiosum and tubular urethra are depicted. The significant discrepancy between the anatomical structure of the corpus spongiosum around the urethra is shown. Tfm mouse is observed to have the wider corpora spongiosum all around the urethra where it is mainly ventral in wild type male and it is mainly dorsal in wild type female.

Figure 8
Selected static images of genital tubercle 3D reconstructions in three different phenotypes (A to D, wild type male; E to H, wild type female; I to L, AR deficient mutant (Tfm)) are seen. Only the glans, corpus cavernosum, tubular urethra and urethral plate are depicted. The urethral opening localization and urethral plate structure is identical in wild type female and Tfm mice.

Figure 9
Selected static images of genital tubercle 3D reconstruction sin three different phenotypes (A to D, wild type male; E to H, wild type female; I to L, Tfm) are seen. All anatomical structures, skin, glans, corpus cavernosum, corpus spongisoum, urethral plate and tubular urethra are depicted.

The blind vaginal pouch was more easily observed at 1 week postnatal. Representative hemotoxylin-eosin staining patterns revealed the corporeal bodies, urethra, urethral spongiosum and glans in the fetal GTs as three distinct morphological phenotypes (Figure 2 and 3).

In the wild type male GT, the corpus cavernosum is observed as a well developed structure with two crura extending into the pelvis. In the wild type female, the corpus cavernosum was also well demarcated. Nevertheless, the AR deficient mutant male mouse had significantly shorter and thinner corporeal structures. The glans was also hypoplastic compared to the female counterpart. The localization of the urethral opening was comparable in the AR deficient mutant male and wild type female mouse (Figure. 1, 2 and 3). The urethral opening was closer to the tip of the GT in the wild type male unlike in the AR deficient mutant male mouse. The urethral spongiosum patterns were different in the three different phenotypic mice as shown in Figure 4 and 5.

2.1.2. FGF-10 Deficient Mutant Mouse Model

The GTs of the FGF-10 deficient mutant male mouse exhibited abnormalities of the urethral seam and surrounding mesenchyme when compared to both normal male and female newborns (Figure 10- 14).

Urethral spongiosum in the wild type male covered the urethra mainly ventrally with a relatively thinner spongiosal tissue between the corporal bodies and urethra. Urethral spongiosum was characterized with abundant vascular lacunas in the male type. Wild type female urethral spongiosum was densely composed of elastic fibers occupying mainly the dorsal side of the urethra. Vascular structures were found between the urethra and corporal bodies in the spongiosal tissue. AR deficient mutant male mouse revealed a different type of urethral spongiosum by covering the urethra ventrally and dorsally in a symmetric fashion. In contrast to wild type female in which thinner spongiosal tissue was seen between the anterior vaginal wall and the urethra. The AR deficient mutant male mouse had a stronger spongiosal tissue between these two structures. The vascular distribution in the urethral spongiosum in AR deficient mutant male was similar to the wild type female. Three dimensional (3D) reconstructed images emphasize the structural differences between the GT of wild type male, female and the Tfm mouse. The size and shape differences between the corporal bodies are significant as shown (Figure 6 - 9).

The urethral folds failed to form a midline seam line and remodel into a tubular urethra.

The supporting mesenchyme also failed to surround the epithelial cells of the advancing urethral plate. This failure resulted in a relatively flat ventral aspect of the GT. The urethral opening in FGF-10 deficient mutant male mouse assumed a hypospadiatic position but clearly had developed more than the female specimens.

Newborn WT Male

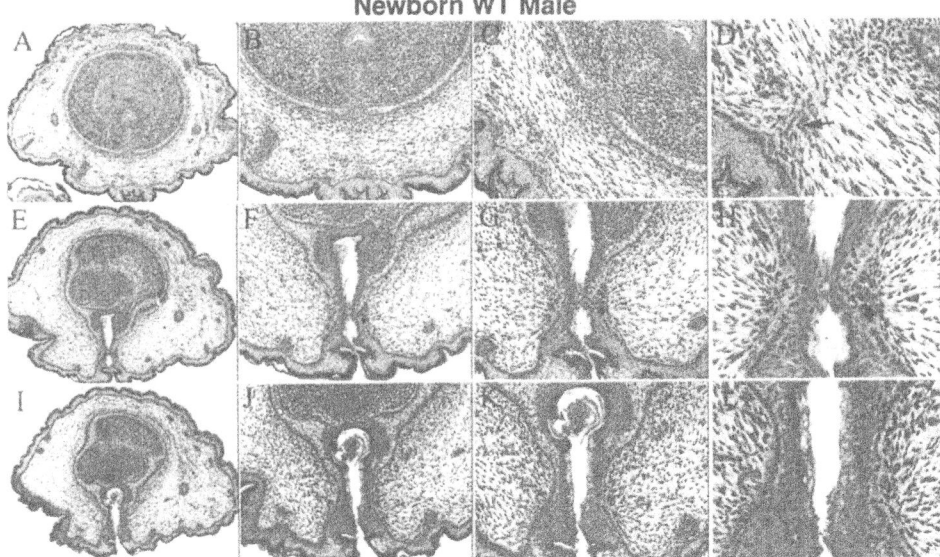

Figure 10
Normal Newborn Male Mouse Genital Tubercle. FGF 10 Wild Type (littermate).   Serial section histologic
analysis.   Note the closed urethra (A-C) and the remodeling urethral seam (D black arrow).   The distal open
urethra is shown in Figures E - L.   Note in Figures G and H (high power) the skin appears to touch, yet two
epithelial surfaces are clearly visualized.   Compare the normal male newborn to the normal female (Figure 11)
and the mutant FGF 10 mouse with hypospadias (Figure 12).

Newborn WT Female

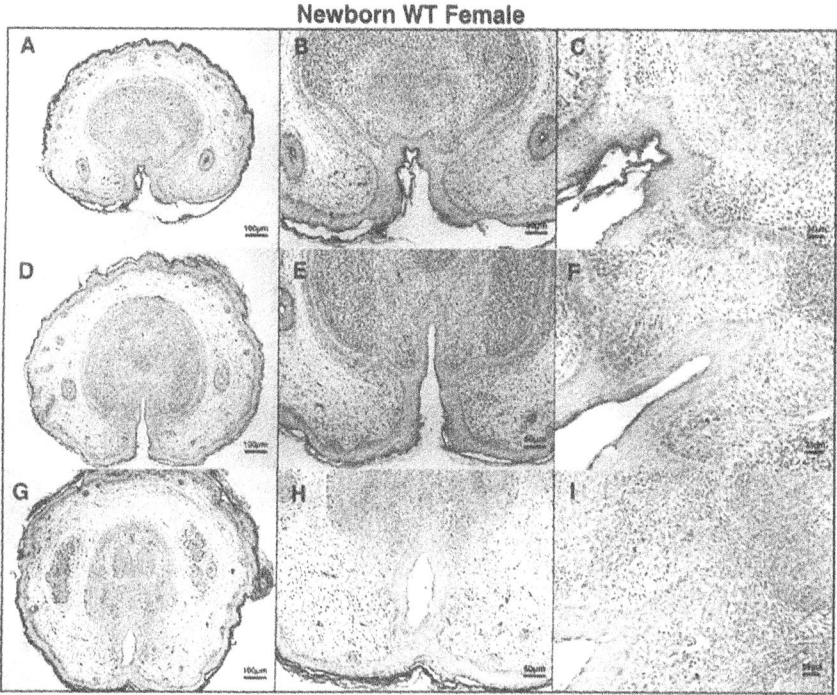

Figure 11
Normal Newborn Female Mouse Genital Tubercle. FGF 10 Wild Type (littermate).   Note the open urethra (A-F).   The epithelial edges fuse proximally although they do not remodel into a mesenchymal seam (G-I). (C, F, I high power) Compare to normal male (Figure 10) and the mutant FGF 10 mouse with hypospadias (Figure 12).

Newborn FGF-10 KO

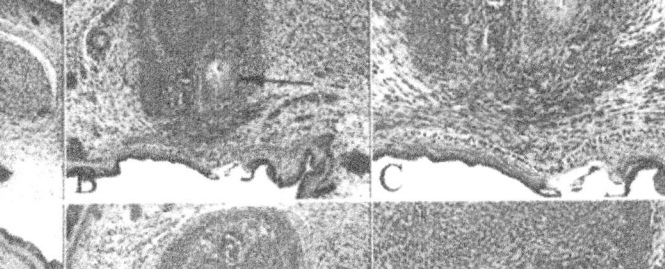

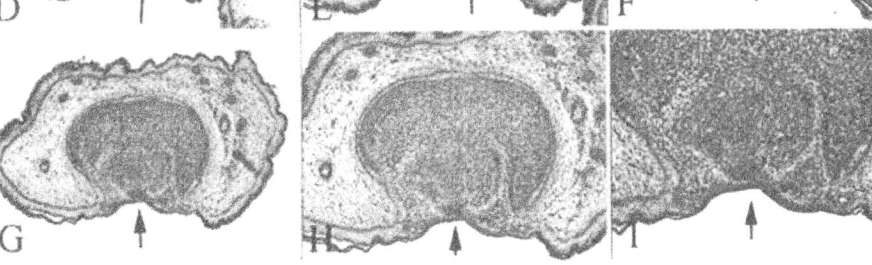

Figure 12

FGF 10 mutant with hypospadias.   Serial section histologic analysis.   Note the closed urethra (A-C, black arrow).   The urethral folds failed to form a midline seam and remodel into a tubular urethra (D-I, black arrow). The supporting mesenchyme also failed to surround the epithelial cells of the advancing urethral plate. This resulted in a relatively flat ventral aspect of the genital tubercle. The urethral opening has assumed a hypospadiac position but clearly has developed more than the female (Figure 11).

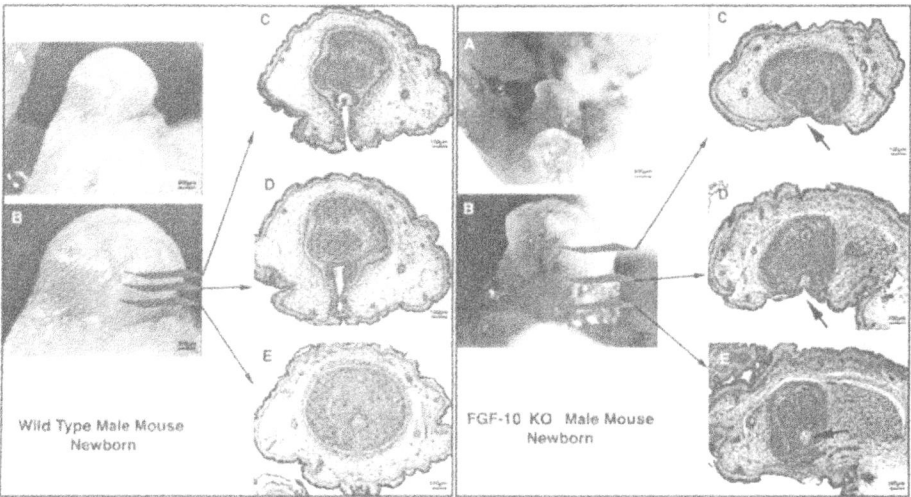

Figure 13

Gross and histologic analysis of hypospadias in the fibroblast growth factor (FGF-10) gene deletion "null" mouse. The male animals have "hypospadias" with a proximal urethral meatus. Urethral seam formation is abnormal when compared to the wild type animals. The panel on the left shows the genital tubercle of the wild type FGF 10 newborn mouse. Note the opening of the urethral meatus close to the end (A and B) of the genital tubercle which is normal for the mouse. The corresponding histologic sections show the normal urethra (D and E) and the open urethral meatus (C). In contrast, the panel on the right reveals the FGF 10 gene deletion knockout mouse genital tubercle (newborn). Note the more proximal urethral opening (A and B) i.e. hypospadias. The histology of the urethra is also quite different. The proximal urethra is intact (E), however the distal urethra is open (C and D (arrows)) with abnormalities of the urethral seam (flattened without adhesion of epithelial edges and mesenchymal seam remodeling).

Newborn FGF-10 KO Male

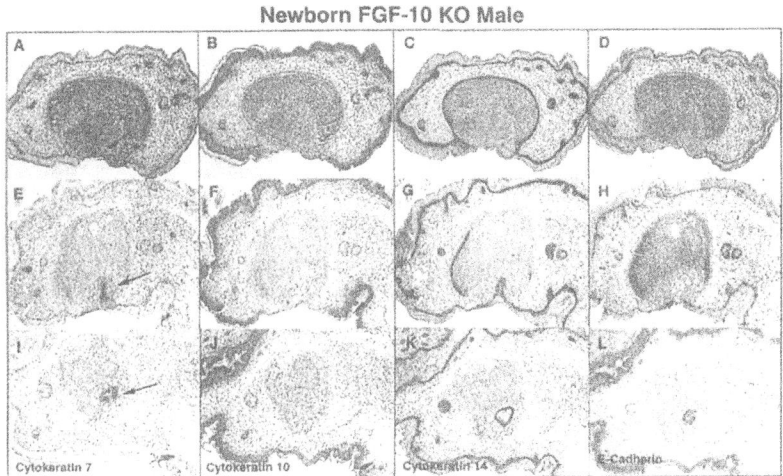

Figure 14
Normal Newborn Male Mouse Genital Tubercle. FGF 10 Wild Type (littermate).   Serial section immunohistochemical analysis for epithelial markers.   Figure A, E, I cytokeratin 7 (marker for urothelium). Figure B, F, J cytokeratin 10 (marker for skin).   Figure C, G, K cytokeratin 14 (marker for basal layer of skin). Figure D, H, L E-cadherin (non-specific marker for epithelial adhesion). Note the variable staining of each epithelial marker.   Specifically cytokeratin 7 stains the urethral area whereas cytokeratin 10 and 14 stain the epidermal/skin areas.   At the distal end of the genital tubercle there is an overlap in staining. Compare to the FGF 10 mutant with hypospadias (Figure 15).

Newborn WT Male

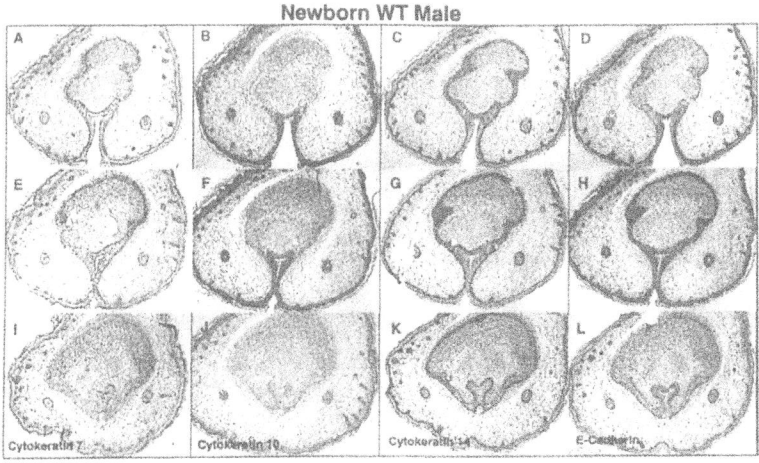

Figure 15
FGF 10 mutant with hypospadias. Serial section immunohistochemical analysis for epithelial markers. Figure A, E, I, cytokeratin 7 (marker for urothelium). Figure B, F, J, cytokeratin 10 (marker for skin).   Figure C, G, K, cytokeratin 14 (marker for basal layer of skin). Figure D, H, L, E-cadherin (non-specific marker for epithelial adhesion). Compare to the normal male Figure 5. Note the flattened and abnormal urethral development with lack of seam formation. Note cytokeratin 7 preferential staining of the proximal urethra and urethral plate (A, E, I) compared to cytokeratin 10 (Figure B, F, J).

Immunohistochemical staining in the FGF-10 deficient mutant male mouse showed that the urethral plate, which expressed Cytokeratin 7 advanced to the tip of the GT as in normal male. However, the urethral plate never formed urethral folds that would normally remodel into the epithelial seam, nor were the urethral folds supported by surrounding mesenchyma. The epidermal skin cells which expressed Cytokeratin 10 remained in contact with the advancing urethral plate and were not excluded from the urethra since seam formation did not take place in the normal males (Figure 15).

3D reconstruction images revealed the relationship of the developing urethra to the surrounding mesenchyme in the normal male, female and affected FGF-10 knockout (Figures 16-18).

2.1.3. Sonic Hedgehog Deficient Mutant Mouse Model

Shh deficient mutant animals demonstrated a persistent common cloaca characterized by failure of the cloacal separation into anterior and posterior compartments characterized by imperforate anus and GT agenesis (Figure 19).

A single opening on the perineum could anatomically be traced into the bladder. As seen in Figure. 20, immunohistochemical staining revealed localization of Cytokeratin 7 from the single perineal opening to the bladder.

Cytokeratin 10 localized only to the epithelial skin cells of ectodermal origin. Remarkably pyknotic nuclei were noteworthy in the mesenchyma surrounding the urogenital sinus epithelium in Figure 21.

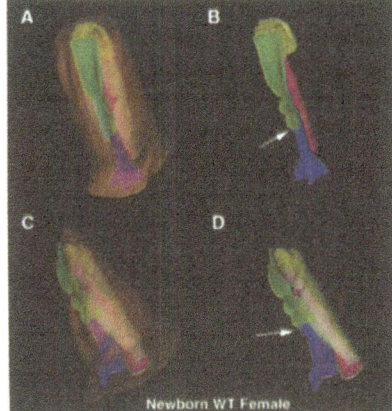

Figure 16

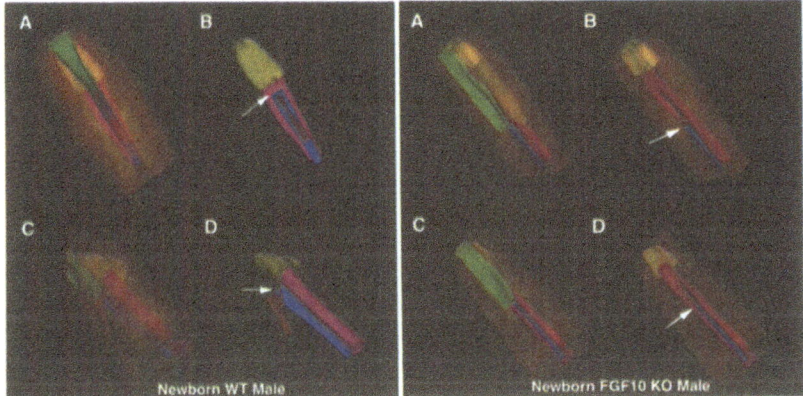

Figure 17   Figure 18

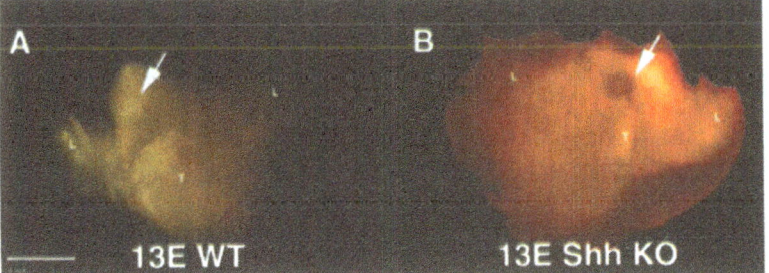

Figure 19

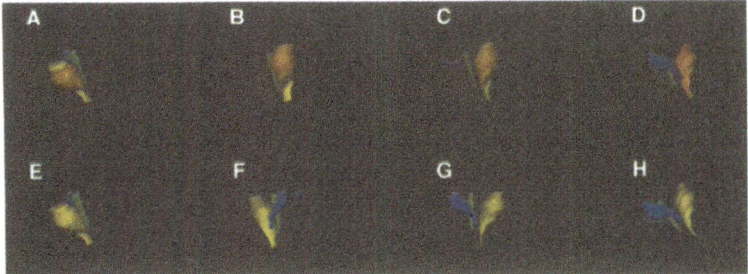

Figure 22

Figure 16
Three dimensional reconstruction of the normal female newborn mouse genital tubercle. Note the true opening of the urethra (arrow) which is below the male (Figure 7). Compare to Figure 17 and 18.

Figure 17
Three dimensional reconstruction of the normal female newborn mouse genital tubercle. Note the true opening of the urethra (arrow) which is below the male (Figure 7). Compare to Figure 16 and 18.

Figure 18
Three dimensional reconstruction of the FGF 10 mutant newborn mouse genital tubercle with hypospadias. Note the true opening of the urethra (arrow) which is more proximal than the male and more distal than the female. The genital tubercle is also flattened compared to the normal male and female.

Figure 19
Genital tubercles of 13 day gestation of (A) wild type, (B) Shh Knockout (KO) mouse embryo. In Shh KO embryo limbs (L) and tail (T) are dysmorphic. The arrowhead shows the urethral plate on the well described GT of wild type embryo whereas in the Shh KO model the common cloaca is observed at the same place. Bar represents 1 mm.

Figure 22
Three dimensional reconstruction of the Shh mutant 13 day mouse pelvis with common cloaca. Note the true opening of the cloaca is at the perineum. The endodermal lining of urogenital sinus covers the bladder and continues to the common cloacal opening. Hindgut connects to this cavitation with a thin fistula. Pubic bones do not meet at the anterior midline. Genital tubercle is completely absent.

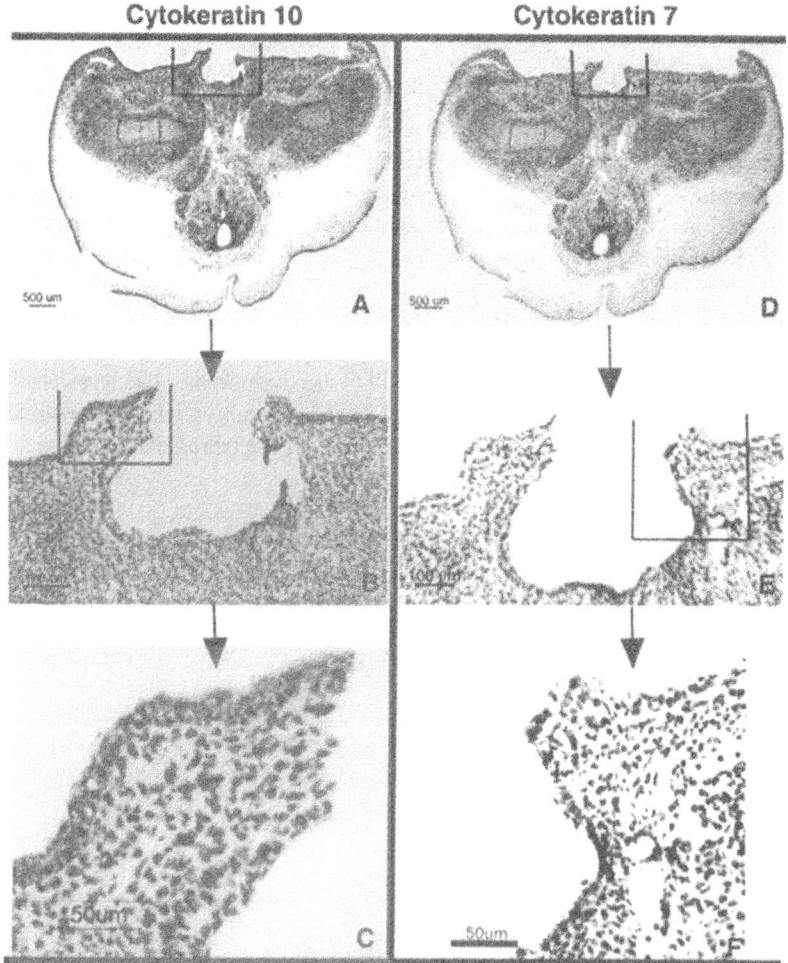

Figure 20
Cytokeratin 7 and 10 immunostaining of 13 day gestation Shh KO model. In A, B and C, the epitelium of the skin expresses cytokeratin 10 whereas in D, E and F inner lining of common cloaca shows endodermal characteristics by showing Cytokeratin 7 expression.

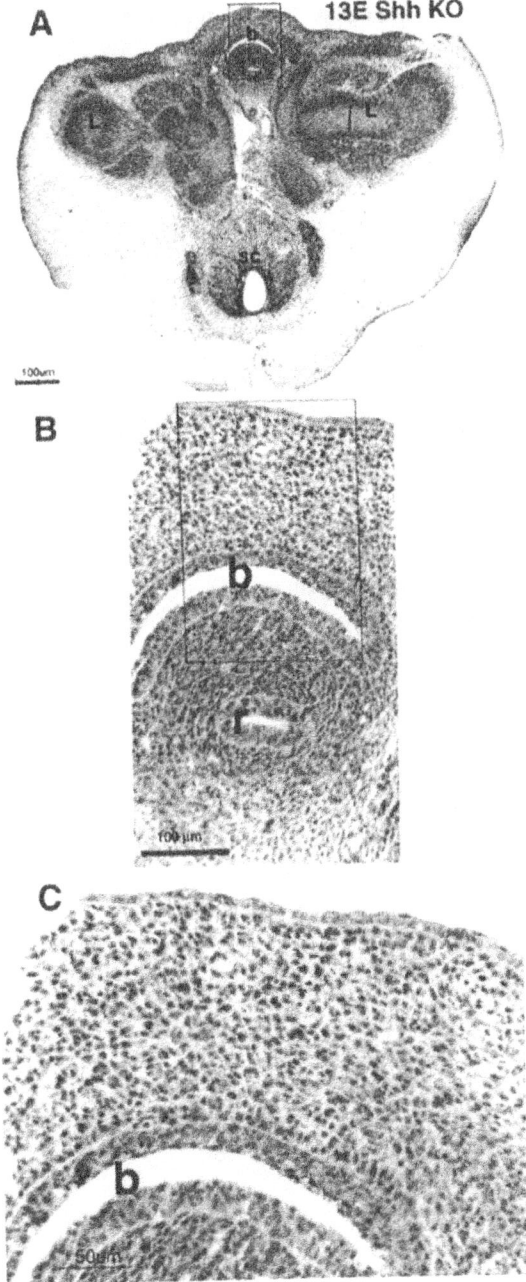

Figure 21
Hematoxylin and Eosin staining of 13 day gestation Shh Knockout (KO) mouse model. Note the pyknotic cell nuclei in mesenchyma around the urogenital endoderm.

Three dimensional construction images revealed that the urinary bladder was morphologically normal at the urachal portion entirely covered with the endodermal layer. Cytokeratin 7 localized to the lining the common cloaca walls without visualization of GT outgrowth. An atretic anal canal localized directly behind the urinary bladder base connecting as a fistula to the common urogenital sinus with a wide open pubic bone (Figure 22).

3. Discussion

In this study, the hypothesis that the male and female GT are indistinguishable without the presence of androgens secondary to a mutant androgen receptor has been tested. We described the distinctive structural anatomical differences between the GTs of wild type female and AR deficient mutant male in the mouse model.   Complete androgen insensitivity does not result in the female phenotype GT but rather in a unique morphological phenotype GT carrying special anatomical structure unique to both sexes. The presence of ARs in the anatomical compartments differentiating into the GT supports the hypothesis that androgenic activity is crucial in the differentiation of urogenital tract into male and female counterparts. However, absence of the androgen effect does not default into the female phenotype (Figure. 6 - 9).

ARs have been found to be abundant in GTs of many animals. Murakami et al. showed that almost all the mesenchymal cells in the GT of fetal rat and mice posses androgen binding capacity (Murakami, 1987). Androgens are believed to be crucial for the induction, growth and differentiation of the male GT (Rey and Picard, 1998) through binding and activation of the AR, a ligand-dependent transcription factor that belongs to the superfamily of the steroid nuclear receptor (Brinkmann et al., 1999). Abnormalities in the synthesis and metabolism of androgens and defects in the androgen receptor have been demonstrated to cause abnormalities in phenotypic sexual development. Complete androgen insensitivity syndrome is the human condition that results when a complete defect in the andreogen receptor occurs (Wiener et al., 1997; Imperato-McGinley et al., 1985; Unoerati-McGinley et al., 1992; Wikf et al., 20000).

Androgenic activity has not been yet elucidated in the female GT development in humans and animals. Conflicting data about the localization of the ARs and the morphological results of possible activities has been reported. Drews et al. showed a lower AR expression in the differentiating corporal bodies than the surrounding mesenchyme in mouse (Drews et al., 2001). They also failed to show the presence of the androgen receptors in the Mullerian duct epithelium reported by Shapiro et al., 2000. The clitoromegaly or phallus-like appearance of the clitoris in animal models exposed to exogenous testosterone can be explained by either the stromal expansion around the corporal bodies or the inductive activity of AR in stroma to develop the corporal bodies. In human, Baskin et al. previously proposed that supralevels of testosterone cause expansion of extracellular matrix rather than the cellular hyperplasia (Baskin et al., 1997). To our knowledge, there no data has been presented to support any of these mechanisms in female GT development.

Kim et al. pointed out the significant presence of   ARs and 5 α-reductase activity in corpus cavernosum, urethral spongiosum, glans penis, urethra and genital skin in human male fetal penile specimens (Kim et al., 2002). In another report studying both male and female human fetal specimens, androgen receptors were demonstrated with similar staining intensities in both sexes (Cohn et al., 1997). We found poorly developed corpus cavernosum and glans in the AR deficient mutant male mouse model compared to wild type male and female mouse. Our finding reveals that androgen activity is essential in the development of corpora cavernosa and glans. Nevertheless, even a minor development of erectile tissue in AR deficient mutant mouse model arises some other developmental mechanisms independent of androgens in the GT development.

Kim et al. illustrated the weak signal of androgen fibers in the   stromal cells of the urethral spongiosum in human fetal male specimens at the gestational ages of 12 to 20 weeks (Kim et al., 2002). This finding differ from the Kalloo et al.' s study in which all corporal bodies were found to be stained equally for AR (Kalloo et al., 1993). In the present study, AR deficient mutant male mouse revealed a well developed urethral spongiosum. Interestingly, while the wild type female mouse was observed to have a thinner urethral spongiosum close to the anterior vaginal wall, AR deficient mutant mouse had a thicker urethral spongiosum close to the anterior wall of the vaginal pouch. This type of urethral spongiosum structure was observed in the wild type male mice. If the development of the urethral spongiosum was strongly dependent on the androgenic effect, it would not develop well in AR deficient mutant male mouse. This result seems to be parallel with the observation of Kim et al. about the presence of androgen receptor in the urethral spongiosum. The male type urethral spongiosum development in the AR deficient mutant mouse may not be solely dependent on androgens.

Limb development may also be a useful paradigm for understanding GT development. The field of limb development stresses the importance of specific cellular and molecular interactions that determine morphology and function during embryogenesis (Cohn et al., 1997; Cohn and Bright, 1999; Cohn and Tickle, 1999). The GT develops in three dimensions 1) proximal-distal; 2) dorsal-ventral and 3) medial-lateral, although unlike the limb, in the GT the medial–lateral dimension is symmetric. The dhand gene encodes a protein that is required for limb bud initiation by aiding or inducing Shh (Tanaka et al., 2000). Through complex epithelial-mesenchymal interactions, genes such as Shh and Fgf-8 signal from the AER (apical ectodermal ridge) to the surrounding limb mesenchyme initiating the expression of Fgf-10, Hoxd13, Msx1 and the Bmp's (as well as other gene products) (Cohn et al., 1997; Cohn and Bright, 1999; Cohn and Tickle, 1999; Tanaka et al., 2000; Niswander et al., 1994). Many other genes are involved in specific morphologic patterning such as digit formation (Shh, Hoxd13) and differentiation of forelimbs from hindlimbs (Homeobox and T-box). We propose that the molecular control of GT developmental is regulated by many of the same genes involved in limb development. Although an AER-like organizer of GT development has not been previously documented, we propose the novel concept that the urethral plate in the developing GT acts as the GT organizer. We and other investigators have documented the importance of epithelial-mesenchymal signaling in the developing GT and have documented the expression of genes important in limb development in the urethral plate (Fgf-8, Shh) and surrounding mesenchyme (Bmp4, Hoxd13, Fgf-10,) (Haraguchi et al., 2000; Bitgood and McMahon, 1995; Kurzrock et al., 1999)

It has been shown that in the developing mouse GT, FGF10 is expressed in the outgrowing GT mesenchyme while urethral expression of FGF-8 remained stable (Haraguchi et al., 2000). FGF-10 beads are known to initiate growth, even though less prominent compared to those of FGF-8 beads. FGF signaling seems to be one of the key systems involved in controlling outgrowth and in regulating various developmental processes of mammalian GT (Haraguchi et al., 2000). The inductive effects of epithelium on the differentiation and growth of tissues through complex epithelial–mesenchymal interactions have been reported recently. Also it has been suggested that initial GT growth before urethral plate formation might be achieved by applying FGF proteins (Haraguchi et al., 2000).

Interestingly, FGF knockout mouse has not been found to show drastic changes in the development of the GT. On the contrary, in FGF-10 expression location in former studies, only the most distal structures of the GT have been shown to perturbed during development. As seen in the histological sections, the urethral plate was observed to continue to the tip of GT. This surprising finding could be explained by possible gene compensatory mechanisms via other functionally or structurally related genes during the GT development.

In normal development of the GT, the   urethral plate becomes separated from the surface epithelium of the glans penis in a proximo-distal sequence and forms an epithelial tube that runs through the glans penis (Baskin et al., 2001). Urethral folds have been observed to fuse at a seam area to form a continuous urethral lumen to the tip of the glans. Endodermal differentiation theory elucidated the distal urethral development in the mammalian species (Kurzrock et al., 1999). The uninterrupted continuation of the endodermal layer to the tip of GT has been proven by immunohistochemichal studies (Kurzrock et al., 1999). The proximal to distal fusion of urethral plates has been described in the mouse GT (Baskin et al., 2001). As the two epithelial surfaces of the urethral folds fuse, an epithelial seam is formed that is subsequently remodeled into the urethra. We have proposed that normal seam formation would result in normal urethral development. In contrast, if the seam area were not to fuse or be properly remodeled, then the urethra would be open, consistent with hypospadias (Baskin et al., 2001). Our hypothesis is consistent in FGF-10 deficient mutant mouse model.   As seen in Figures 16-18, even though the urethral plate continues to the tip, the plate fails to fuse to form the normal urethra and results in a hypospadiac appearance. We had concluded that any arrest in seam formation or remodeling anywhere along the GT will result in an abnormal urethral opening. FGF-10 signaling appears to be one of the most important seam area regulators. It has been shown that FGF-10 is expressed in mesenchyma adjacent to the midline at the later stages of GT differentiation as urethral tube formation occurs (Haraguchi et al., 2000). This suggest the crucial role of FGF-10 is in the morphogenesis of the distal most urethral plate fusion.

Shh is known to mediate the signaling activities of many organizing and polarizing centers in mammalian embryos, including the polarizing region of the limb bud and the floor of the neural tube (Riddle et al., 1993; Echelard et al., 1993). An elegant study by Perriton et al. indicated that urethral epithelium has polarizing activity, can regulate tissue movements characteristic of genital morphogenesis and expresses Shh (Perriton et al., 2002). They also concluded that Shh expression is continuous throughout the endodermal urogenital sinus and urethral epithelium suggesting that the entire urethra develops from Shh-expressing endoderm (Perriton et al., 2002). This finding also supports the recent endodermal differentiation theory in the development of glandular urethra (Kurzrock et al., 1999). In addition, the dramatic increase in apoptosis observed in GT of Shh deficient mutants implicates Shh as a cell survival factor (Perriton et al., 2002; Sanz-Ezquerro and Tickle, 2000). In this study, similar observation has been documented by remarkable pyknotic cell nuclei in the mesenchyme around the cloacal endodermal layer. In the GT, Shh signaling from the endoderm appears to regulate the cell survival in the surrounding mesenchyma.

In this study, we found that   Shh deficient mutant mouse model reveals complete absence of GT development with a common cloaca appearance. A persistent common cloaca is a well described human congenital birth defect characterized by absence of external genitalia and imperforate anus (Kim et al., 2001) . As seen in the early stages of embryogenesis before the separation of the cloacal membrane, the hindgut and urogenital system remains open into the same perineal aperture. This anomaly is generally associated with vertebral, gut, respiratory system and limb anomalies (Kim et al., 2001). Striking similarity between VACTERL syndrome and Shh knockout mouse is noteworthy. Also the endodermal lining of the common cloacal aperture revealed endodermal cytokeratin expression characteristics. It may be speculated that the lack of Shh signaling in the urogenital sinus epithelium prevents the outgrowth of the GT.

The wide spectrum of birth defects of the GT involves a broad range of molecular signaling elements. Each element seems to be responsible for certain parts of GT morphogenesis. The precise description of the functional anatomy of GT in animal models created by knockout technology is of paramount importance in understanding the molecular events and their functional end results in normal and abnormal embryogenesis.

4. Conclusion

Even though the sexually indifferent GT stage is a morphologic reality in mammalian embryogenesis, androgenic impact does not appear to be the only participant in the differentiation of the sex specific GT. Ventral and dorsal components of the GT have different morphological molecular regulators which are only partially dependent on androgens. Androgen receptor mutation tissue recombination studies could answer many undisclosed mechanisms in the sex specific GT development. Urothelium is the most important element in the morphogenesis of normal GT development. Shh signaling appears to play a major role in the epithelial-mesencyhmal interaction of GT development. Normal urethral tube formation requires fusion of urethral folds on the whole urethra including the glandular urethra. Any molecular event perturbing seam area formation or remodeling can cause the abnormal opening or hypospadias which is the most common congenital abnormality of the GT. The entire urethra develops from the urogenital sinus without any contribution of ectoderm. FGF-10 appears to be important in the seam formation leading to hypospadias.

Supported by a grant from the National Institute of Health # RO1PK058105.

REFERENCES

Anderson, C.A., and Clark, R.L., 1990, External genitalia of the rat: Normal development and histogenesis of 5 alpha-reductase inhibitor –induced abnormalities, *Teratology*. **42**, 483-496.

Baskin, L. S., Sutherland, R. S., DiSandro, M. J., Hayward, S., Lipshultz, J., Cunha, G. R., 1997, The effect of testosterone on androgen receptors and human penile growth, *J. Urol.* **158:** 1113.

Baskin L.S, Colborn, T. et al., 2001, Hypospadias and endocrine disruption: is there a connection? *Environmental Health Perspectives*. **109:** 1175-1183.

Baskin, L. S., Erol, A., Li, Y.W., Cunha, G.R., 1998, Anatomical studies of hypospadias, *J. Urol.* **160(3 Pt 2):** 1108-15; discussion 1137.

Baskin, L.S., Sutherland, R.S., DiSandro, M.J., Hayward, S.W., Lipshultz, J., and Cunha, G.R., 1997, The effect of testosterone on androgen receptors and human penile growth, *J. Urol.* **158,** 1113-1118.

Baskin L.S., Erol, A., Jegatheesan, P., Li, Y., Liu, W., Cunha, G.R., 2001, Urethral seam formation and hypospadias, *Cell Tissue Res* **305,** 379-387.

Bellusci, S., Grindley , J., Emoto, H., Itoh, N. and Hogan, B.L., 1997, Fibroblast growth factor 10 (FGF10) and branching morphogenesis in the embryonic mouse lung, *Development*. **124,** 4867-4878.

Bellusci, S., Furuta, Y., Rush, M.G., Henderson, R., Winnier, G. and Hogan, B.L., 1997, Involvement of sonic hedgehog (Shh) in mouse embryonic lung growth and morphogenesis, *Development*. **124,** 53-63.

Bitgood, M.J., McMahon, A.P., 1995, Hedgehog and Bmp genes are coexpressed at many diverse sites of cell-cell interaction in the mouse embryo, *Dev Biol*. Nov: **172(1):**126-38.

Brinkmann, A. O., Blok, L. J., et al., 1999, Mechanisms of androgen receptor activation and function. *J. Steroid Biochem. Mol. Biol.* **69:** 307.

Brinkmann, A.O., 2001, Molecular basis of androgen insensitivity, *Mol Cell Endocrinol*. **179,** 105-109.

Chiang, C., Litingtung, Y., Lee, E., Young, K.E., Corden, J.L., Wesphal, H. and Beachy, P.A., 1996, Cyclopia and defective axial patterning in mice lacking Sonic hedgehog gene function, *Nature*. **383,** 407-413.

Cohn, M.J., Izpisua-Belmonte, J.C., Abud, H., Heath, J.K. and Tickle, C., 1995, Fibroblast growth factors induce additional limb development from the flank of chick embryos, *Cell*. **80,** 739-746.

Cohn, M.J., Patel, K., Krumlauf, P., Wilkinson, D.G., Clarke, J.D., Tickle, C., 1997, Hox9 genes and vertebrates limb specification, *Nature*. **387,** 97-101.

Cohn, M.J., Bright, P.E., 1999, Molecular control of vertebrate limb development, evolution and congenital malformations, *Cell Tissue Res*. **296(1):**3-17.

Cohn, M.J., Tickle, C., 1999, Developmental basis of limblessness and axial patterning in snakes, *Nature*. **399(6735):**474-9.

Dolle, P., Izpisua-Belmonte, J.C., Brown, J.M., Tickle, C. and Duboule, D., 1991, HOX-4 genes and the morphogenesis of mammalian genitalia, *Genes Dev*. 5, 1767-1768.

Dolle, P., Dierich, A., LeMeur, M., Schimmang, T., Schuhba, B., Chambon, P. and Duboule, D., 1993, Disruption of the Hoxd-13 gene induces localized heterochrony leading to mice with neotenic limbs, *Cell* **75,** 431-441.

Drews, U., Sulak, O., Oppitz, M., 2001, Immunohistochemical localisation of androgen receptor during sex-specific morphogenesis in the fetal mouse, *Histochem. Cell Biol.* **116:** 427.

Echelard, Y., Epstein, D.J., St-Jacques, B., Shen, L., Mohlere, J., Mcmahon, J.A. and McMahon, A.P., 1993, Sonic hedgehog, a member of a family of putative signaling molecules, is implicated in the regulation of CNS polarity, *Cell*. 75, 1417-1430.

Goodman, F.R., Bacchelli, C., Brady, A.F., Brueton, L.A., Fryns, J.P., Mortlock, D.P., Innis, J.W., Holmes, L.B., Donnenfeld, A.E., Feingold, M., Beemer, F.A., Hennekam, R.C. and Scambler, P.J., 2000, Novel HOXA 13 mutations and the phenotypic spectrum of hand-foot-genital syndrome, *Am J Hum Genet*. 67, 197-202.

Haraguchi, R., Suzuki, K., Murakami, R., Sakai, M., Kamikawa, M., Kengaku, M., Sekine, K., Kawano, H., Kato, S., Ueno, N. and Yamada, G., 2000, Molecular analysis of external genitalia formation: the role of fibroblast growth factor (Fgf) genes during genital tubercle formation, *Development*. 127, 2471-2479.

Haraguchi, R., Mo, R., Hui, C., Motoyoma, J., Makino, S., Shiroshi, T., Gaffield, W. and Yamada, G., 2001, Unique functions of Sonic hedgehog signaling during external genitalia development, Development. 128, 4241-4250.

Imperato-McGinley, J., Binienda, Z., Arthur, A., Mininberg, D. T., Vaughan, E. D., 1985, The development of a male pseudohermaphroditic rat using an inhibitor of the enzyme 5 alpha-reductase, *Endocrionology*. **116:** 807.

Imperato-McGinley, J., Sanchez, R. S., Spencer, J. R., Yee, B., Vaughan, E. D., 1992, Comparison of the effects of t he 5 alpha- reductase inhibitor finasteride and the antiandrogen flutamide on prostate and genital differentiation: dose response studies, *Endocrinology*. 131:1149.

Kalloo, N.B., Gearhart, J.P. and Barrack, E.R., 1993, Sexually dimorphic expression of estrogen receptors, but not of androgen receptors in human fetal external genitela, *J Clin Endocrinol Metab.* 77, 692-698.

Kim, J., Kim, P., Hui, C.C., 2001, The VACTERL assosciation: lessons from the Sonic Hedgehog pathway, *Clin Genet.* **59**: 306-15.

Kim, K. S., Liu, W., Cunha, G. R., Russell, D. W., Huang, H., Shapiro, E., Baskin, L. S., 2002, Expresion of the androgen receptor and 5 alpha- reductase type 2 in the developing human fetal penis and urethra, *Cell Tissue Res.* 307: 145.

Kondo, T.J., Innis, J.W. and Duboule, D., 1997, Of fingers, toes and penises, *Nature.* 390, 29.

Kurzrock, E.A., Jegatheesan, P., Cunha, G.R. and Baskin, L.S., 2000, Urethral development in the fetal rabbit and induction of hypospadias: A model for human development, *J Urol.* 164, 1786-1792.

Kurzrock, E.A., Baskin, L.S,, Li, Y., Cunha, G.R., 1999, Epithelial-mesenchymal interactions in development in development of the mouse fetal genital tubercle, *Cells Tissues Organs* 164, 125-130.

Kurzrock, E.A., Baskin, L.S., Cunha, G.R., 1999, Ontogeny of the male urethra: theory of endodermal differentiation, *Differentiation.* **64**: 115-122.

Martin, G.R., 1998, The roles of FGFs in the early development of vertebrate limbs, *Genes Dev.* 12, 1571-1586.

Mortlock, D.P. and Innis, J.W., 1997, Mutation of HOXA 13 in hand-foot-genital syndrome, *Nat Genet.* 156, 179-180.

Motoyoma, J., Liu, J., Mo, R., Ding, Q., Post, M. and Hui, C.C., 1998, Essential function of Gli2, and Gli3 in the formation of lung, trachea and eosophagus, *Nat. Genet.* 20, 54-57.

Murakami, R. and Mizuno, T., 1986, Proximo-distal sequence of development of the skeletal tissues in the penis of rat and the inductive effect of epithelium, *J Embryol Exp Morphol.* 92, 133-143.

Murakami, R., 1987, A histological study of the development of the penis of wild-type and androgen-insensitive mice, *J. Anat.* 153, 223-31.

Niswander, L., Tickle, C., Vogel, A., Martin, G., 1994, Function of FGF-4 in limb development. *Mol Reprod Dev.* Sep, **39(1)**:83-8.

Ohuchi, H., Nakagawa, T., Yamamato, A., Araga, A., Ohata, T., Ishimaru, Y., Yoshioka, H., Kuwana, T., Nohno, T., Yamasaki, M., Itoh, N., Noji, S., 1997, The mesenchymal factor, FGF10, initiates and maintains the outgrowth of the chick limb bud through interaction with FGF8 , an apical ectodermal factor, *Development.* 124, 2235-2244.

Ornitz, D.M., 2000, FGFs, heparan sulfate and FGFRs: complex interactions essential for development, *Bioessays.* 22, 108-112.

Paulozzi, L., D. Erickson, et al., 1997, Hypospadias Trends in Two US Surveillance Systems, *Pediatrics.* **100(5)**: 831-834.

Paulozzi, L. J., 1999, International trends in rates of hypospadias and cryptorchidism, *Environ Health Perspect.* **107(4)**: 297-302.

Perriton, C.L., Powles, N., Chiang, C., Maconochie, M.K., Cohn, M.J., 2002, Sonichedgehog signaling from the urethral epithelium controls external genital development, *Developmental Biology.* 247, 26-46.

Podlasek, C.A., Barnett, D.H., Clemens, J.Q., Bak, P.M. and Bushman, W., 1999, Prostate development requires Sonic Hedgehog expressed by the urogenital sinus epithelium, *Dev Biol.* 209, 28-39.

Ramalho-Santos, M., Melton, D.A., McMahon, A.P., 2000, Hedgehog signals regulate multiple aspects of gastrointestinal development, *Development.* 127, 2763-2772.

Rey, R., Picard, J. Y., 1998, Embryology and endocrinology of genital development, *Baillieres Clin. Endocrinol. Metab.* **12**: 17.

Riddle, R.D., Johnson, R.L., Laufer, E., Tabin, C., 1993, Sonic Hedgehog mediates the polarizing activity of the ZPA, *Cell.* 75, 1401-1416.

Roberts, D.J., Johnson, R.L., Burke, A.C., Nelson, C.E., Morgan, B.A, and Tabin, C., 1995, Sonic Hedgehog is an endodermal signal inducing Bmp-4 and Hox genes during induction and reginalization of the chick hindgut, *Development.* 121, 3163-3174.

Sanz-Ezquerro, J.J. and Tickle, C., 2000, Autoregulation of Shh expression and Shh induction of cell death suggest a mechanism for modulating polarising activity during chick limb development, *Development.* 127, 4811-4823.

Sekine, K., Ohuchi, H., Fujiwara, M., Yamasaki, M., Yoshizawa, T., Sato, T., Yagishita, N., Matsui, D., Koga, Y., Itoh ,N., Kato, S., 1999, Fgf10 is essential for limb and lung formation, *Nat. Genet.* 6, 348-356.

Shapiro, E., Huang, H., Wu X., 2000, Uroplakin and androgen receptor expression in the human fetal genital tract: insights into the development of the vagina, *J. Urol.* **164**: 1048.

Tanaka, M., Cohn, M.J., Ashby, P., Davey, M., Martin, P., Tickle, C., 2000, Distribution of polarizing activity and potential for limb formation in mouse and chick embryos and possible relationships to polydactyly, *Development.* Sep, **127(18)**:4011-21.

Wall, N.A. and Hogan, B.L., 1995, Expression of bone morphogenetic protein –4 (BMP-4), bone morphogenetic protein – 7 (BMP-7), fibroblast growth factor –8 (FGF-8) and sonic hedgehog (SHH) during branchial arch development in the chick, *Mech Dev.* 53, 383-392.

Warot ,X., Fromental-Ramain, C., Fraulob, V., Chambon, P. and Dolle, P., 1997, Gene dosage-dependent effects of the Hoxa-13 and Hoxd-13 mutations on morphogenesis of the terminal parts of the digestive and urogenital tracts, *Development.* 124, 4781-4791.

Wiener, J. S., Teague J. L. et al., 1997, Molecular biology and function of the androgen receptor in genital development, *J. Urol.* **157**: 1377,.

Wolf, C. J., LeBlanc, G. A., Ostby. J. S., Gray, L. E. Jr., 2000, Characterization of the period of the sensitivity of fetal male sexual development to vinclozolin, *Toxicol. Sci.* **55**: 152.

Yamaguchi, T.P, Bradley, A., McMahon, A.P. and Jones, S., 1999, A Wnt5a pathway underlies outgrowth of multiple structures in the vertebrate embryo, *Development.* 126, 1211-1223.

Yamasaki, M., Miyake, A., Tagashira, S. and Itoh, N., 1996, Structure and expression of the rat mRNA encoding a novel member of the fibrobalst growth factor family, *J Biol Chem.* 271, 15918-15921.

Yucel, S. and Baskin, L. S., in press, Identification of Communicating Branches Between the Dorsal, Perineal and Cavernosal Nerves of the Penis, *J Urol.*

Zakany, J., Fromental-Ramain, C., Warot, X. and Duboule, D., 1997, Regulation of number and size of digits by posterior Hox genes: A dose-dependent mechanism with potential evolutionary implications, *Proc Natl Acad Sci USA*, 94, 13695-13700.

DEVELOPMENTAL GENETICS OF THE EXTERNAL GENITALIA

Martin J. Cohn*

1. INTRODUCTION

The incidence of congenital malformation of the urogenital system is second only to that of the cardiovascular system, yet comparatively little is known about the cellular and molecular mechanisms that regulate urogenital organogenesis. In this chapter, I review recent advances in the developmental biology of the external genitalia, and discuss the implications of this work for our understanding of hypospadias. The majority of research into external genital development and hypospadias has focused on the endocrine system, particularly on the role of androgens (see accompanying chapters in this volume). A relatively unexplored area of genital morphogenesis is the early, genetically controlled process of pattern formation, when genital tubercle outgrowth and three-dimensional patterning occurs (Figure 1). These processes occur in the absence of endocrine signals, and identification of the molecular mechanisms of early genital development is crucial to our understanding of congenital anomalies.

One of the surprises of comparative developmental studies is that evolution has been relatively conservative; the same genetic cassettes are involved in development of eyes, limbs and nerves, for example, in animals as diverse as flies and humans. The knowledge that genetic circuits have been repeatedly co-opted during the evolution of embryonic development provides a springboard for investigating the urogenital system. We have used the vertebrate limb as a paradigm for investigating the mechanisms involved in external genital development. Limbs and external genitalia undergo many similar morphogenetic processes, and we have hypothesized that the same molecular mechanisms may operate during development of the limb bud and the genital tubercle. Here I report on some of the initial tests of this hypothesis.

2. BUILDING A THIRD AXIS: BUDDING AND PROXIMODISTAL OUTGROWTH

Appendages, such as limbs and external genitalia, develop at particular positions along the anteroposterior and dorsoventral axes of the embryo. The molecular mechanisms that

*Martin J. Cohn, Department of Zoology, University of Florida, 223 Bartram Hall, Box 11858, Gainesville, FL 32611-8525; E-mail: cohn@zoo.ufl.edu

Hypospadias and Genital Development, edited by
L. Baskin, Kluwer Academic/Plenum Publishers, 2004

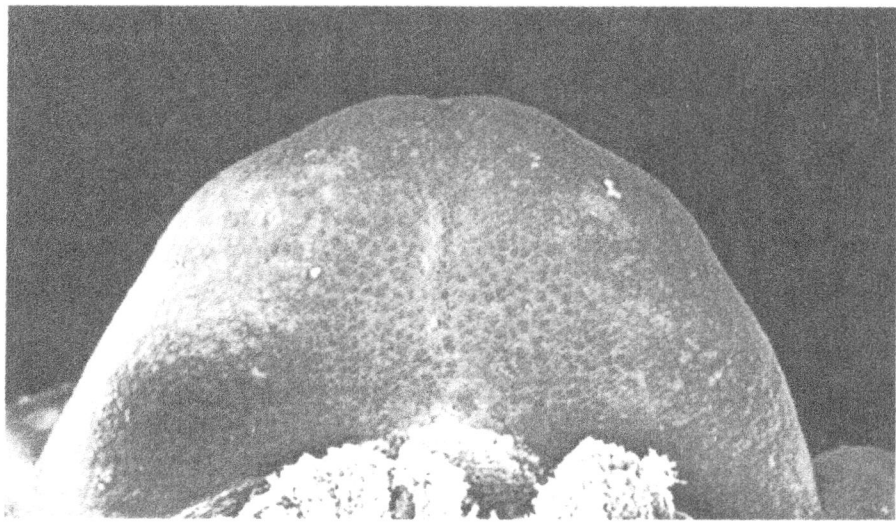

Figure 1. Scanning electron micrograph of the ventral side of a mouse genital tubercle at embryonic day 12.5.

determine the position at which limbs form are beginning to be understood (Cohn et al., 1997; Altabef and Tickle, 2002), and the nature of the signals involved in initiating outgrowth of the limbs is now known (Cohn et al., 1995; Crossley et al., 1996; Min et al., 1998; Sekine et al., 1999). Local expression of secreted proteins known as fibroblast growth factors (FGFs) initiates the process of limb budding at two positions on either side of the body. These limb buds consist of undifferentiated mesenchyme encased in an ectodermal jacket. Outgrowth of these buds is sustained by a specialized epithelial structure at the distal tip of the limb bud known as the *apical ectodermal ridge* (AER), which is itself a source of secreted Fgf (Figure 2). The AER may be thought of as a growth factor factory, and the underlying mesenchymal cells respond to these growth factors by continuing to divide, thereby extending the limb bud further from the primary body axis. If the AER is removed surgically, chemically or pathologically, outgrowth of the limb arrests and distal limb structures do not develop. The earlier in development the ridge is removed, the more severe the truncation, indicating that the limb is laid down in a proximal to distal sequence (Summerbell, 1974). For example, a very early excision of the ridge may result in loss of all structures distal to the humerus, whereas a comparatively late removal of the ridge may lead to loss of only distal phalanges. Does a similar structure regulate proximodistal outgrowth of the external genitalia?

The external genitalia begin as a pair of swellings on either side of the   cloacal plate. These swellings grow out in a coordinated manner to form the single genital eminence or tubercle (Figure 1). The mechanisms regulating the initiation and maintenance of genital budding are not well-understood. The genital tubercle does not appear to have a morphologically distinct AER-like structure, however, experimental manipulations suggest that there may be a functionally equivalent tissue. In 1986, Murakami and Mizuno demonstrated that the epithelial component of the rat genital tubercle is required for growth and differentiation of the adjacent mesenchyme (Murakami and Mizuno, 1986). A similar epithelial requirement has been demonstrated in the mouse genital tubercle (Kurzrock et al., 1999; Haraguchi et al., 2000). Murakami and Mizuno found that removal of epithelium resulted in stage-dependent truncation of the penile tissues, with later removals resulting in loss of progressively more distal structures. Thus, outgrowth of both the genital tubercle and the limb bud depends on epithelial signaling, and the connective tissue and skeleton of these appendages are laid down in a proximal to distal sequence. The parallel between the AER of the limb and the distal urethral epithelium of the genital tubercle may extend to the molecular level; both tissues are site of FGF expression (Figure 2), and both can be replaced experimentally by beads loaded with FGF protein (Niswander *et al.*, 1993; Haraguchi *et al.*, 2000).

A. LIMB BUD   B. GENITAL TUBERCLE

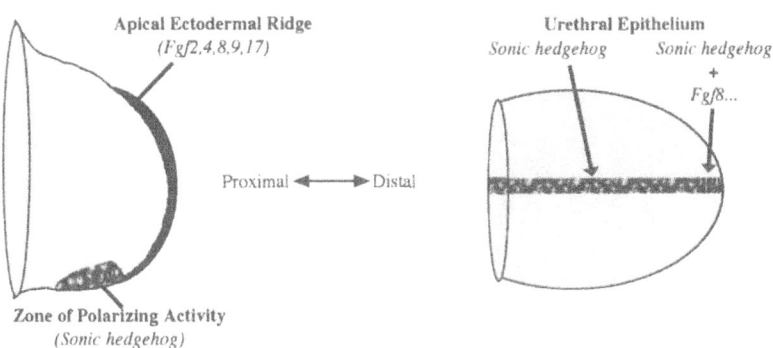

Figure 2. Schematic diagram of signaling centers in the limb bud (A) and genital tubercle (B). Both are depicted in ventral view.

Loss-of-function mutations in mice thus far have implicated a small number of genes in the regulation of outgrowth in both limbs and external genitalia. Some of these genes are simply "outgrowth" genes, in that they are expressed in all organs that grow distally, whereas others appear to regulate outgrowth specifically in limb buds and genital tubercles, but not other appendages. Of the former group, *Wnt5a,* a member of the Wnt gene family (vertebrate orthrologs of the fly *wingless* gene), is expressed in a large number of embryonic outgrowths, including the limb, genital tubercle, tongue, mandibular arch and distal lung (Yamaguchi et al., 1999; Li et al., 2002). Expression of *Wnt5a* is graded from distal to proximal and, as Wnt5a regulates cell proliferation, loss of Wnt5a function impairs distal outgrowth of these structures (Yamaguchi et al., 1999; Li et al., 2002). Of the latter group, members of the Hox paralogy group 13 have been most intensively studied. *Hoxd13* and *Hoxa13* are involved in the patterning of structures at the terminus of the limbs (i.e., the digits) and the terminus of the primary body axis (i.e., posterior vertebrae, anus and genitalia (Figure 3). *Hoxd13* expression in the genitalia and limbs is controlled by a single enhancer (Kmita et al., 2002). Such striking conservation of genetic regulation suggests a molecular mechanism in support of the idea that limbs and genitalia have a common evolutionary history (van der Hoeven *et al.*, 1996; Kondo *et al.*, 1997). Loss of *Hoxd13* and *Hoxa13* leads to agenesis of the genital tubercle and digits, and heterozygosity for either causes malformation of the phallus and limbs (Dollé et al., 1993; Warot et al., 1997; Zákany et al., 1997). In humans, mutations in the *HOXA13* gene are responsible for Hand-Foot-Genital Syndrome, which affects the distal aspects of the limb and external genitalia (Goodman *et al.*, 2000). Thus, FGF, Wnt and Hox genes have important roles in early genital development, and their functions in the genital tubercle appear to mirror their functions in the limb bud.

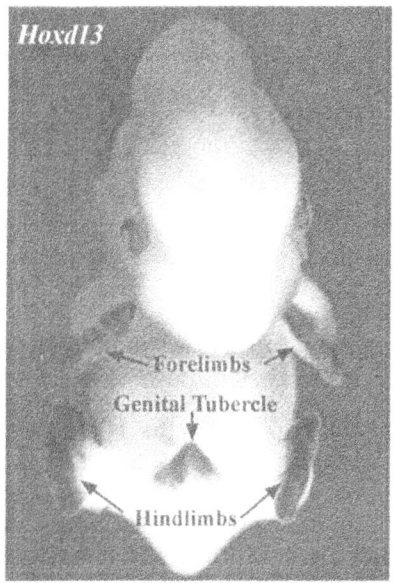

Figure 3. Whole-mount *in situ* hybridization of a mouse at embryonic day 13.5 showing *Hoxd13* mRNA localization (dark staining) to the genital tubercle and the developing digits of the forelimb and hindlimb buds.

3. COORDINATING OUTGROWTH WITH PATTERN FORMATION

The external genitalia are clearly polarized along three axes ; proximodistal, dorsoventral (or anteroposterior), and mediolateral. Establishment of these axes, and the polarization of developmental processes along them, is likely to be a complex business. Again, the limb is a useful starting point for unravelling these developmental mechanisms, as we have a relatively detailed picture of the cells and molecules responsible for polarizing the limb

A specialized group of mesenchymal cells, collectively known as the *zone of polarizing activity* (ZPA) or *polarizing region*, acts to polarize the limb bud along the anterior to posterior (or thumb to small finger) axis (Figure 2). The polarizing or organizing activity of these cells was originally demonstrated by classical experimental embryology, in the form of a 'cut-and-paste' experiment. When a limb bud receives a graft of posterior mesenchymal cells to its anterior margin, the result is a "posteriorization" of anterior cells (reviewed in Tickle, 2002). In other words, cells at the anterior margin of the limb bud, responding to a signal emitted by the graft, acquire posterior character and give rise to posterior digits. The ultimate effect is development of a limb with a mirror-image duplication of the digits along the anteroposterior axis. Cells in the polarizing region express the *Sonic hedgehog (Shh)* gene, which codes for a powerful secreted signaling molecule. The Shh signal is responsible for the specification of "posterior" positional identities in the limb bud (Riddle *et al.*, 1993). As such, when exposed to Shh, cells situated at the anterior of the limb bud are reprogrammed follow a posterior program of differentiation. Cells with organizing properties are found throughout the embryo, where they orchestrate developmental processes such as gastrulation, specification of motor neurons in the ventral spinal cord, and anteroposterior regionalization of the brain. Moreover, many of these organizing tissues utilize the Shh signal to polarize neighboring cells. Could a similar organizer exist in the genital tubercle, and if so, what signal or signals might it utilize to pattern the tubercle?

The pattern of digits that develop within the chick limb bud has long been used as an assay for polarizing signals. Tissues with organizing activity, such as the notochord, the floor plate of the neural tube and the node of the gastrula, can lead to anterior-posterior duplication of the digits. We have previously shown that the urethral epithelium, but not the adjacent mesenchyme, of the mouse genital tubercle could also result in a mirror-image duplication of the digits, when grafted to the anterior margin of the chick limb bud (Perriton *et al.*, 2002). Although this finding demonstrated that urethral plate epithelium could polarize the limb, further work was required to test its function during external genital development. Urethral plate epithelial cells express *Shh*, which may account for the ability of these cells to polarize the limb (Figure 2). The function of the *Shh*-expressing urethral epithelial cells in the genital tubercle was determined by two independent analyses of genital development in mice with a loss-of-function mutation in *Shh*. Both studies revealed that *Shh-/-* mice exhibit agenesis of the external genitalia (Haraguchi *et al.*, 2001; Perriton *et al.*, 2002). Interestingly, Shh is not required for the initial outgrowth of the genital swellings (Perriton *et al.*, 2002), however sustained outgrowth, polarized gene expression and cell survival all depend on the presence of Shh. Thus, the Shh signal from the urethral plate is essential for external genital development. These findings raise new questions concerning the relationship of

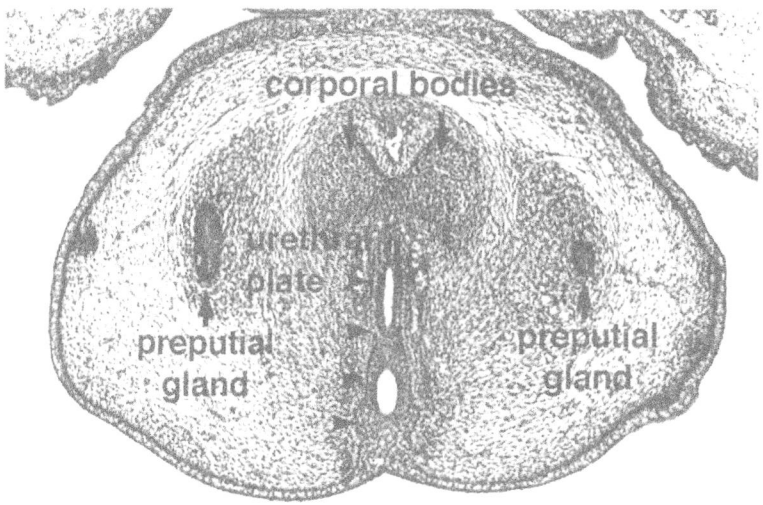

Figure 4.. Transverse histological section through the genital tubercle of a mouse embryo at embryonic day 16.
Note that mesenchymal condensation and differentiation occurs in arc-like pattern, which is centered around
the urethral plate epithelium

outgrowth to axial patterning. Does the urethral plate also determine dorsoventral
polarity of structures within the genital tubercle? The connective and erectile tissues of
the genital tubercle develop with a radial symmetry focused around the urethral plate
(Figure 4), which is suggestive of an organizing influence of the urethral epithelium on
the surrounding cells. In addition to causing digit duplications, transplantation of the
urethral epithelium to the chick limb bud results both in the formation of an epithelial
tube within the limb and in the induction of ectopic limb muscle around the tube (for
details see Perriton *et al.*, 2002). This result raises the possibility that, within the
tubercle, the urethral epithelium may function in organizing tubulogenesis and in
patterning the smooth muscle cells of the corporal bodies. A number of genes, such a
bone morphogenetic proteins (BMPs), Patched, Gli and FGFs, are expressed in
dorsoventrally restricted patterns. Analysis of the function of these genes in genital
development is presently underway in a number of laboratories.

4.   BUILDING THE URETHRA

In addition to having to coordinate axial patterning and outgrowth of the genital
tubercle, the embryo must simultaneously orchestrate complex tissue movements in order
to build a tubular urethra in the appropriate position. The alarmingly high incidence of
urethral malformations, in particular hypospadias, in humans should direct our attention
towards the cellular and molecular biology of urethragenesis. As mentioned previously
in this chapter and elsewhere in this volume, most of the work in this area has concerned
the role of androgen signaling (including androgen agonists and antagonists) in genital
development. The sequencing of the human and mouse genomes has created a new

opportunity for us to study the genetics of urethral development, and the early indications are that highly conserved developmental genetic programs play a key role in morphogenesis of urethral canal. In a screen for genes expressed in tissues affected by hypospadias, we identified a growth factor receptor, *FgfR2*, that is expressed in the prepuce and urethral plate of mouse embryos. Moreover, a loss of function mutation in this receptor results in gross hypospadias (Perriton, Petiot, Dickson and Cohn, unpublished data). FGFs and FGFRs have been suggested to act as andromedins, or mediators of androgen signaling, and it is tempting to speculate that this may account for the similar genital phenotype in antiandrogen-treated and Fgfr2 -/- mice. Work on prostate development and disease, however, suggests that the relationship between FGF and androgen signaling is not quite so clear, with some studies arguing against and others in favor of the andromedin hypothesis (Thomson *et al.,*1999, Thomson, 2001). Given the timing of expression of FGFs and *Fgfr2* relative to the production of androgen by the embryo, it seems likely that the FGF pathway operates independently of androgens during early genital development. Any andromedin-like function could occur only after the synthesis of testosterone by the embryo. Our findings suggest that the function of growth factor signaling in urethral development may be multiphasic. At early stages, a genetically "wired", functionally autonomous morphogenetic program regulates urethral development in the absence of hormones, however after development of the endocrine system, the embryo's hormonal milieu provides a new context within which local signals operate. Understanding the interactions between the local and global signaling systems that operate within the embryo, and between these endogenous signals and exogenous environmental agents, will be critical to unravelling the etiology of hypospadias.

ACKNOWLEDGEMENTS

I thank Larry Baskin for the invitation to participate in the Hypospadias and Genital Development Symposium, Claire L. Perriton, and Philippa Bright for stimulating discussion and critical comments on this chapter, and C.L.P. for providing the data shown in Figures 1 and 4.

REFERENCES

Altabef, M. and Tickle, C., 2002, Initiation of dorso-ventral axis during chick limb development. *Mech Dev* **116**, 19-27.

Cohn, M. J., Izpisúa-Belmonte, J. C., Abud, H., Heath, J. K. and Tickle, C., 1995, Fibroblast growth factors induce additional limb development from the flank of chick embryos. *Cell* **80**, 739-46.

Cohn, M. J., Patel, K., Krumlauf, R., Wilkinson, D. G., Clarke, J. D. and Tickle, C., 1997, *Hox9* genes and vertebrate limb specification. *Nature* **387**, 97-101.

Crossley, P. H., Minowada, G., MacArthur, C. A. and Martin, G. R., 1996, Roles for FGF8 in the induction, initiation, and maintenance of chick limb development. *Cell* **84**, 127-36.

Dollé, P., Dierich, A., LeMeur, M., Schimmang, T., Schuhbaur, B., Chambon, P. and Duboule, D., 1993, Disruption of the Hoxd-13 gene induces localized heterochrony leading to mice with neotenic limbs. *Cell* **75**, 431-41.

Goodman, F. R., Bacchelli, C., Brady, A. F., Brueton, L. A., Fryns, J. P., Mortlock, D. P., Innis, J. W., Holmes, L. B., Donnenfeld, A. E., Feingold, M. et al., 2000, Novel HOXA13 mutations and the phenotypic spectrum of hand-foot-genital syndrome. *Am J Hum Genet* **67**, 197-202.

Haraguchi, R., Mo, R., Hui, C., Motoyama, J., Makino, S., Shiroishi, T., Gaffield, W. and Yamada, G., 2001, Unique functions of Sonic hedgehog signaling during external genitalia development. *Development* **128**, 4241-50.

Haraguchi, R., Suzuki, K., Murakami, R., Sakai, M., Kamikawa, M., Kengaku, M., Sekine, K., Kawano, H., Kato, S., Ueno, N. et al., 2000, Molecular analysis of external genitalia formation: the role of fibroblast growth factor (Fgf) genes during genital tubercle formation. *Development* 127, 2471-9.

Kmita, M., Fraudeau, N., Herault, Y. and Duboule, D., 2002, Serial deletions and duplications suggest a mechanism for the collinearity of Hoxd genes in limbs. *Nature* **420**, 145-50.

Kondo, T., Zakany, J., Innis, J. W. and Duboule, D., 1997, Of fingers, toes and penises. *Nature* **390**, 29.

Kurzrock, E. A., Baskin, L. S., Li, Y. and Cunha, G. R., 1999, Epithelial-mesenchymal interactions in development of the mouse fetal genital tubercle. *Cells Tissues Organs* **164**, 125-30.

Li, C., Xiao, J., Hormi, K., Borok, Z. and Minoo, P., 2002, Wnt5a participates in distal lung morphogenesis. *Dev Biol* **248**, 68-81.

Min, H., Danilenko, D. M., Scully, S. A., Bolon, B., Ring, B. D., Tarpley, J. E., DeRose, M. and Simonet, W. S., 1998,. Fgf-10 is required for both limb and lung development and exhibits striking functional similarity to Drosophila branchless. *Genes Dev* **12**, 3156-3161.

Murakami, R. and Mizuno, T., 1986, Proximal-distal sequence of development of the skeletal tissues in the penis of rat and the inductive effect of epithelium. *J Embryol Exp Morphol* **92**, 133-43.

Niswander, L., Tickle, C., Vogel, A., Booth, I. and Martin, G. R. 1993. FGF-4 replaces the apical ectodermal ridge and directs outgrowth and patterning of the limb. *Cell* **75**, 579-87.

Perriton, C. L., Powles, N., Chiang, C., Maconochie, M. K. and Cohn, M. J., 2002, Sonic hedgehog signaling from the urethral epithelium controls external genital development. *Dev Biol* **247**, 26-46.

Sekine, K., Ohuchi, H., Fujiwara, M., Yamasaki, M., Yoshizawa, T., Sato, T., Yagishita, N., Matsui, D., Koga, Y., Itoh, N. et al., 1999, Fgf10 is essential for limb and lung formation. *Nat Genet* **21**, 138-41.

Riddle, R. D., Johnson, R. L., Laufer, E., Tabin, C., 1993, Sonic hedgehog mediates the polarizing activity of the ZPA. *Cell* **75**, 1401-16

Summerbell, D., 1974, A quantitative analysis of the effect of excision of the AER from the chick limb-bud, *J Embryol Exp Morphol.* **32**, 651-660

Thomson, A. A., 2001, Role of androgens and fibroblast growth factors in prostatic development. *Reproduction* **121**,187-195

Thomson, A. A. and Cunha, G. R , 1999, Prostatic growth and development are regulated by FGF10. *Development* **126**, 3693-3701.

Tickle, C., 2002, The early history of the polarizing region: from classical embryology to molecular biology. *Int J Dev Biol* **46**: 847-52

van der Hoeven, F., Zákány, J. and Duboule, D., 1996, Gene transpositions in the HoxD complex reveal a hierarchy of regulatory controls. *Cell* **85**, 1025-35.

Warot, X., Fromental-Ramain, C., Fraulob, V., Chambon, P. and Dollé, P., 1997, Gene dosage-dependent effects of the Hoxa-13 and Hoxd-13 mutations on morphogenesis of the terminal parts of the digestive and urogenital tracts. *Development* 124, 4781-91.

Yamaguchi, T. P., Bradley, A., McMahon, A. P. and Jones, S., 1999, A Wnt5a pathway underlies outgrowth of multiple structures in the vertebrate embryo. *Development* 126, 1211-23.

Zákany, J., Fromental-Ramain, C., Warot, X. and Duboule, D., 1997, Regulation of number and size of digits by posterior Hox genes: a dose-dependent mechanism with potential evolutionary implications. *Proc Natl Acad Sci U S A* **94**, 13695-700.

DEVELOPMENT OF THE MOUSE EXTERNAL GENITALIA: UNIQUE MODEL OF ORGANOGENESIS

Kentaro Suzuki[1], Kohei Shiota[2], Yanding Zhang[3], Lei Lei[1], and Gen Yamada[1]*

1. INTRODUCTION

The mammalian external genitalia develop highly specialized structures suitable for efficient internal fertilization. The advent of molecular developmental biology has led to the accumulation of large amounts of data regarding mammalian organogenesis. However, there is not much research on the development of external genitalia (Cunha and Lung, 1979, Dolle et al., 1991; Kondo et al., 1997; Warot et al., 1997). Even recent molecular analyses of this organ utilize embryological observations often dated several decades ago.

Molecular developmental analysis of several appendages, e.g., limbs, fins and external genitalia, has revealed some indication for shared developmental programs for organogenesis.

The genital tubercle (GT), an anlage of the external genitalia, differentiates into a penis in male and a clitoris in female. The first sign of urogenital differentiation is observed in the genital ridge during the murine embryonic development. The structural difference between the male and female genital ridge can be first recognized at 12.5 d.p.c. (Brennan et al., 1998; Ikeda et al., 1994), at which stage the development of external genitalia is under the influence of steroid hormones.

Although there have been some suggestions regarding the similarity of developmental programs between external genitalia and other bud-type anlage, e.g., limb buds, an important characteristic point for external genitalia development is this sexual dimorphism that occurs at late stage in its development. The development of the

[1] Center for Animal Resources and Development (CARD) and Graduate School of Molecular Genomic Pharmacy, Kumamoto University, Honjo 2-2-1, Kumamoto 860-0811, Japan;
[2] Congenital Anomaly Research Center and Department of Anatomy and Developmental Biology, Graduate School of Medicine, Kyoto University, Kyoto, 606-8501, Japan;
[3] College of Bioengineering, Fujian Teachers University, Fuzhou, Fujian 350007, China.
*Author for correspondence (e-mail: gen@kaiju.medic.kumamoto-u.ac.jp)
Tel: 81 (Japan) 96 373 6569, Fax: 81 (Japan) 96 373 6560

Hypospadias and Genital Development, edited by
L. Baskin, Kluwer Academic/Plenum Publishers, 2004

fetal external genitalia can be divided into two distinct processes. The first process involves the initial outgrowth of the GT, from the main embryonic body trunk, and has been proposed to be independent of the action of steroid hormones, such as androgens. The second process eventually generates the sexually dimorphic formation of the external genitalia through the differentiation of the mesenchyme and the urethra of the GT.

Previous studies on human hereditable diseases have indicated the possibility that some genes might be involved in various organogenesis possibly in the basic developmental processes which underlie such organgenesis affected in the above diseases. In fact, pioneering studies have suggested that GT development shows some similarities with limb development, considering that both structures exhibit budding type outgrowth from the main embryonic trunk (Dolle, et al., 1991; Yamaguchi et al., 1999). A congenital anomaly with developmental defects both in limbs and external genitalia in humans has been reported (Goff and Tabin, 1996; Kondo, et al., 1997; Mortlock and Innis, 1997). Double gene knockout muntantion of *Hox* genes (*Hox a13* and *Hox d13*) has been reported to display a similar hypoplasic phonetype both in external genitalia and limbs (Dolle et al. 1991; Kondo et al., 1997).

Given the possible similarities between external genitalia and limb organogenesis, it would be worthwhile to investigate functional similarities or diversities of various developmental signaling genes. Such signaling genes are also expressed in functionally important embryonic epithelial or mesenchymal regions, e.g., in embryonic limbs or external genitalia (Haraguchi et al., 2000; 2001). It has been reported that various signaling genes are expressed in signaling epithelia in a complex and reiterated manner during development. Vertebrate limb development depends on the establishment and maintenance of the apical ectodermal ridge (AER), a specialized ectodermal structure at the distal tip of the limb bud (Duboule, 1993; Johnson and Tabin, 1997; Tickle and Eichele, 1994). It is well recognized that proliferation of limb mesenchymal cells is controlled by the AER or factors emanating from the AER (Figure 1 I).

Among the growth factor genes, FGFs (fibroblast growth factors) have been identified as essential signaling molecules for embryogenesis (Bellusci et al., 1997; Cohn et al., 1995; Martin, 1998; Ornitz, 2000; Wall and Hogan, 1995; Yamasaki et al., 1996; Yonei-Tamura et al., 1999). One of the FGFs, FGF8, is important for the growth and patterning of embryonic limbs (Crossley and Martin, 1995; Crossley et al., 1996; Johnson and Tabin, 1997; Ohuchi et al., 1994). More recent studies using conditional knockout mice for several of the FGF genes revealed that there are complex and reiterated functions of several FGFs in addition to other growth factors during limb development (Sun et al., 2002). It is possible that loss of function studies of the Fgf8 gene failed to identify limb phenotypes due to compensation by other FGFs (Moon and Capecchi, 2000). Thus, it is of particular interests to examine the possible functions of various FGFs during other embryonic bud analge including the external genitalia formation. We have previously investigated the expression pattern and functions of various FGFs during mouse embryonic external genitalia development (Haraguchi et al. 2000).

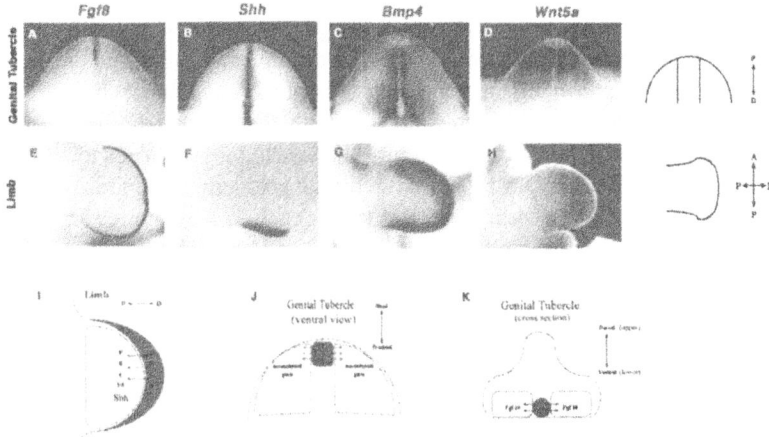

Figure 1. *In situ* hybridization for gene expression (FGF*8*, *Shh*, *Bmp4*, *Wnt5a*)

Whole-mount *in situ* hybridization was performed with digoxigenin-labeled Fgf8, Shh, Bmp4, and Wnt5a probes with by standard procedures.
A~D: embryonic GT at 12.5 d.p.c..
E~H: embryonic limb at 11.5 d.p.c..
I: a simplified model for FGF8-FGF10 interaction during limb formation.
J: Fgf 8 expression as a marker for DUE during GT formation.
K: midline urethral plate epithelium (UPE) versus bilateral Fgf10 expression in the cross section of the developing GT.

2. EXPRESSION OF THE Fgf8, Shh, Bmp4, Wnt5a GENES DURING MURINE GENITAL TUBERCLE DEVELOPMENT

To better understand the complex and reiterated expression and functions of developmental genes in organogenesis, the expression pattern of several growth and transcription factors was examined.

The expression pattern of the Fgf8 gene was examined at various developmental stages. At 10.5 d.p.c., prior to the onset of GT outgrowth, the Fgf8 and Fgf10 genes were both expressed in the urogenital sinus as well as in the epithelium of the distal portion of urogenital sinus (data not shown; Haraguchi et al., 2000). At 11.5 d.p.c., after initial expression and prior to GT protrusion, the Fgf8 gene was expressed at the distal signaling epithelia, the Distal Urethral Epithelia (DUE) of the GT (Figure 1A,J). The region was also termed as solid urethral plate epithelium in some publications. FGF8 mRNA levels increased to a maximum level at 11.5 d.p.c., and then decreased gradually around 14.5 d.p.c. (data not shown). In contrast to the rather wide spread exprssion of FGF8 in AER, the distal FGF8 expression in the DUE may constitute an intriguing contrast to the expression of FGF8 in AER (Figure 1A,E; data not shown).

In contrast, the FGF10 gene was expressed in the mesenchyme during the outgrowth phase of GT (Haraguchi et al., 2000; data not shown). Hence, both the Fgf8 gene expression in the distal tip of the urethral epithelium and mesenchymal FGF10 gene expression continued through the outgrowth period of the GT from 11.5 to 14 d.p.c.

It has been known that multiple FGF genes are expressed in an overlapping manner during the formation of several organs. The FGF4 gene, one of the other key FGF genes, was not expressed prominently during GT development (data not shown). Several other FGF genes, including the FGF receptors, are also dynamically expressed during external genitalia formation (Satoh et al., in preparation; Perriton et al., 2002).

It may be noteworthy to comment that in both the GT and limbs, mesenchymal Bmp4 is expressed, at least, partly in ventral mesenchymes adjacent to the signaling epithelia (Figure 1C, G; Bmp4 is expressed in the subectoderm region to AER in limb at 11.5 d.p.c.). It is also expressed in the dorsal (upper) GT mesemchyme (Suzuki et al., in preparation). Expression of the other key signaling gene for limb patterning, the Sonic Hedgehog (Shh) gene, was found to be initially expressed in the cloaca epithelium and then in the distal regions of the urethral epithelium where the FGF8 gene was also expressed (Haraguchi et al., 2001, data not shown). The Shh gene was also expressed at the site of the urethral epithelium (Haraguchi et al., 2001; Perriton et al., 2002; Figure 1B, K). It has been well known that posterior-localized expression of Shh in the limb bud offers a cue for anterior-posterior (A-P) development of the developing limb (Figure 1F).

During limb development, the epithelial-mesenchymal interaction is considered essential to regulate the expression of the mesenchymal genes. Therefore, expression of several markers, such as the Bmp4 and Msx1 genes, were examined. Intriguingly, they were expressed in the mesenchyme of the outgrowing GT from 10.5 to 14.5 d.p.c. (Figure 1C; Suzuki et al., in preparation). An intriguing "growth-promoting" gene derived from mesenchyme is *Wnt5a*, originally proposed by Yamaguchi et al. (Yamaguchi et al., 1999), and also observable in the distal limb bud mesenchyme (Fig.1H). *Wnt5a* is expressed in the distal GT mesenchymes (Fig.1D). The expression pattern of these genes led us to investigate whether similar regulatory gene expression may occur in GT formation.

3. SCANNING ELECTRON MICROSCOPE (SEM) ANALYSIS OF GENITAL TUBERCLE (GT) DEVELOPMENT

Morphological analysis of the caudal region of the developing embryos before and during the GT development was performed by SEM scanning electron microscope (Figure 2).

The cloacal membrane is formed at the ventral midline in the caudal regions of developing embryos. It is partly recognized that the development of the GT occurs in coordination with the development of the caudal embryonic region, including the formation of the cloaca and its surrounding mesenchyme and tail region (Gofflot et al.,. 1997; Goldman et al., 2000). Thus, the functional and morphological analysis of cloaca formation associated with GT development is essential to our understanding of the coordinated development of the embryonic caudal region.

Regions bilaterally adjacent to the cloacal membrane swell slightly to generate the genital folds, forming the cloacal membrane as a shallow fold (Figure 2A). Parts of the cloaca are connected caudally into a blind-ended tailgut. Although not yet well understood, the tailgut, cloaca, and VER (ventral ectodermal ridge) are transient embryonic structures, which are considered to affect each development leading to the regulation of various aspects of caudal embryonic morphogenesis (Gofflot et al., 1997; Goldman et al., 2000). After 10.5 d.p.c., the GT starts to display protrusion at the cloacal membrane (Figure 2B).

Structures of the GT become prominent at 11.5 d.p.c., with its bilaterally located ventral swellings gradually emerging as a protruding bud. This further grew out proximo-distally (P-D) at 12 d.p.c. (Figure 2C). The urethral plate was formed in the ventral (lower) median plane, which is continuous with the cloacal membrane on its caudal side. The cloaca is subdivided into the urogenital sinus and anus, concomitant with the development of the urorectal septum. The fusion of urorectal septum to the cloaca membrane or "degeneration of the cloaca membrane" associated with the septum development has been discussed in recent publications. The orifice of the urogenital sinus is surrounded by the urogenital folds.

Of particular note is the presence of the distal-most region at the tip of the GT, emerging at approximately 12.0 d.p.c. (white arrowheads; Fig.2C). It has been shown that the distal urethral epithelium (DUE), a region defined by various criteria, plays an essential role in the control of GT outgrowth (Haraguchi, et al., 2000, 2001; Ogino, et al., 2001).

Formation of such a distal-most protrusion in the developing external genitalia was also described several decades ago by Glenister (1954). Formation of the distal end of the urethra should be discussed in relation with the development of the distal protrusion, also discussed as against the possibility of ectodermal ingrowth for the anterior end of the urethra (reviewed in Kurzrock et al., 1999).

From the viewpoint of molecular developmental biology, recent findings have suggested that the distal signaling epithelia may play an important developmental role in the coordinated outgrowth of the external genitalia (Haraguchi et al., 2000). However, the possible role of this protrusion in human GT development remains to be clarified. There have been various reports describing the functional and morphological character of the signaling epithelia, e.g., AER in limbs. Thus, the roles of the DUE during GT morphogenesis would be particularly intriguing from the molecular and morphological perspectives.

The outgrowth of the GT continues further with the urethral plate gradually forming on its ventral side. This development of the urethral plate to the urethra has been discussed in several publications. This process is believed to involve the fusion of the urethral folds and prepuce, as well as the canalization of the urethral plate (Figure 2D). The urogenital folds develop into urethral folds on their distal side, and into anal folds at their more proximal side. Baskin et al. (2001) described that urethral folds and groove in the mouse are not exactly analogous to that of human because the tips of the urethral folds fuse to generate the tubular human urethra. The mechanisms of seam formation after the urethral fold fusion requires further investigation. The prepuce starts to grow P-D along the outgrowing GT. This is a bilateral fold of skin covering the GT, which first forms as a small swelling at the proximal base of the GT from 13.5 d.p.c. (also seen in Figure 2E,F). The differentiation of the labioscrotal swelling and the elevation of the prepuce up to the tip of the GT proceeds dramatically following GT outgrowth. The male and female external genitalia appear similar until 15.5 d.p.c.

The first morphological difference between male and female external genitalia is

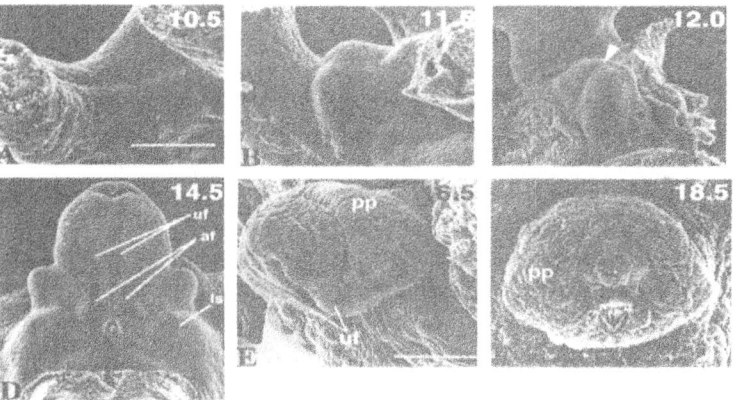

Figure 2. Scanning electron microscopy (SEM) analysis

Mouse genital tubercles were fixed in 4% paraformaldehyde (PFA) and dehydrated for scanning electron microscopy (SEM) analysis. Embryos were sexed at the time of fixation by examining the gonads located at the anterior side of the kidneys (Morphological criteria for typing sexes of embryos is feasible from 12.5 d.p.c.). Tails were removed from embryos (12.0 d.p.c. to 18.5 d.p.c.) for better morphological presentation of the external genitalia.

Embryonic specimens dated 10.5 d.p.c. (A), 11.5 d.p.c. (B), 12.0 d.p.c. (C), 14.5 d.p.c. (D), 16.5 d.p.c. (E), and 18.5 d.p.c. (F) are shown.

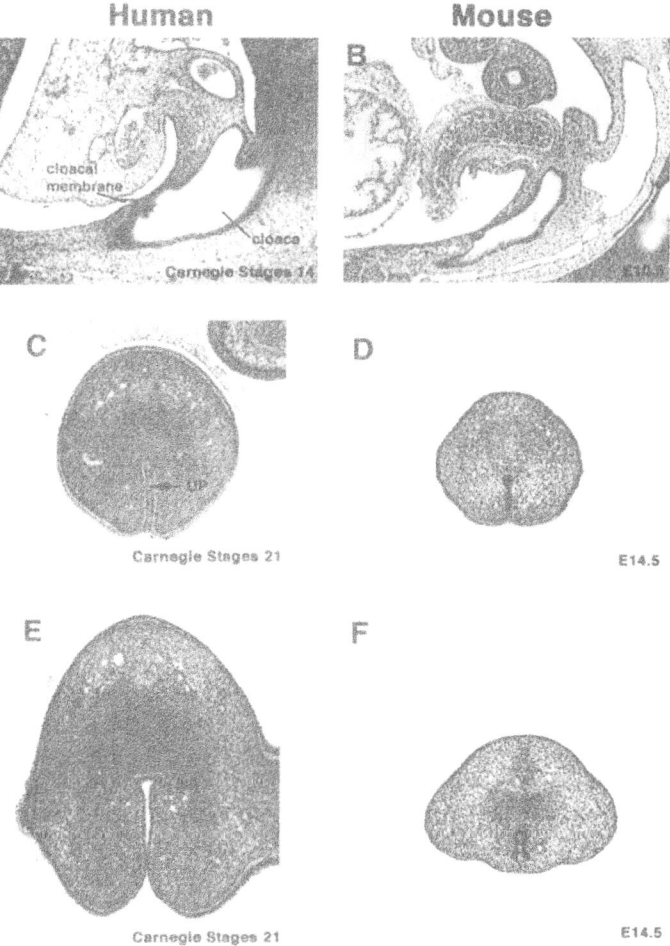

Figure 3. Histological analysis of human and mouse GT

Mouse embryos were fixed in 4% PFA, dehydrated in graded ethanol, embedded in paraffin and sectioned with 5.0 micron thickness. Sections were stained with hematoxylin/eosin (H&E) by standard procedures. Human embryonic samples were photographed for samples of the Congenital Anomaly Research Center, Koyto University Graduate School of Medicine.

evident after 16.5 d.p.c. (Figure 2E,F). The prepuce grows rapidly and engulfs the proximal region of the glans penis or clitoridis, although the ventral side of the prepuce develops more prominently in males than in females. The glans lamellae, a bilayer of adhered epithelial cells that separates the prepuce from the glans penis in males and from the clitoris in females, begins to develop concomitantly with the growth of the prepuce in both sexes. The fusion of the urethral folds in the ventral midline was prominent in males between 16.5 and 17.5 d.p.c. The glans lamellae completely encircles the glans penis, except for a narrow region at the ventral midline of the glans at 18.5 d.p.c. (Figure 2F). The canalization of the urethral plate, which is initially connected to the surface epithelium, proceeds P-D in the glans of the male at 18.5 d.p.c. In the female, the urethral folds were not completely fused and it remained in the ventral midline separating the glans lamellae at 18.5 d.p.c. (data not shown). The mechanism of this canalization process, which displays a transition from the urethral plate to urethral fold and further to urethra, has been an issue of discussion among many scientists (see reference in Kurzrock et al., 1999; Baskin et al., 2001).

4. HISTOGENESIS OF HUMAN AND MOUSE EXTERNAL GENITALIA DEVELOPMENT

We compared the histogenesis of the human and mouse embryonic external genitalia. The cloacal membrane was observed in sagittal sections from human embryos at Carnegie stage 14 (Figure 3A) and from mouse at E10.5 (Figure 3B). Different aspects of morphogenesis between human and murine tailgut, the caudal endodermal organ connected to cloaca, have been reported. Molecular biological characters of the clocal membrane between these two species are now under investigation (in preparation).

In later development, human and mouse GT exhibit somewhat similar morphology in coronal sections (Figure 3C,D and Figure 3E,F). There have been reports about the differences underlying the process of seam formation in human and mouse GT development (Baskin 2000, Baskin et al., 2001). In the early stages, there are no prominent differences in urethral plate formation observed at lower magnification (Figure 3C,D). Upon further differentiation of the urethral plate and mesenchymes, some cytodifferentiation occurs that can be detected by histological alteration, e.g., epithelial cell differentiation or mesenchymal condensation (data not shown). In the case of the mouse GT, initial urethral plate adhesion is observed at later stages and two distinct sites of seam formation in the urethral folds are observed at later stages of development (data not shown; Baskin et al., 2001).

5. FUNCTIONAL ANALYSIS OF THE DISTAL SIGNALING EPITHELIA, DISTAL URETHRAL EPITHELIUM (DUE) AND SEVERAL GROWTH FACTORS ASSOCIATED WITH DUE

Recently, we investigated the molecular mechanisms of GT formation and showed that the DUE, a region marked by FGF8 expression, regulates the initial outgrowth process of the murine GT (Haraguchi et al., 2000). Based on several functional assays, we have proposed that the DUE of the developing GT may be a signaling epithelia

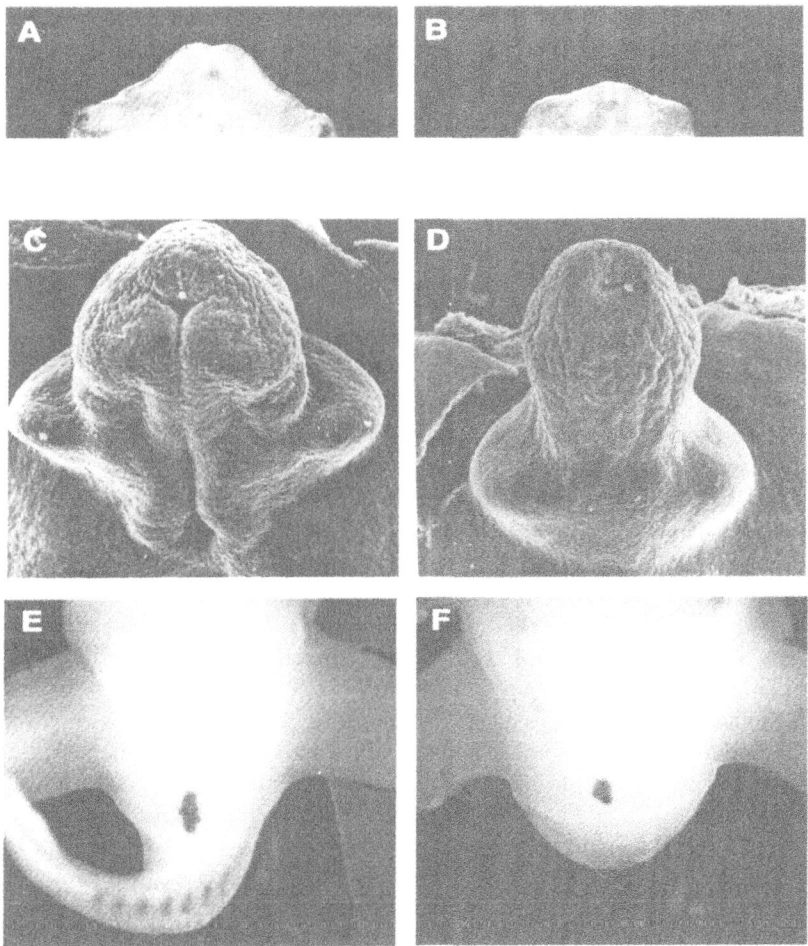

Figure 4. Role of DUE and effect of RA exposure on GT formation

Ablation of DUE by micro-surgery inhibited embryonic GT outgrowth (Figure 4A,B). Pregnant mice of 9.0 d.p.c. were treated with all-*trans* RA (Sigma) at a concentration of 100 mg/kg body weight by oral gavage (Kochhar, 1973; 1985). Controls were given an appropriate volume of the vehicle alone. SEM analysis of RA-treated and control GTs at 14.5 d.p.c. Figure 4C shows control, and Figure 4D shows RA-treated GT. Whole mount in situ gene expression analysis for FGF8. FGF8 was expressed in the DUE of control GT (Figure 4E), and in the abnormal GT by RA treatment at 11.5 d.p.c. (Figure 4F)

orchestrating the initial development of the external genitalia (Haraguchi et al., 2000; Ogino et al., 2001). Ablation of the DUE with a micro-needle inhibited GT outgrowth (Figure 4A,B), as did addition of an anti-FGF8 antibody (Haraguchi et al., 2000). In addition to several FGFs, expression of multiple regulatory genes involved in the formation of other appendages have been detected in the developing GT (Suzuki et al., in preparation). These similarities, as well as some degree of differences in gene expression and function related to the DUE and its adjacent mesenchymes might suggest the existence of a shared type molecular developmental program for some vertebrate appendages.

6. INVOLVEMENT OF THE RA SYSTEM AND GROWTH FACTOR SIGNALING

Interaction between retinoic acid (RA) and Fgf signaling has been recently shown to be essential for limb development (Mercader et al., 2000). In this context, the roles of regulatory factors and pathways, such as FGFs, Bmps, RA system, and Sonic Hedgehog (Shh), require further analysis. In order to analyze the teratogenic actions of RA exposure and the possible interactions of RA and growth factors, we examined the histology and embryonic development of the GT in RA-treated mice.

In the RA-treated mice, GT outgrowth was almost normal, although the morphology of the ventral portion of the GT was severely affected (Figure 4C,D). Tubular structures, which possibly represent the urethral epithelium were observed after RA treatment. Expression of FGF8, one of the marker FGFs expressed in the DUE, was still detected in the distal GT regions of RA-treated embryo (Figure 4E,F). By SEM analysis, we had observed a protrusion at the distal tip of the developing GT in RA-treated embryos (data not shown, Suzuki et al., 2002). This distal-most "epithelial tag", had been partially reported in a previous histological analysis decades ago for the human GT (Glenister, 1954).

Previous histological analyses had only suggested that this region is highly proliferative, but careful embryological and morphorogical studies of the functionally defined DUE had not been undertaken. Accordingly, more detailed studies on the functions of the protrusion (tag) are required, especially to determine an overlap or distinction with the functionally defined DUE.

In summary, these results suggest the intriguing possibility that the outgrowth of the GT in RA-treated mice was maintained possibly due to the presence of DUE. Further molecular studies are required to clarify the teratogenic action of RA during GT development, including its influence on the DUE and urethral plate morphogenesis. Given the complex and dynamic interaction between the RA and several growth factors in the development of several organs, the function of RA-induced pathways and its teratogenic action should be studied further for urogenital system. Cascades, including RA and Shh, have been suggested to function to generate an A-P axis in limb formation.

7. ABERRANT EXTERNAL GENITALIA FORMATION IN FGF10 KNOCKOUT MICE

Following the initial outgrowth of the GT, distal portions differentiate into the penis or clitoridis together with the preputial glands, urethrae, prepuces and the corpus cavernosum. Upon mesenchymal differentiation of the GT, FGF10 was expressed in the mesenchyme adjacent to the midline urethral plate at 13.5 d.p.c., prior to the stage at which sexual dimorphism is displayed (Haraguchi, et al., 2000). It has been suggested that mesenchymal FGF10 plays an important role in the genesis of several distinct organs (Sekine et al., 1999, Min et al., 1998). Judged by the fundamental role of FGF10 in limb initiation and its mesenchymal functions, investigation of the role of FGF10 in external genitalia development would be quite intriguing.

Recently, it has been shown that the external genitalia of FGF10 mutant mice have striking defects in urethral tube formation of the glans regions (Haraguchi, et al., 2000). Such dysmorphogenesis was clearly seen by gross morphology and by SEM analysis (Figure 5A,B,C,D). Characteristic morphological defects were found in the ventral side of the glans penis and the clitoridis, especially in the prepuce. In normal females at 18.5 d.p.c., morphogenesis of the urethra and prepuce in the glans clitoridis is very similar to that of the penis, though the incorporation of the tubular urethral epithelium is incomplete. In FGF10 mutants, tubular urethrae was not formed at the glans penis or clitoridis and failure of the prepuce fusion at the ventral midline of the glans was evident (Haraguchi et al., 2000).

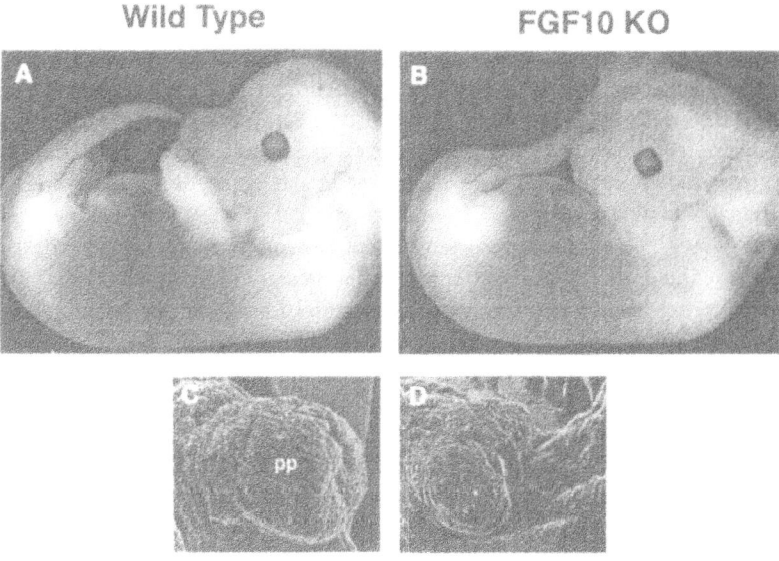

Figure 5. Phenotypes of Fgf10 mutantion in the GT and limbs.

Normal ICR mice (6 week-old) and Fgf10 gene knockout mice are shown. The FGF10 knockout mice and genotyping have been previously described (Sekine et al., 1999). Embryos were aseptically dissected and GTs were microsurgically dissected and processed for SEM analysis.
Wild type embryo at 12.5 d.p.c. (Figure 5 A) and FGF10 knockout embryo at 12.5 d.p.c. (Figure 5 B). SEM analysis of normal GT (Figure 5 C), and FGF10 knockout GT (Figure 5 D).

More recent analysis has shown that the interaction of the growth factor cascade between mesenchymal FGF10 expression and midline epithelial Shh expression plays an important role in external genitalia formation (Haraguchi et al., 2001; Figure 1K). Another intriguing characteristic of the FGF10 knockout mice is the defects in the development of various organs and appendages, including malformation of growing buds, e.g., limb and GT (Figure 5A,B). The FGF10 gene has been shown to be an essential for limb development (Ohuchi et al., 1997; 1999). In contrast with the drastic limb agenesis phenotype displayed in, hypoplasia of the GT of FGF10 knockout mice, the hypoplasia defects in GT development may be considered more subtle .

8. SEXUAL DIMORPHISM OF EXTERNAL GENITALIA DEVELOPMENT

In contrast to the early sex-independent differentiation, the later development of external genitalia involves a characteristic androgen-dependent dimorphic process.

The involvement of AR-mediated signaling during external genitalia development has been also revealed by genetic analysis. Several investigators have reported that in *Testicular feminization (Tfm)* mice, which are insensitive to androgens due to a mutation in the AR gene, the external genitalia develop into a feminized urethra very different from that of normal males (Murakami, 1987).

Baskin et al. has described series of studies on the mechanism of several types of human birth defects, including the histology and clinical problems associated with hypospadias (Baskin, 2000; Kurzrock et al., 2000; van der Werff and Ultee, 2000), and cryptorchidism (Jegou et al., 1983; Weidner et al., 1998; 1999). However, few morphological studies in relation with recent molecular embryological findings for murine external genitalia development have been reported. It has been widely accepted that genetically engineered mice are useful model systems to study organogenesis. It has been also noted that there are some differences in the human and mouse developmental processes. In particular, external genital formation displays significant morphogenic variation among species. These variations include the formation or absence of penile bones in the penis (e.g., OS penis), the process of urethra formation and corpus cavernous formation. It would be worth mentioning that one of the important and frequently observed malformations in human infants are in the urogenital systems, e.g., hypospadias. As such, a significant number of publications have been devoted to models of urethra development. It would be useful to investigate further murine external genitalia development in the molecular regulatory programs controlling morphogenesis.

ACKNOWLEDGMENTS

We thank Drs. Laurence Baskin, Gerald Cunha, Makoto Ishibashi, Takashi Miura, Toshiya Takigawa, Chigako Uwabe, Juan Carlos Izapisua-Belmonte, Pascal Dolle, Wade Bushman, Carol Podlasek, Marty Cohn, Scott Stadler, Hironori Kato, Yukiko Ogino, Mr. Hidenao Ogi, Yoshihiko Satoh and Marc Lamphier for their help. This work was supported by General promotion of cancer research in Japan, the Grants-in Aid from the Ministry of Education Scientific Research for Priority Areas (A) and by 21th Century COE Research.

REFERENCES

Baskin, L.S., 2000, Hypospadias and urethral development. *J. Urol.* **163**: 951-956.

Baskin, L.S., Erol, A., Jegatheesan, P., Li, Y.W., Liu, W.H., Cunha, G.R., 2001, Urethral seam formation and hyposapadisa. *Cell. Tissue. Res.* **305**: 379-387.

Bellusci, S., Grindley, J., Emoto, H., Itoh, N. and Hogan, B. L., 1997, Fibroblast growth factor 10 (FGF10) and branching morphogenesis in the embryonic mouse lung. *Development* **124**: 4867-4878.

Brennan, J., Karl, J., Martineau, J., Nordqvist, K., Schmahl, J., Tilmann, C., Ung, K., and Capel, B., 1998, Sry and the testis: molecular pathways of organogenesis. *J. Exp. Zool.* **281**: 494-500.

Cohn, M. J., Izpisua-Belmonte, J.C., Abud, H., Heath, J. K. and Tickle, C., 1995, Fibroblast growth factors induce additional limb development from the flank of chick embryos. *Cell* **80**: 739-746.

Crossley, P. H., and Martin, G. R., 1995, The mouse FGF8 gene encodes a family of polypeptides and is expressed in regions that direct outgrowth and patterning in the developing embryo. *Development* **121**: 439-451.

Crossley, P. H., Minowada, G., MacArthur, C.A., and Martin, G.R., 1996, Roles for FGF8 in the induction, initiation, and maintenance of chick limb development. *Cell* **84**: 127-136.

Cunha, G. R., Lung, B., 1979, The importance of stroma in morphogenesis and functional activity of urogenital epithelium. *In Vitro* **15(1)**: 50-71.

Dolle, P.,.Izpisua-Belmonte, J. C., Brown, J. M., Tickle, C., and Duboule, D., 1991, HOX-4 genes and the morphogenesis of mammalian genitalia. *Gene. Dev.* 5:1767.

Duboule, D., 1993, The function of Hox genes in the morphogenesis of the vertebrate limb. *Ann. Genet.* **36**: 24-29.

Glenister, T. W., 1954, The origin and fate of the urehral plate in man. *J. Anat.* **88**: 413-425.

Goff, D. J. and Tabin, C. J. , 1996, Hox mutations au naturel [news]. *Nat. Genet.* **13**: 256-258.

Gofflot, F., Hall, M., and Morriss-Kay, G.M., 1997, Genetic patterning of the developing mouse tail at the time of posterior neuropore closure *Dev. Dyn* **210**: 431-445.

Goldman, D.C., Martin, G.R., and Tam, P.P.L., 2000, Fate and function of the ventral ectodermal ridge during mouse tail development. *Development* **127**: 2113-2123.

Haraguchi, R., Suzuki, K., Murakami, R., Sakai, M., Kamikawa, M., Kengaku, M., Sekine, K., Kawano, H., Kato, S., Ueno, N., and Yamada, G., 2000, Molecular analysis of external genitalia formation: the role of fibroblast growth factor (Fgf) genes during genital tubercle formation. *Development* **127**: 2471-2479.

Haraguchi, R., Mo, R., Hui, C.C., Motoyama, J., Makino, S., Shiroishi, T., Gaffield, W., and Yamada, G., 2001, Unique functions of Sonic hedgehog signaling during external genitalia development. *Development* **128**: 4241-4250.

Ikeda, Y., Shen, W.H., Ingraham, H.A., and Parker, K. L., 1994, Developmental expression of mouse steroidogenic factor-1, an essential regulator of the steroid hydroxylases. *Mol. Endocrinol.* **8**: 654-662.

Jegou, B., Laws, A. O., and de Kretser, D. M., 1983, The effect of cryptorchidism and subsequent orchidopexy on testicular function in adult rats. *J. Reprod. Fertil.* **69**: 137-145.

Johnson, R. L., and Tabin, C. J., 1997, Molecular models for vertebrate limb development. *Cell* **90**: 979-990.

Kochhar, D. M., 1973, Limb development in mouse embryos. 1. Analysis of teratogenic effects of retinoic acid. *Teratology* **7**: 289-295.

Kochhar, D. M., 1985, Skeletal morphogenesis: comparative effects of a mutant gene and a teratogen. *Prog. Clin. Biol. Res.* **171**: 267-281.

Kondo, T. , Zakany, J., Innis, J. W., and Duboule, D., 1997, Of fingers, toes and penises. *Nature* **390**: 29.

Kurzrock, E. A., Baskin, L. S., and Cunha, G. R., 1999, Ontogeny of the male urethra: theory of endodermal differentiation. *Differentiation* **64**: 115-122.

Kurzrock, E. A., Jegatheesan, P., Cunha, G. R., and Baskin, L. S. , 2000, Urethral development in the fetal rabbit and induction of hypospadias: a model for human development. *J. Urol.* **164**: 1786-1792.

Martin, G. R., 1998, The roles of FGFs in the early development of vertebrate limbs. *Gene. Dev.* **12**: 1571-1586.

Mercader, N., Leonardo, E., Piedra, M. E., Martinez, A. C., Ros, M. A., and Torres, M., 2000, Opposing RA and FGF signals control proximodistal vertebrate limb development through regulation of meis genes. *Development* **127**: 3961-3970.

Min, H., Danilenko, D. M., Scully, S. A., Bolon, B., Ring, B. D., Tarpley, J. E., DeRose, M., and Simonet, W. S., 1998, FGF10 is required for both limb and lung development and exhibits striking functional similarity to Drosophila branchless. *Genes. Dev.* **12**: 3156-3161.

Moon, A. M., and Capecchi, M. R., M. R., 2000, FGF8 is required for outgrowth and patterning of the limbs [In Process Citation]. *Nat. Genet.* **26:** 455-459.

Mortlock, D. P., and Innis, J. W., 1997, Mutation of HOXA13 in hand-foot-genital syndrome. *Nat. Genet.* **15:**179-180.

Murakami, R., 1987, A histological study of the development of the penis of wild-type and androgen-insensitive mice. *J. Anat.* **153:** 223-231.

Ogino, Y., Suzuki, K., Haraguchi, R., Satoh, Y., Dolle, P., and Yamada, G., 2001, External genitalia formation: Role of fibroblast growth factor, retinoic acid signaling and distal urethral epithelium (DUE). *Ann.N.Y.Acad.Sci.* **948:** 13-31.

Ohuchi, H., Yoshioka, H., Tanaka, A., Kawakami, Y., Nohno, T., and Noji, S., 1994, Involvement of androgen-induced growth factor (FGF8) gene in mouse embryogenesis and morphogenesis. *Biochem. Biophys. Res.Commun.* **204:** 882-888.

Ohuchi, H., Nakagawa, T., Yamamoto, A., Araga, A., Ohata, T., Ishimaru, Y., Yoshioka, H., Kuwana, T., Nohno, T., Yamasaki, M., Itoh, N., and Noji, S., 1997,The mesenchymal factor, FGF10, initiates and maintains the outgrowth of the chick limb bud through interaction with FGF8, an apical ectodermal factor. *Development.* **124:** 2235-2244.

Ohuchi, H., Noji, S., 1999, Fibroblast-growth-factor-induced additional limbs in the study of initiation of limb formation, limb identity, myogenesis, and innervation. *Cell. Tissue. Res.* **296:** 45-56.

Ornitz, D. M., 2000, FGFs, heparan sulfate and FGFRs: complex interactions essential for development. *Bioessays* **22:** 108-112.

Perriton, C. L., Powles, N., Chiang, C., Maconochie, M. K., and Cohn, M. J., 2002, Sonic hedgehog signaling from the urethral epithelium controls external genital development. *Dev Biol.* Jul 1;**247(1):** 26-46.

Sekine, K., Ohuchi, H., Fujiwara, M., Yamasaki, M., Yoshizawa, T., Sato, T., Yagishita, N., Matsui, D., Koga, Y., Itoh, N., and Kato, S., 1999, FGF10 is essential for limb and lung formation. *Nat. Genet.* **21:** 138-141.

Sun, X., Mariani, F. V., and Martin, G. R., 2002, Functions of FGF signaling from the apical ectodermal ridge in limb development. *Nature.* **418:** 501-508.

Suzuki, K.,. Ogino, Y., Murakami, R., Satoh, Y., Bachiller, D., and Yamada, G., 2002, Embryonic development of mouse external genitalia: insights into a unique mode of organogenesis. **4:2** 133-141.

Tickle, C. and Eichele, G., 1994, Vertebrate limb development. *Annu. Rev. Cell Biol.* **10:** 121-152.

van der Werff, J. F., Ultee, J., 2000, Long-term follow-up of hypospadias repair. *Br J Plast Surg.* 53(7): 588-592.

Wall, N. A., and Hogan, B. L., 1995, Expression of bone morphogenetic protein-4 (BMP-4), bone morphogenetic protein-7 (BMP-7), fibroblast growth factor-8 (FGF-8) and sonic hedgehog (SHH) during branchial arch development in the chick. *Mech. Dev.* **53:** 383-392.

Warot, X., Fromental Ramain, C., Fraulob, V., Chambon, P., and Dolle, P., 1997, Gene dosage-dependent effects of the *Hoxa-13* and *Hoxd-13* mutations on morphogenesis of the terminal parts of the digestive and urogenital tracts. *Development* **124:** 4781-4791.

Weidner, I. S., Moller, H., Jensen, T. K., and Skakkebaek, N. E., 1998, Cryptorchidism and hypospadias in sons of gardeners and farmers. *Environ Health Perspect* **106:** 793-796.

Weidner, I. S., Moller, H., Jensen, T. K., and Skakkebaek, N. E., 1999, Risk factors for cryptorchidism and hypospadias. *J. Urol.* **161:** 1606-1609.

Yamaguchi, T. P., Bradley, A., McMahon, A. P., and Jones, S., 1999, A *Wnt5a* pathway underlies outgrowth of multiple structures in the vertebrate embryo. *Development* **126:** 1211-1223.

Yamasaki, M., Miyake, A., Tagashira, S., and Itoh, N., 1996, Structure and expression of the rat mRNA encoding a novel member of the fibroblast growth factor family. *J. Biol. Chem.* **271:** 15918-15921.

Yonei-Tamura, S., Endo, T., Yajima, H., Ohuchi, H., Ide, H., and Tamura, K., 1999, FGF7 and FGF10 directly induce the apical ectodermal ridge in chick embryos. *Dev. Biol.* **211(1):** 133-143.

NEW CONCEPTS ON THE DEVELOPMENT OF THE VAGINA

Ellen Shapiro MD*, Hongying Huang, MD, and Xue-Ru Wu, MD

1. INTRODUCTION

The female reproductive tract develops from the mullerian ducts. Mullerian duct development will not proceed in the absence of the mesonephric duct (Freedman and Shapiro, 1997; Gruenwald, 1943). The mullerian ducts adhere to each other just before they project into the dorsal wall of the urogenital sinus (UGS) causing an elevation termed the sinus tubercle. When the fused tips contact the sinus tubercle, the ducts fuse cranially, forming a tube with a single lumen called the uterovaginal primordium or canal (O'Rahilly, 1977). This tube will ultimately become the superior aspect of the vagina and uterus (Larsen, 1993). The cranial portion of the mullerian ducts which are unfused, develop into the fallopian tubes (George and Wilson, 1988). The original coelomic epithelial ostium remains as the abdominal opening of the fallopian tube. A septum initially divides the uterus into two cavities. This septum between the fused ducts disappears after 9 weeks, forming a single uterine cavity. The mesonephric ducts regress one week later, leaving a remnant in the female which is termed Gartner's duct (Arey, 1974). Muscularization of the uterus is completed by 17 weeks gestation and forms from the mesenchyme surrounding the mullerian ducts.

Vaginal development begins as early as 9 weeks with this Y-shaped solid mass of cells. The tissue of the sinus tubercle thickens forming endodermal evaginations termed the sinovaginal bulbs. These sinovaginal bulbs give rise to the lower third of the vagina. The inferior portion of the uterovaginal canal becomes occluded by tissue termed the vaginal plate. The etiology of this plate is unclear. It may form from the sinovaginal bulbs, the adjacent caudal end of the wolffian ducts, the mullerian ducts, or a combination of these structures Gruenwald, 1943; Acien et al., 1987; Bok and Drews, 1983; Bulmer, 1957). During the 3rd to the 5th month of gestation, the plate elongates, and the central cells of the plate break down by a process of desquamation with the lower vaginal lumen appearing at 11 weeks. The lining of the vagina and cervix is derived from the endodermal epithelium of the definitive UGS (Larsen, 1993). By 20 weeks,

* Ellen Shapiro, M.D., Department of Urology, 150 East 32nd Street, 2nd Floor, New York, N.Y. 10016, Phone: (646) 825 6326, FAX: (646) 825 6397, e-mail: ellen.shapiro@msnyuhealth.org

Hypospadias and Genital Development, edited by
L. Baskin, Kluwer Academic/Plenum Publishers, 2004

canalization of the plate is complete (Koff, 1933). Although the dual origin of the vagina has been proposed, other theories suggest a müllerian duct (MD) or wolffian (WD) origin or various combinations of these structures and the UGS (O'Rahilly, 1977; George and Wilson, 1988; Bulmer, 1957; Koff, 1933; Gray et al., 1994; Cunha, 1975; Hunter, 1930; Hart, 1911; Mijsberg, 1924). Prior studies on the development of the vagina were limited since they were performed prior to the advent of tissue selective antigens detectable by immunohistochemistry. Therefore, differential staining of the UGS, MDs, and WDs during the early development of the vagina would provide definitive evidence for the origin of the lower vaginal segment.

Uroplakins (UPs) are specialized membrane proteins of the urothelial plaque constituting the asymmetrical unit membrane of the bladder and represent specific molecular markers of urothelial differentiation (Wu et al., 1990; Wu, et al., 1994). Acid phosphatase activity has been localized to the urothelium of the fetal UGS during weeks 8-10, but is absent in urothelium of the dorsal wall of the UGS in the region of the MDs and WDs (Kellokumpu-Lehtinen, 1980). We hypothesize that the epithelium of the dorsal wall of the UGS involved in the formation of the sinovaginal bulbs (SVB) will express UPs, implying that the lower vaginal segment is of UGS origin.

It has been shown in animal models, that testosterone inhibits formation of the lower vagina (Bok and Drews, 1983). Congenital adrenal hyperplasia (CAH) is a form of female pseudohermaphroditism due to prenatal exposure to adrenal androgens during gestational weeks 9-12. The timing of the exposure to androgens is critical (8-13 weeks) (Grumbach and Ducharme, 1960). Exposure at progressively earlier stages of differentiation results in persistence of the UGS and labioscrotal fusion. If exposure occurs sufficiently early, the labia fuse to form a penile urethra. If exposure to androgens occurs after 12 weeks, only clitoromegaly occurs (Grumbach and Ducharme, 1960).

Since the phenotype in CAH is dependent upon the timing of the exposure to androgens, we examined the localization and temporal expression of the androgen receptor (AR) during development of the vagina and uterus. This may, in part, explain the varying phenotypes observed in CAH .

2. METHODS

2.1. Tissues and Antibodies

Lower genitourinary (GU) tracts were obtained from 4 human female fetuses gestational age 9 - 18 weeks (Shapiro et al., 2000). Approval for use of these specimens was obtained from the New York University Institutional Board Review Association. Each specimen was paraffin-embedded and serially sagittally-sectioned (3 microns thick). Representative sections were stained with hematoxylin and eosin (H and E) and examined with light microscopy. A uroplakin antibody raised in rabbit was reactive strongly with uroplakin III, moderately with uroplakin I and weakly with uroplakin II (Wu et al., 1990). Antibody to androgen receptor was obtained from Santa Cruz Biotechnology, Inc. (Santa Cruz, CA), and antibodies to acid phosphatase from DAKO Corporation (Carpinteria, CA).

2.2 Immunohistochemistry

An indirect peroxidase staining method was employed for the detection of various differentiation markers in urogenital tissues (Zhang et al., 1999). Representative tissue sections were deparaffinized followed by quenching of endogenous peroxidase with 3% hydrogen peroxide. Antigen retrieval was performed with 0.5% trypsin at 37^0 C for 15 minutes (for uroplakins and acid phosphatase) or using a microwave for 15 minutes (for androgen receptor). The sections were then incubated with primary antibodies followed by a corresponding secondary antibody conjugated with horseradish peroxidase. Slides were developed in a solution containing diaminobenzidine (DAB) from DAKO Corporation (Carpinteria, CA) and hydrogen peroxide.

3. RESULTS

The WDs and MDs were observed at 9 weeks. (Figure 1 a and b).

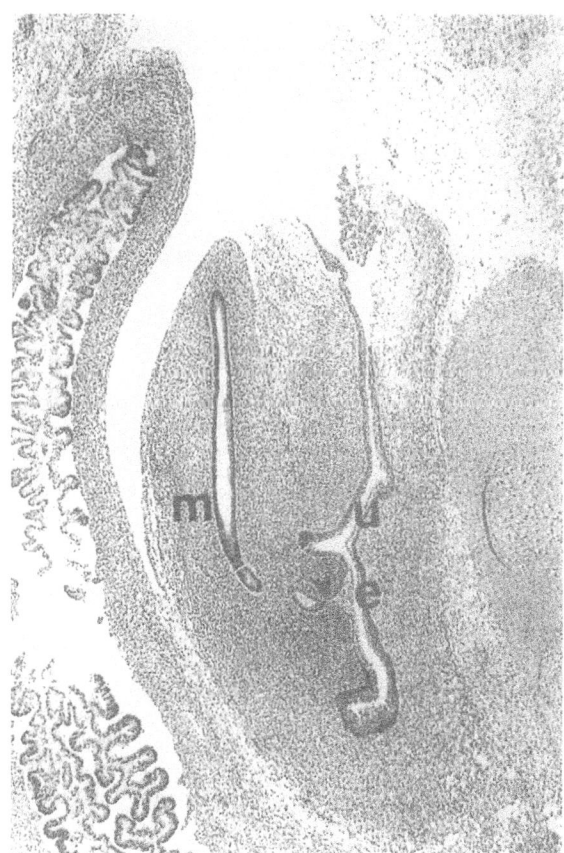

Figure 1a. Sagittal section of 9 wk fetus stained with H and E shows a) the MD (m) point of contact (v) between MD and area of evagination (e) of the UGS (u), and entrance site of WD (w) (40x) and b) (200x).

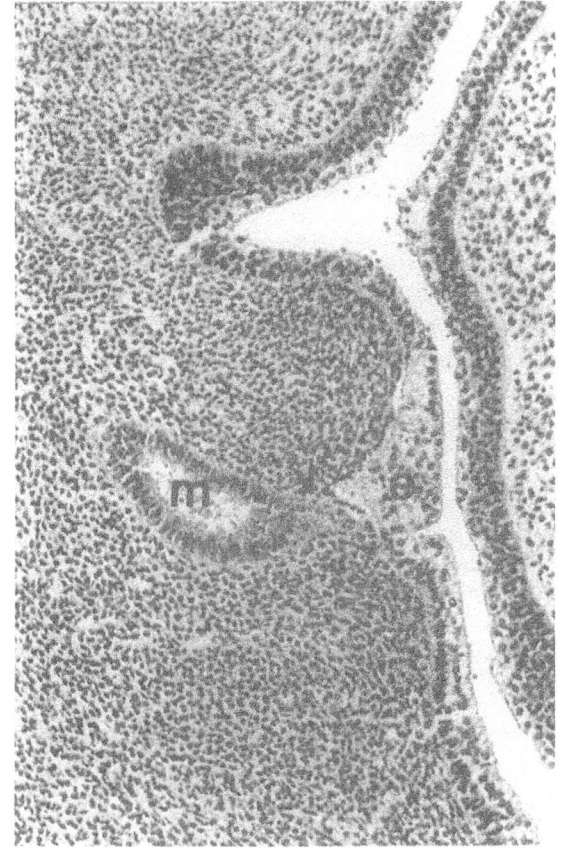

Figure 1b.

The UGS showed evagination and the subsequent formation of the sinovaginal bulbs (SVBs) (Figure 1 b and 2 a).

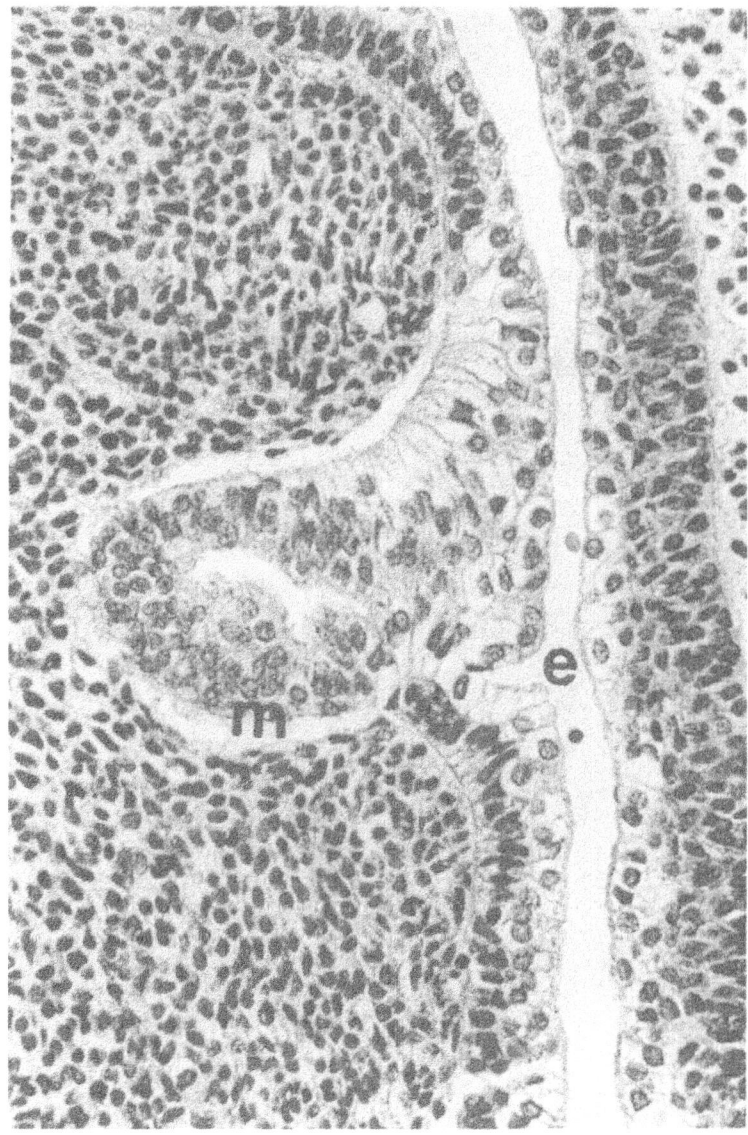

Figure 2a. Serial sagittal sections of 9 wk fetus a) stained with H and E showing point of contact between MD (m) and area of evagination of UGS (e) (400x), b) uroplakin expression (brown stain) in the UGS at the site of contact. (400x) and c) uroplakin staining of the UGS region involved in evagination 12 microns from serial sections in a and b No uroplakin staining is present in the MDs The entrance site of WD (w) is shown. (200x).

The urothelium of the entire UGS expressed UPs including the region of evagination. (Figure 2 b and c).

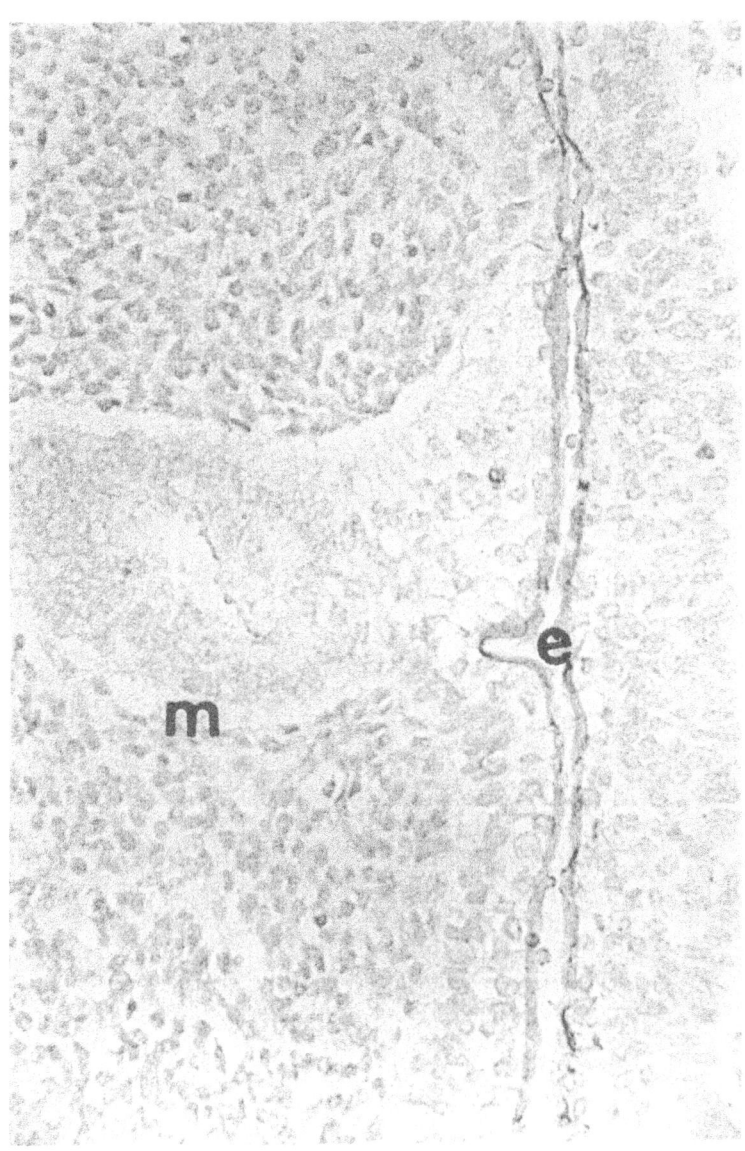

Figure 2b.

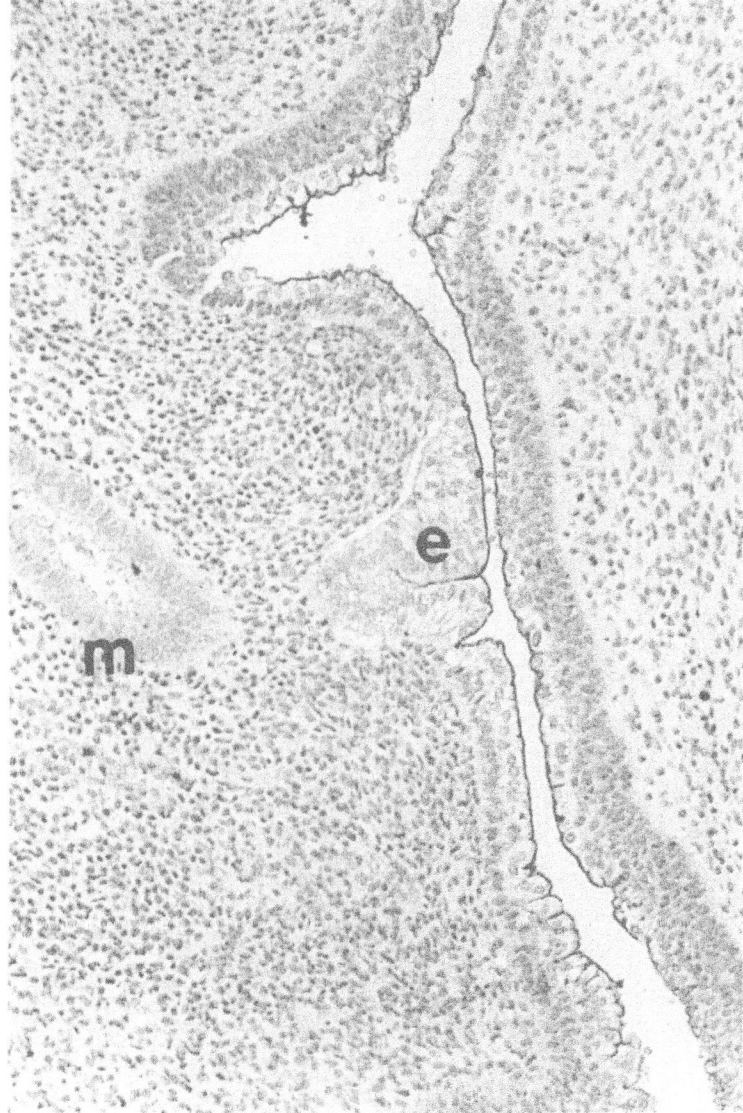

Figure 2c.

The MDs were in direct contact with the region of UGS evagination, but were not contiguous with the UGS. (Figure 2 a and b) Acid phosphatase expression was observed in the UGS except in the area of evagination (Figure 3 a) but by 12 weeks, acid phosphatase was expressed throughout the UGS. (Figure 3 b).

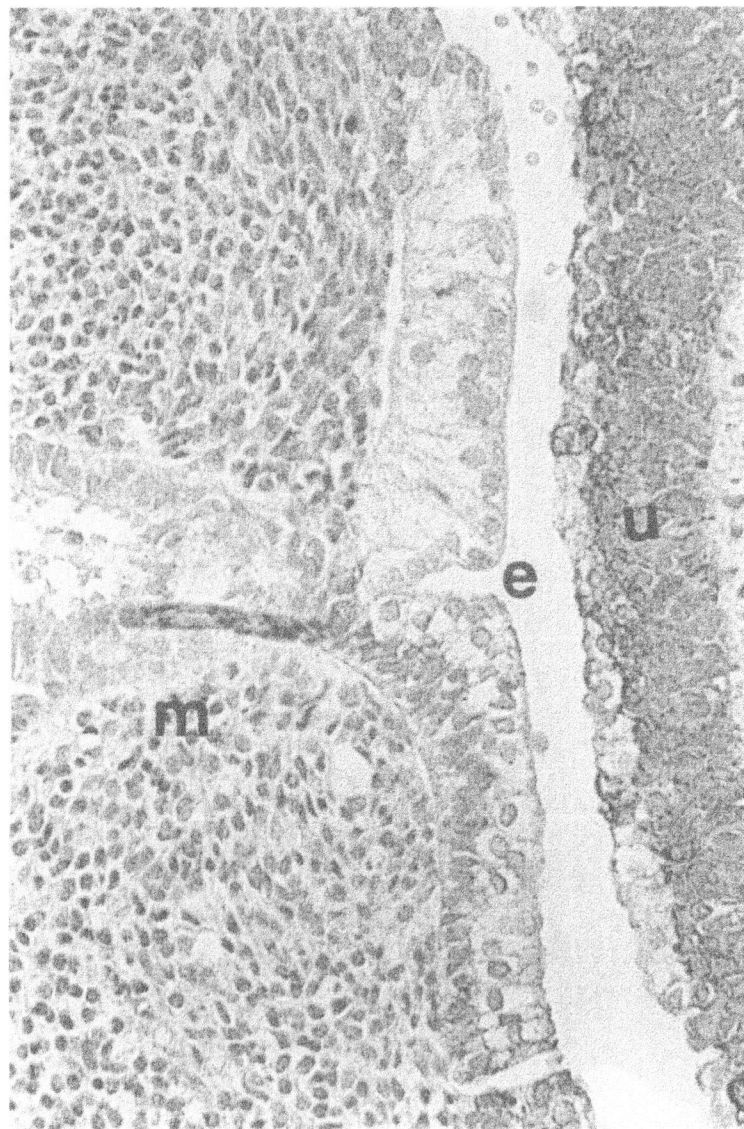

Figure 3a. Sagittal section of 9 wk fetus showing acid phosphatase localization (brown stain) along the UGS (u) except in the region of the MDs (m) and evagination of the dorsal wall (e). (100x) b) Sagittal section of 12 wk fetus in the region of a sinovaginal bulb (s) showing acid phosphatase expression throughout the UGS (u). (200x) and c) Sagittal section of 9 wk fetus showing cytokeratin 7 staining thoughout the UGS. No CK 7 expression is seen in the MD. (200x)

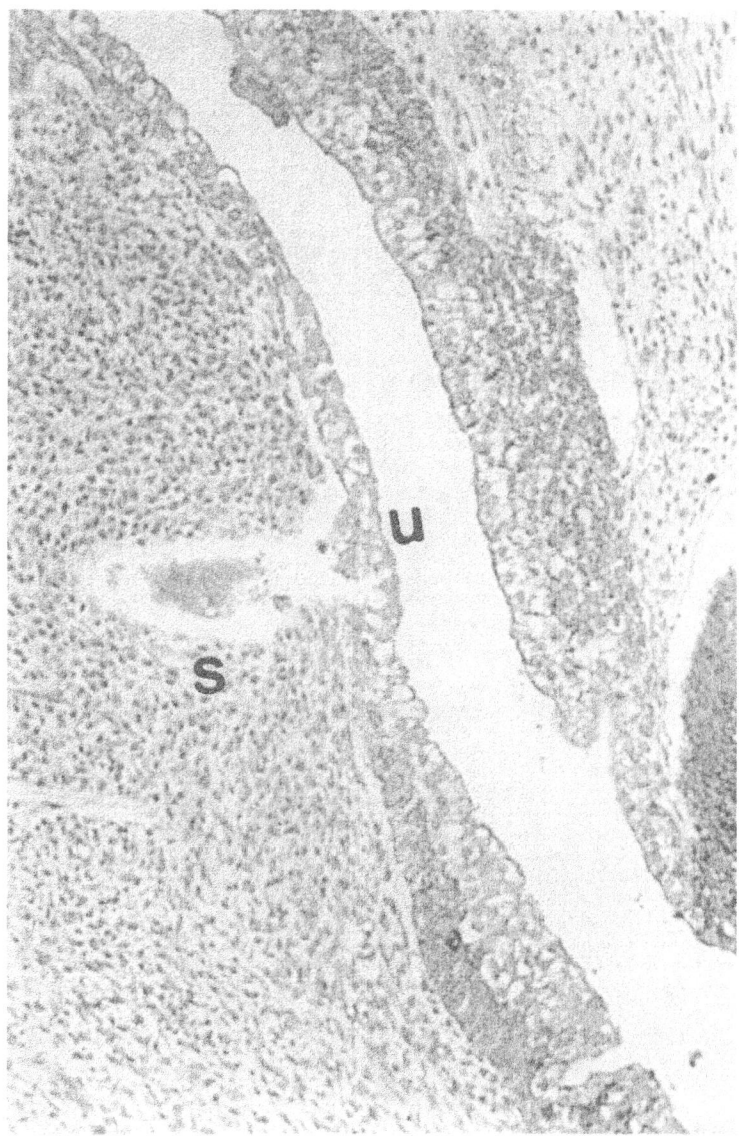

Figure 3b.

By 9 weeks, ARs expression was seen in the epithelium and the stroma of the UGS, SVB, MDs and the WDs. (Figure 4 a and b)

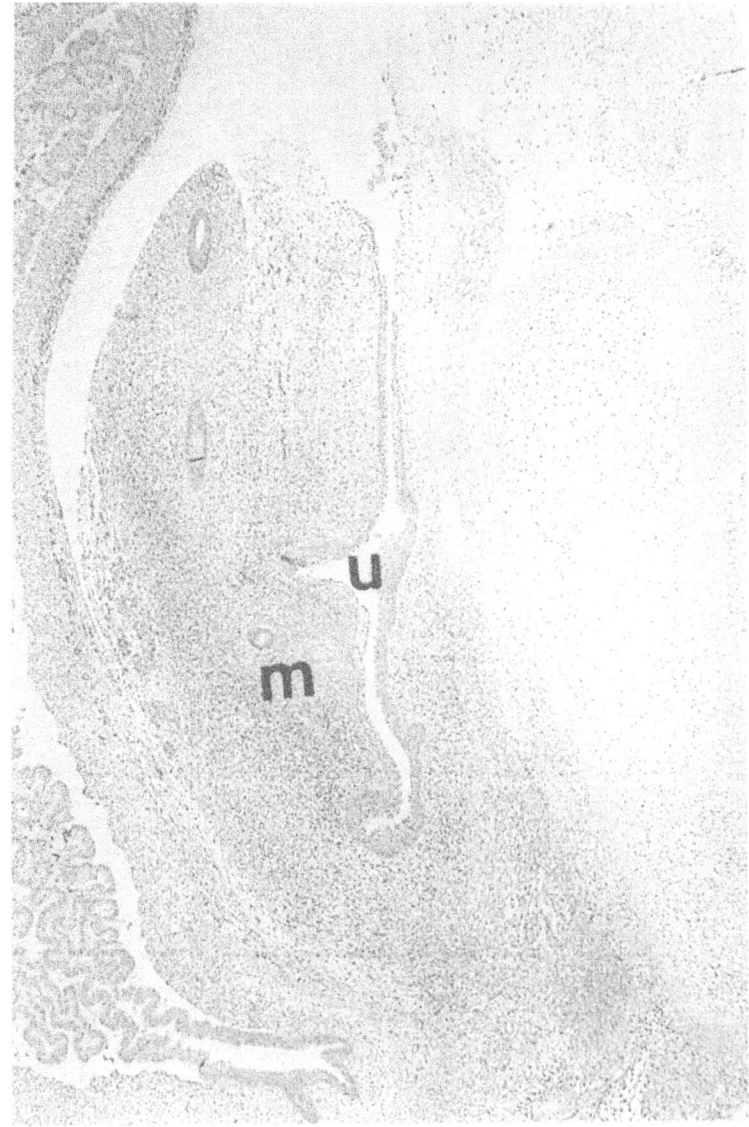

Figure 4a. Sagittal sections of androgen receptor expression (brown stain) in 9wk fetus at a) level of evagination; MDs (m), WDs (w) entrance site, UGS (u), and surrounding stroma stain positive for AR. (40x) b) (400x)

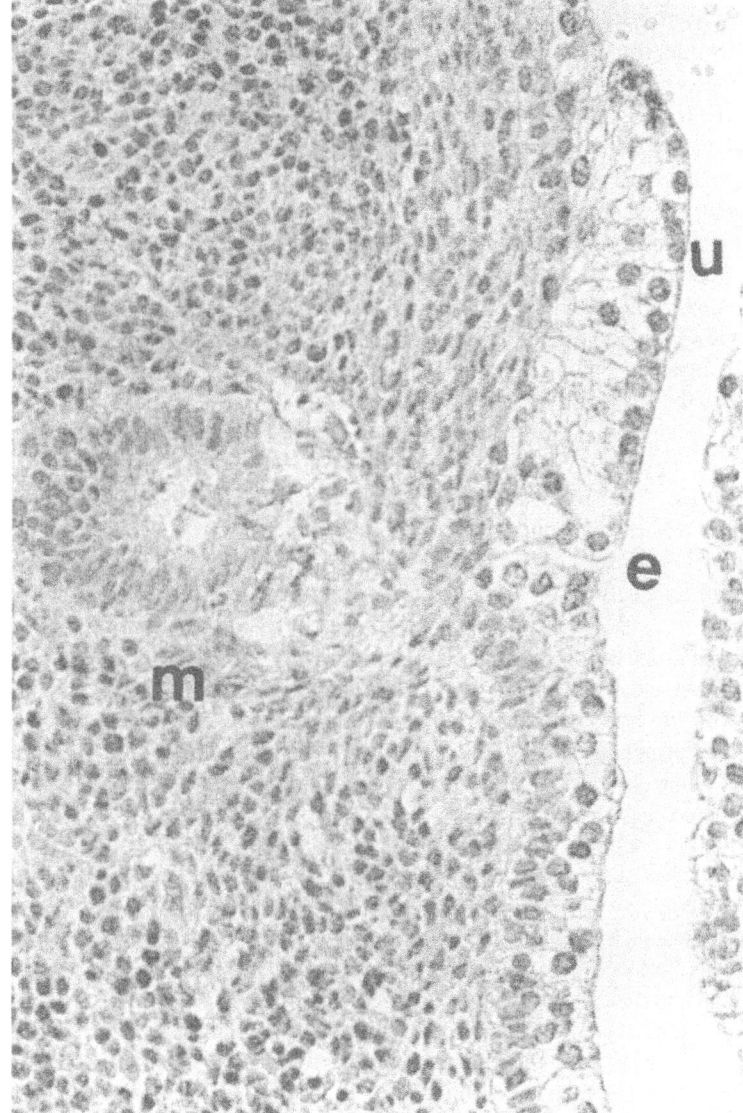

Figure 4b.

By 14 weeks, AR expression was marked by diminished in the urothelium of the UGS, the epithelium of the lower vagina and MDs and in the stroma surrounding these structures.

4. DISCUSSION

The area of evagination of the UGS expresses UPs but not acid phosphatase and is involved in the formation the SVBs which differentiate into the lower vagina by 18 weeks. These immunohistochemical studies provide definitive evidence for the UGS origin of the lower vagina. Testosterone inhibits the development of the lower vagina in male mammalian animal models (Bok and Drews, 1983). Although all females with CAH are exposed to testosterone in-utero, the degree of masculinization and uterovaginal descent vary (Grumbach and Conte, 1992).   Therefore, the timing of the in-utero exposure to testosterone in CAH will determine the phenotype of the external genitalia and the effect on the genital tract. When exposure to testosterone occurs after 12 weeks, only clitoromegaly occurs (Grumbach and Ducharme, 1960). Therefore, earlier exposure to testosterone leads to greater but varying degrees of virilization and varying degrees of incomplete uterovaginal descent. Those individuals with the greatest virilization usually have little or no descent of the uterovaginal complex (Grumbach and Conte, 1992). Since there is no evidence that the lower vagina is not formed in CAH, the effect of androgens is most likely targeted to the UGS and genital tubercle and not the lower vagina per se (Grumbach and Ducharme, 1960; Grumbach and Conte, 1992). We postulate that uterovaginal descent can not proceed normally when there is elongation of the UGS following early androgen exposure. This results in incomplete uterovaginal descent since there is insufficient time for the complex to transcend the increased distance resulting from elongation. When androgen exposure occurs later in gestation, the AR is absent in the urothelium of the UGS and the epithelium of the vagina and the MD by 14 weeks, suggesting that the UGS and its derivatives are androgen insensitive and uterovaginal descent will proceed normally. This is supported by observing acid phosphatase localization throughout the UGS by 12 weeks, implying that the phenotype of the epithelium along the dorsal wall has changed following the initial development of the UGS into the lower vagina.

Kalloo et al. demonstrated the location of the AR in the external genitalia of human male and female fetuses at gestational ages 18-22 weeks and found that the distribution and staining intensity was similar in both groups (Kalloo et al., 1993). In the male, the corpus cavernosum, stroma of the inner prepuce, scrotum, and periphery of the glans expressed AR. In the female, the glans, inner prepuce, stroma of the labia minora, corpus cavernosum, and labia majora also expressed AR. Since the age of the specimens correspond to the time when testosterone levels are declining, fetal penile growth after 16-18 weeks is most likely an androgen-independent event despite the presence of the AR after that time. In females with CAH, it is assumed that the AR distribution in the external genitalia is similar to normal females. Si⸱ ᴣ only clitoromegaly results with androgen exposure after 12 weeks, then the external genitalia have diminished sensitivity to high levels of endogenous testosterone despite the presence of AR. This suggests that development of the genital tubercle in the male is initiated via the androgen receptor before 12 weeks and that other growth factors are responsible for second trimester growth of the external genitalia when testosterone levels are diminished.

5. CONCLUSIONS

The region of evagination of the UGS expresses UPs and is involved in the formation the SVBs supporting the UGS origin of the lower vagina. Since testosterone inhibits the formation of the lower vagina, the timing of exposure to endogenous testosterone in CAH determines the phenotypic appearance of the external genitalia and the effect on lower genital tract development. Clinically, if exposure to testosterone occurs after 12 weeks, only clitoromegaly occurs. Since there is no AR expressed in UGS urothelium, vaginal epithelium and MDs by 14 weeks, this would infer that these tissues become androgen-insensitive and vaginal descent will be unaffected.

REFERENCES

Acien, P., Arminana, E. and Garcia-Ontiveros, E., 1987, Unilateral renal agenesis associated with ipsilateral blind vagina, *Arch. Path.* 240, 1.

Arey, L.B., 1974, Developmental anatomy, 7th edition, W.B. Saunders Co., Philadelphia.

Bok, G. and Drews, U., 1983, The role of the Wolffian ducts in the formation of the sinus vagina: an organ culture study, *J. Embryol. Exp. Morph.* **73**: 275-295.

Bulmer, D., 1957, The development of the human vagina, *J. Anat.* **91**: 490-509.

Cunha, G. R., 1975, The duel origin of vaginal epithelium, *Am. J. Anat.* **143**: 387-392.

Freedman, A. L. and Shapiro, E., 1997, The embryological basis of genitourinary malformations. in: Urological Disease of the Fetus and Infant. D.F.M. Thomas (ed.), Butterworth Heinemann, Oxford, Chapter 2, pp 24-37.

George, F. W. and Wilson, J. D. 1988, Sex determination and differentiation. in: E. Kobel, ed. The Physiology of Reproduction, New York: Raven Press, 3-26.

Gray, S. W., Skandalakis, J. E., and Broecker, B., 1994, The female reproductive tract. IN: J. E. Skandalakis and S. W. Gray, eds. Embryology for Surgeons, 2nd edition. Baltimore: Williams and Wilkins, Chapter 22, 816-847.

Gruenwald, P., 1943, The normal changes in the position of the embryonic kidney, *Anat. Rec.* 85, 163.

Grumbach, M. M. and Ducharme, J. R., 1960, The effects of androgens on fetal sexual development: Androgen-induced female pseudohermaphrodism, *Fertility and Sterility.* **11**: 157-180,.

Grumbach, M. M. and Conte, F. A., 1992, Disorders of sex differentiation. In: Jean D. Wilson and Daniel W. Foster, eds. Williams Textbook of Endocrinology, 8th edition. Philadelphia: W.B. Saunders, Chapter 18, 853-951.

Hart, D. B., 1911, Adenoma vaginae diffusum (adenomatosis vaginae) with a critical discussion of present views of vaginal hymenal development, *Edinb. Med. J.* **6**: 577-590.

Hunter, R. H., 1930, Observations on the development of the human female genital tract, Contrib. Embryol. Carnegie Inst. Wash., **22**: 91-108.

Kalloo, N. B., Gearhart, J. P., and Barrack, E. R., 1993, Sexually dimorphic expression of estrogen receptors, but not of androgen receptors in human fetal external genitalia, *J. Clin. Endocrinol. Metab.* **77**: 692-698.

Kellokumpu-Lehtinen, P., 1980, The histochemical localization of acid phosphatase in human fetal urethral and prostatic epithelium, *Invest. Urol.* **17**: 435-440.

Koff, A. K., 1933, Development of the vagina in the human foetus. Contrib. Embryol. Carnegie Inst. Wash., **24**: 61-90.

Larsen, W. J., 1993, Human Embryology. Churchill Livingstone, New York.

Mijsberg, W. A., 1924, Uber die Entwicklung der Vagina, des Hymen und des Sinus urogenitalis biem Menschen. *Z. Anat. Entw.* **74**: 684-760.

O'Rahilly, R., 1977, The development of the vagina in the human. Birth Defects: Original Article Series, **13**: 123-136.

Shapiro, E., Huang, H-Y, and Wu, X.-R., 2000, Uroplakin and androgen receptor expression in the human fetal genital tract: insights into the development of the vagina, *J. Urol.* **164**: 1048-1051.

Wu, X.-R., Manabe M, Yu, J., Sun, T. T., 1990, Large scale purification and immunolocalization of bovine uroplakins I, II, and III. Molecular markers of urothelial differentiation, *J. Biol. Chem.* **265**(31): 19179.

Wu, X.-R., Lin, J. H., Waltz, T., Haner, M., Yu, J., Aebi, U. and Sun, T.-T., 1994, Mammalian uroplakins. A group of highly conserved urothelial differentiation-related membrane proteins, *J. Biol. Chem.* **269**: 13716-13724.

Zhang, Z. T., Pak, J., Shapiro, E., Sun, T. T., Wu, X.-R., 1999, Urothelium-specific expression of an oncogene in transgenic mice induced the formation of carcinoma in situ and invasive transitional cell carcinoma, *Cancer Res.* **59**(14): 3512.

Endocrine Disruptors &
Sexual Dimorphism
in the Animal Kingdom

ENDOCRINE DISRUPTION OVERVIEW:
ARE MALES AT RISK?

Theo Colborn, Ph.D. [1]

1. INTRODUCTION

The goal of this paper is to encourage physicians and others responsible for public health to start thinking outside of the box -- to consider what has been learned from wildlife that has contributed considerably toward understanding the etiology of a number of human disorders that have increased significantly over the past 50 years in the developed world. Hopefully this will lead to practitioners' thinking in broader terms when the etiology of a disorder is not evident. The increased incidence of hypospadias, for instance, would not have been discovered if it had not been for a troubled population of alligators in Lake Apopka, Florida (Guillette et al., 1994). The reproductive success of the alligators had dropped to 10% and many of the males had undersized phalluses. These findings led to a series of laboratory studies that eventually revealed that there are synthetic anti-androgens in the environment that can interfere with male development and function (Kelce et al., 1994; Kelce et al., 1995). Serendipitous discoveries such as this concerning wildlife health have led to intense scrutiny concerning the etiology of a number of other human disorders related to the endocrine system such as abnormal sexual development and neurological and reproductive impairment.

2. BACKGROUND

In 1991, experts warned that *"A large number of man-made chemicals that have been released into the environment, as well as a few natural ones, have the potential to disrupt the endocrine system of animals, including humans."* (Clement and Colborn, 1992). They also estimated with confidence that *" Unless the environmental load of*

---

[1]World Wildlife Fund 1250 24th St. Washington DC 20037 theo.colborn@wwfus.org 970-527-6548 (phone and fax).

[2]Adapted from Birth Defects Surveillance Data (Teratology, 1997. Vol. 56: 115-175).

[3]Adapted from National Birth Defects Prevention Network (1997).

[4]Adapted from Gray, E. et al. (1999); Gupta, (2000).

Hypospadias and Genital Development, edited by
L. Baskin, Kluwer Academic/Plenum Publishers, 2004

synthetic hormone disruptors is abated and controlled, large scale dysfunction at the population level is possible. The scope and potential hazard to wildlife and humans are great because of the probability of repeated and/or constant exposure to numerous synthetic chemicals that are known to be endocrine disruptors." They also published a list of 51 widely used chemicals that were known endocrine disruptors at that time. Today, the list has grown to over 140 (Brucker-Davis, 1998; Colborn, 1998; Short and Colborn, 1999; European Commission, 2001).

In 1995, the U.S. Environmental Protection Agency adopted the following definition of an endocrine disruptor: *An Endocrine Disruptor is an exogenous agent that interferes with the production, release, transport, metabolism, binding, action, and/or elimination of hormones in the body responsible for the maintenance of homeostasis and the regulation of developmental processes AND FUNCTION. (note, that function was left out of the definition ..author's addition).* An exogenous agent can be a synthetic chemical or a product of the modern, unusual use of concentrated natural foods, such as soy. These chemicals have been described as stealth chemicals that cause silent and delayed toxicity (Weiss, 1998).

The health endpoints that result from exposure to endocrine disruptors *in utero* are unlike those endpoints expressed following postnatal and adult exposure. Also, the health endpoints that are initiated *in utero* are unlike the health endpoints upon which chemical safety is currently determined by governments. Consequently, there has never been a government approved set of testing protocols to systematically test chemicals for their possible adverse effects on development and function.

2.1 The Insidious Latent Activity of Endocrine Disruptors

The question therefore is how did so many chemicals slip through the regulatory process? Because, endocrine disruption is insidious -- it inconspicuously intrudes on physiologic, morphologic, immunologic, neurologic, and reproductive development and function during gestational exposure. Its impact may not be expressed measurably or visibly until adulthood, long after the mother and prenate were exposed. Prenatal exposure to endocrine disruptors has been shown to irreversibly alter cognition, behavior, and immune competency, and to predispose autoimmune and fertility problems and gonadal cancers. The endpoints are so numerous they are difficult to quantify in human populations, let alone wildlife. And for physicians, many of the symptoms can be confused with the symptoms of already defined clinical diseases.

The most sensitive cohort among humans is the unborn (Bern, 1992) which means that at any one time during the year only 1.1% of the U.S. population is at risk. The latency of expression and endless list of irreversible errors of programming that can occur during this early life stage, contribute to the insidious nature of endocrine disruption. Even after a generation time, twenty years later, the impact at the population level might not be discernible because:

(1) Not every individual would have been exposed at levels where impairment would have taken place.

(2) No two exposure scenarios could possibly be similar when one considers the number of chemicals in use today, 87,000 --15,000 of which are produced in high volumes (EDSTAC, 1998). This is not like administering the same pharmaceutical to selected individuals across a population and knowing what to look for. Consider all the different side effects and the range of responses in a population-wide experiment that could

become evident when only one drug had been administered. Then consider how diverse and confusing the effects could be in the uncontrolled study that is going on today with industrial and agricultural chemicals in our environment.

(3) To the above, add in the complexity of the vast mixes of chemicals in the environment and it is easy to imagine the difficulty of recognizing or quantifying damage from endocrine disruptors in a population.

(4) In addition, no two individuals could have been exposed identically at the same critical moment or stage in development.

(5) Then there are the sexually dimorphic differences between boys and girls that do not develop alike.

(6) And finally, variation in human resilience and genetic predisposition would have to be factored in, which would make some individuals more or less susceptible.

Consequently, the range of adverse health responses to endocrine disruptors is endless and even after 20 years a particular endpoint might not reach statistical significance in a nation-wide population study. However, clusters of abnormal developmental and/or functional problems have been reported in discrete geographic regions around the world (Garry, 2000; Mastroiacovo, 1988; Dolk et al., 1998; Kristensen et al., 1997; Eskenazi, 2002). All of the above reasons work in favor of those who argue that exposure to synthetic chemicals that have been demonstrated in the laboratory to be endocrine disruptors are not posing a health problem to humans.

2.2 The Wildlife Scenario

So how can one argue that endocrine disruptors are affecting the health of humans? In order to do this one must go back in time to the literature on wildlife in aquatic systems where persistent chemicals biomagnified to such high concentrations that they were easy to detect in animal tissue.

Fortunately for humans many wildlife species breed in the spring. And in the case of shorter lived species that reach sexual maturity in a year, approximately all the females, or almost 1/2 of the population will be producing offspring during a short period of time each year. And if the adult animals were exposed to endocrine disrupting chemicals prior to, or on their breeding grounds, this would lead to a clustering of health problems. This is exactly what happened in the past where damage was clearly visible among aquatic bird species that nest in colonies (Fry and Toone, 1981; Fry et al., 1987; Fox, 2001). If these same birds were continuous breeders like humans, the impacts would have been far less obvious. As a matter of fact, if the births had been spread out over the year, in all probability the problems would never have been recognized by wildlife biologists. Fortunately, natural selection has in many instances limited wildlife birthing to the most benign season of the year (birds especially)--and also at a time when wildlife biologists and their undergraduate and graduate students can get into the field and see what is happening in the bird colonies.

As early as 1960, biologists began to report that chemical residues in breeding females were causing lesions in the chicks that led to early mortality. The chemicals were also affecting parental behavior. Three egg-switching studies, each one becoming more sophisticated, revealed that lack of parenting was in part causing chick mortality (Peakall et al., 1978; Fox et al., 1978; Kubiak et al., 1989). Whether this was the result of pre-hatching exposure by the parents or post-hatching exposure was not determined. Also in the late 60s and early 70s there were reports of female/female pairing in both Great

Lakes and Pacific coast colonies of gulls (Hunt and Hunt 1977; Fry et al., 1987). Males especially seemed to be vulnerable. They were growing ovo-testes and forming fraternities rather than attending to reproductive activities. Biologists reported skewed sex ratios in bird populations with more phenotypic females than males. There were no sex probes available to determine genetic sex at that time. However, it has been speculated that perhaps some of the females were genetic males with female phenotypes.

By the 1970s, fisheries biologists were reporting strange thyroid disturbances across the Great Lakes in top predator salmonid species. One team found enlarged thyroids and aberrant triiodothyronine/thyroxine (T3/T4) ratios in every top predator fish they examined (Leatherland, 1992). About that time Environment Canada started a herring gull monitoring program throughout the four Great Lakes with a common border between the United States and Canada looking at thyroid hormone levels and thyroid histology in the gulls that eat the fish. Today, the thyroids in the Lake Erie fish are so enlarged that they often rupture and are visible (Leatherland, 1998). No clear link with a specific chemical that could be causing the goiters has been discovered to date.

Also, in the 1970s biologists from across the Great Lakes were reporting visible damage in avian species and/or complete extirpation of local populations. After close scrutiny they realized that in almost every instance only the youngsters were having visible problems if they hatched. A large number of chicks were dying before hatching, shortly after hatching, or at the time of fledging, suggesting that death was only postponed in the lesser contaminated animals (Kubiak et al., 1989). It was among terns and gulls nesting in large numbers in relatively small geographic areas that the passing of persistent chemicals from one generation to the next was documented. From this, the phenomenon of endocrine disruption began to emerge as wildlife biologists formed collaborations with scientists from other disciplines and began to think outside the box.

2.3 The Human Scenario

The expression of effects following exposure to synthetic chemicals in wildlife with generation times of one to five years is far more robust than those expressed in humans with a long generation time of twenty years. Consequently, it takes years for chemically-induced effects or prenatal exposure to be recognized among human populations (see Table 1). Nonetheless, a pattern of increased prevalence of human disorders seemingly follows the increases in production and release of man-made chemicals that are known to interfere with development. For example, in the 1920s the plastic monomer, bisphenol A, was introduced into commerce, and the unintentional release of dioxin led to its building up in the Great Lakes as the result of the production and use of free chlorine. In 1929, the production of polychlorinated biphenyl, dielectric, fire retardants made possible the long distance distribution of electricity and other technological advances. In 1938, 1,1,1-trichloro-2,2-bis(p-chlorophenyl) ethane (DDT) was first used as an insecticide. Widespread postnatal exposure to increasing numbers of new man-made chemicals commenced in the mid 1940s during, and shortly after WWII. By 1950, a generation of adults who, although they had not been exposed prenatally, had accumulated the new synthetic chemicals in their bodies and were producing the first generation of offspring exposed to numerous synthetic chemicals during gestation. By 1970, these post-WWII babies were adults and began having babies of their own. It was during the 1970s that what appears to be increases in unusual, previously rare neurodevelopmental health problems such as autism, ADD-like symptoms, manic depression, and schizophrenia were catching the attention of health professionals (Berkow and Fletcher, 1978; Gershon and Reider, 1992). Parental support groups and non-profit organizations dedicated to these problems began to emerge, and in 2000, a presidential initiative led to the establishment of children's health centers nationwide to develop treatments and cures for these problems.

The list has now grown to include increases in juvenile arthritis, juvenile cancers, juvenile diabetes, juvenile delinquency, juvenile violence, early puberty, hypospadias, incidence and mortality of early testicular cancer, incidence of breast, testicular, and prostatic cancers, other secondary sex organ problems, and loss of fertility, signaled by a growing industry in *in vitro* fertilization (Herman-Giddens et al., 1997; Paulozzi et al., 1997; King and Schottenfeld, 1998; Shibata et al., 1998; Guillette, 1999; Mckiernan et al., 1999; van Maanen et al., 2000; Kaipiainen-Seppanen and Savolainen, 2001; Jensen et al., 2002).

Table 1. Chronology of Human Exposure

1920s-30s	BPA, PCBs and DDT released
1940s-WWII	First wide-scale exposure to man-made chemicals
1940s-50s	First generation exposed prenatally
1950s-70s	First generation born that was exposed in the womb
1970s-90s	First generation that was exposed in the womb reached reproductive age
1980s-present	Second generation born that was exposed in the womb

2.4 Sensitivity of the Male

Recently more and more publications in the science/health literature provide evidence that there is a synthetic chemical-related challenge to the integrity of the male. For example, starting from conception to birth, in Japan since the 1970s the ratio of the male/female fetal deaths reached more than 2 in 1996. When comparing the male/female ratio of fetal deaths with the male/female ratio of infant deaths and neonatal deaths there is a striking difference. The ratios of male/female deaths at birth and deaths shortly after birth have remained the same albeit with a great deal of variance. This paper points out that in Japan something is weeding out males before birth (Mizuno, 2000).

Two studies from the U.S. and one from Canada reviewed the male/female live birth ratio from approximately 1970 to 1995 and reported a decline in male/female birth ratios (Davis et al., 1998; Marcus et al., 1998; Allan et al., 1997). The authors of the Canadian study searched back to 1930 and found that the decline did not start in Canada until 1970 (Allan et al., 1997). This agrees with the timeline in Table 1 when one would expect to see an effect at the population level. The Canadian team found a cumulative loss of 2.2 males per 1000 births across Canada; in the eastern Atlantic region of Canada the loss was 5.6 male births per 1000 ($p<0.001$) between 1970 and 1990. The Canadian team also analyzed the U.S. data and found a highly significant decrease in the male/female birth ratio at 1.0 male per 1000 live births ($p<0.001$).

An Atlanta study using the Metropolitan Atlanta Congenital Defects Program data found major birth defects in males were 3.9% compared with 2.8% for females from 1968 to 1995 (Lary and Paulozzi, 2001). Defects of sex organs were 8.5 times more

Table 2. Percent Change in Hypospadias and Epispadias Rates for Caucasians from 18 U.S. States, 1985 to 1995[2]

U.S. state	Percent change
Arizona	+ 9.2
Arkansas	+ 9.6
California	- 33.5
Colorado	+ 6.6
Georgia	- 12.45
Hawaii	+ 50.0
Iowa	+ 18.4
Kansas	+ 29.3
Maryland	- 11.1
Missouri	- 7.4
Nebraska	+ 2.0
New Jersey	+ 57.2
New York	- 1.8
North Carolina	+ 20.1
Oklahoma	+ 5.7
Virginia	- 20.7
Washington	- 2.4
West Virginia	+ 56.6

prevalent among males than females. Urinary tract defects were 62% more and gastrointestinal tract defects were 55% more prevalent in males than in females.

Another study suggested that the prevalence of hypospadias doubled since 1968 and estimated that 1 in 125 boys in the U.S were born with the anomaly in 1989 (Paulozzi et al., 1997). Severe cases increased more than the milder forms. This anomaly is the result of disturbances around the 8th week of gestation when under normal conditions fetal testosterone is converted to dihydrotestosterone (DHT) by the enzyme 5 alpha reductase. DHT initiates the simultaneous growth of the penis and the urethra. Only recently was it discovered that certain synthetic chemicals could interfere with the masculinization of the external genital and development of internal male gonads.

Several studies suggest that the increase in hypospadias has slowed down (Toppari et al., 2002). Table 2 was developed from the most recent statistics available on hypospadias published in the Journal of Teratology. Table 3 provides a simple analysis of what the data reveal.

For a comprehensive review of the literature on hypospadias that covers prevalence, etiology, risk factors from genetics to environmental factors, exposure, and the wildlife experience see Baskin et al. (2001).

A seminal paper in the field of environmental health introduced the term "environmental anti-androgen" in 1994 (Kelce et al., 1994). In this study, a widely used fungicide, vinclozolin, was given to pregnant rats. Their male pups experienced significant increases in undescended testicles, permanent nipples, hypospadias, vaginal pouchs and atrophy in seminal vesicles and ventral prostate glands. The most sensitive endpoint in this study was anogenital (AG) distance, measurable at birth.

A year later in 1995, a paper in Nature received a great deal of attention because of the discovery that 1,1-dichloro-2,2-bis(p-chlorophenyl)ethylene (DDE), the persistent metabolite of DDT was a powerful antiandrogen (Kelce et al., 1995). Until then it was a mystery how DDT caused the feminizing effects seen in wildlife --where genetic male birds grew female plumage, behaved like females, and grew ovo-testes. Damage from lab exposure to o,p' -DDT (the estrogenic component of DDT) never was as severe as the damage seen in wildlife. Because of this, in early reports, cautious researchers called the feminizing effects in the animals "estrogen like effects" (Burlington and Lindeman, 1950). This 1995 study (Kelce et al., 1995) revealed that DDE, which is found in almost all human and wildlife tissue around the world, is 1/10th as potent as flutamide, a pharmaceutical used to chemically castrate men with prostate cancer. It is found in umbilical cord blood and amniotic fluid within the range where it causes hypospadias in male rats in the laboratory. Similar to the results in the vinclozolin study, AG distance was the most sensitive endpoint for detecting the anti-androgen in this study as well.

As one can see from Table 4, the classes of chemicals that are known anti-androgens and can cause hypospadias range from agricultural to industrial chemicals.   The mechanism of action that leads to hypospadias of most of the chemicals that cause hypospadias is diverse. For example, the parent compound of the fungicide vinclozolin is biologically inert. It has two active metabolites that competitively bind the androgen receptors (ARs) thus shifting the number of active ARs available for endogenous androgens. One metabolite is more active than the other and depending on conditions in the tissue the results vary. Thus making it is difficult to predict the outcome of exposure. Another fungicide, Procymidon, inhibits DHT production. The metabolite of the insecticide DDT, DDE, blocks ARs. In a not yet fully understood mechanism, the

Table 3. Number of States with Changes in Hypospadias and Episadias Rates for Caucasians in 18 U.S. States, 1985 to 1995[3]

Change	# of U.S. states
Increased >15%	6
Decreased > 15%	2
Any increase	11
Any decrease	7

phthalates interfere with fetal testosterone production. They do not interfere with the AR. The Centers for Disease Control and Prevention (CDC) in a nationwide survey of parent compound and metabolites in human urine reported in 2000 that the cohort of women between the ages of 20 and 40 (child bearing age) were excreting 9 times more of two phthalates (monobutyl phthalate and monobenzyl phthalate) than any other cohort of individuals (Blount et al., 2000). Upon further delineation the CDC discovered that these particular phthalates are found in nail polish, cosmetics and perfumes.

Table 4. Environmental Endocrine Disruptors that Cause Hypospadias in Laboratory Animals[4]

Agricultural and Public Health Chemicals

Industrial Chemicals

Insecticide
p,p′-DDE +
(DBP) +

(DEHP) +
Fungicides
Vinclozolin +
Procymidone +
Organochlorines

Herbicide
Linuron +

Plastic Components
Di-(n-butyl)   phthalate

Diethylhexyl   phthalate

Bisphenol A

Persistent

Dioxin (TCDD) *
PCB 169 *

+ Reduced anogenital distance. The most sensitive endpoint.
* Females. In all other cases, females have not been examined to date.[1]

Traditionally, estrogenic substances were suspected of causing hypospadias. This could very well be true because most of the lab studies on estrogens used diethylstilbestrol (DES) as a model and focused only on the females; the male rodents were discarded at birth. According to a leading authority on DES, "No one looked at the males" (McLachlan, 2000). However, there is one study not in the peer literature but in a registrant's submission to the government where all the rat pups exposed to DES on gestation days 13-20 had hypospadias at all doses used. And another paper from the Netherlands suggests that the initial damage that initiates hypospadias might be from estrogenic activity (Weidner et al., 1998).

In what appears to be a second generation effect, the sons of DES daughters have a significantly increased risk of developing hypospadias compared with boys whose mothers were not exposed to DES prenatally (Klip et al., 2002). The mothers (26,428) in this study were all having fertility problems between 1980 and 1995. Of these, 205 mothers were exposed to DES prenatally and four of them had sons with severe cases of hypospadias (1 in 50 boys). However, the study was not designed to look at hypospadias. One open-ended question in the instrument asked the mothers to report serious health problems in their children. The four cases and eight others (among the remaining 26,223 cases) required surgery for penoscrotal or distal shaft openings. In all probability hypospadias was underreported in this study. The problem of monitoring disturbed development as the result of *in-utero* exposure to xenobiotics in second generation populations will be even more difficult than monitoring disruption in first generation populations. Unlike the study above, where the second generation mothers knew that their mothers were exposed to DES, most mothers will not know what they were exposed to prenatally or throughout their gestational years that may have caused a health problem in their children.

2.5. Male Dysgenesis

Some of the most provocative papers on the male predicament in recent years have come from Denmark, starting in 1994 with the release of a meta-analysis of sperm count on a worldwide basis (Carlsen et al., 1992). It revealed a 50% drop in sperm count and reduced semen quality between 1939 and 1990. Several years later, in response to criticism of the Danish team's findings, a reanalysis confirmed the earlier results (Swan et al., 1997).

The Danish team in its more recent papers describe what they call the testicular dysgenesis syndrome (TDS) (Boisen et al., 2001). They provide a list of abnormalities in males that are not generally found alone and use this to defend what they describe as *"adverse trends in male reproductive health"*. In essence, the Danish team is describing a population level effect. They list the following as conditions of this syndrome:

Increased incidence in testicular cancer

Decreased semen quality

Increased incidence of undescended testicles and hypospadias

Increased need for assisted reproduction

They note that the trend toward increases in these anomalies has been overlooked because of specialization in medicine. For example, pediatricians, endocrinologists, urologists, andrologists, and oncologists work independently and therefore the big picture is missed. The conditions the team lists are not expressed simultaneously but over a span of 20 or 30 years as the patient ages and as he progresses from one doctor to another. The Danish team concludes that TDS is increasingly common and state that "Testicular Dysgenesis Syndrome is a result of disruption of embryonal programming and development during fetal life". The authors suggest that future epidemiology not focus on just one symptom alone, but should be more comprehensive and take into consideration co-morbidity.

3. Stepping outside the box

The vulnerability of males was recently confirmed again in terms of their susceptibility to neurodevelopmental problems in two separate reports. In the first instance, a North Carolina epidemiological study reported that boys are more likely to develop ADHD-like symptoms than girls (Rowland et al., 2002). In line with the Danish team's statement about disruption of embryonal programming and development, the question arises "Why would not the other developing organ systems be equally vulnerable to a xenobiotic in the womb environment during a critical stage of development?" In the North Carolina study, boys from the first grade through the fifth grade were significantly more susceptible to ADHD-like symptoms compared with girls. In the 4th and 5th grade 20% of the children were affected -- 15% were boys and 5% were girls. Cases were those who were on prescribed medication for the problem prompting the authors to state that these are underestimates of the problem. In the second instance, disorders of autism spectrum in preschool children has increased 100% since 1970 (Gillberg and Wing, 1999; Schettler et al., 2000) with boys 4 times more likely to express the symptoms than girls (Gillberg and Wing, 1999).

Needless to say, the economic and social costs are prohibitive. More men will need medical attention throughout their lives. Many may never reach their fullest intellectual and social potential. It has been estimated that a loss of 5 IQ points means between $275 to $326 billion per year in lost earnings in the US (Muir and Zegarac, 2001). The authors of the study argue that there are enough data to support the premise that chemicals are causing from 10% to 50% of the neurodevelopmental problems that would lead to this loss. Taking their most conservative estimate of 10%, that still means $27.5 billion per year impact on income alone. This does not take into consideration all the other impacts lowered IQ can have on a society nor the costs of treating hypospadias, reduced fertility, and gonadal cancers.

4. Conclusion

Over the past decade, rapid advances in integrated molecular, cellular, physiologic, biochemical, medical, and toxicologic research have revealed several stages of male urogenital development that are vulnerable to endocrine disrupting chemicals. Some of these findings might explain non-genetic hypospadias for which there have been few explanations. The feminizing effects of xenoestrogens on developing males do not completely explain the idiopathic cases of hypospadias. However, since environmental antiandrogens were first discovered in 1994, several critical stages of male urogenital development have been revealed where specific synthetic chemicals can impede normal molecular and biochemical activity leading to frank expression of hypospadias and other male gonad anomalies in laboratory animals. Despite these new discoveries, a disconnect between a putative causal agent and hypospadias in humans continues to pose a problem. This will become more of a problem if the list of antiandrogenic synthetic chemicals continues to grow and if practitioners do not become aware of the hazards some chemicals can pose during sexual differentiation to the male fetus. It will also increase the difficulty of protecting reproductive age females and males from conception through adulthood from antiandrogenic substances. Hopefully, as the CDC nationwide urine sampling effort broadens to include more environmental contaminants it will continue to provide guidance for reducing maternal, embryo, and fetal exposure. In light of the human suffering associated with hypospadias, determining the etiology and exercising prevention should be major goals for public health authorities and clinicians.

REFERENCES

Allan, B. B., Brant, R., Seidel, J. E., and Jarrell, J. F., 1997, Declining sex ratios in Canada. *Can Med Assoc J,* **156(1):** 37-41. B.

Baskin, L. S., Himes, K., and Colborn, T., 2001, Hypospadias and endocrine disruption: Is there a connection? *Environ Health Perspect,* **109(11):** 1175-1183.

Berkow, R. and Fletcher, A. J., eds., 1978, Attention deficit disorder (ADD), in: The Merck Manual of Diagnosis and Therapy. 15th ed., Merck and Co. Inc., pp. 1978-1980.

Bern, H. A., 1992, The fragile fetus, in: Colborn, T. and Clement, C., eds. Chemically-Induced Alterations in Sexual and Functional Development: The Wildlife/Human Connection. Princeton Scientific Publishing, Princeton. pp. 9-15.

Blount, B. C., Silva, M. J., Caudill, S. P., Needham, L. L., Pirkle, J. L., Sampson, E. J., Lucier, G. W., Jackson, R. J., and Brock, J. W., 2000, Levels of seven urinary phthalate metabolites in a human reference population, *Environ Health Perspect.* **108(10):** 972-982.

Boisen, K. A., Main, K. M., Rajpert-De Meyts, E., and Skakkebaek, N. E., 2001, Are male reproductive disorders a common entity? The testicular dysgenesis syndrome. *Ann N Y Acad Sci.* **948:** 90-99.

Brucker-Davis, F., 1998, Effects of environmental synthetic chemicals on thyroid function, *Thyroid.* **8(9):** 827-856.

Burlington, H. and Lindeman, V. F.,1950, Effect of DDT on testes and secondary sex characters of white leghorn cockerels, *Proc Soc Exp Biol Med.* **74:** 48-51.

Carlsen, E., Giwercman, A., Keiding, N., and Skakkebaek, N. E., 1992, Evidence for decreasing quality of semen during past 50 years, *BMJ.* **305:** 609-613.

Clement, C. R. and Colborn, T., 1992, Herbicides and fungicides: A perspective on potential human exposure, in: Colborn, T. and Clement, C., eds. Chemically-Induced Alterations in Sexual and Functional Development: The Wildlife/Human Connection. Princeton Scientific Publishing, Princeton. pp. 347-364.

Colborn, T., 1998, Endocrine disruption from environmental toxicants, in: Rom, W. N., ed., Environmental and Occupational Medicine. 3rd ed. Lippencott-Raven Publishers, Philadelphia. pp. 807-816.

Davis, D. L., Gottlieb, M. B., and Stampnitzky, J. R., 1998, Reduced ratio of male to female births in several industrial countries - a sentinel health indicator, *J Am Med Assoc.* **279(13):** 1018-1023.

Dolk, H., Vrijheid, M., Armstrong, B., Abramsky, L., Bianchi, F., Garne, E., Nelen, V., Robert, E., Scott, J. E., Stone, D., and Tenconi, R., 1998, Risk of congenital anomalies near hazardous-waste landfill sites in Europe: The EUROHAZCON study, *Lancet.* **352(9126):** 423-427.

EDSTAC, 1998, Endocrine Disruptor Screening and Testing Advisory Committee Final Report. Vol 1. United States Environmental Protection Agency, Washington DC.

Eskenazi, B., Mocarelli, P., Warner, M., Samuels, S., Vercellini, P., Olive, D., Needham, L. L., Patterson, D. G. Jr., Brambilla, P., Gavoni, N., Casalini, S., Panazza, S., Turner, W., and Gerthoux, P. M., 2002, Serum dioxin concentrations and endometriosis: A cohort study in Seveso, Italy, *Environ Health Perspect.* **110(7):** 629-634.

European Commission, 2001, Annex 7: Human health and wildlife relevant data on endocrine disruption included in the database on 146 substances evaluated in the Expert Meeting, in: Communication from the Commission to the Council and the European Parliament on the implementation of the Community Strategy for Endocrine Disruptors – A range of substances suspected of interfering with the hormone systems of humans and wildlife. COM 261.

Fox, G. A., 2001, Effects of endocrine disrupting chemicals on wildlife in Canada: Past present and future, *Water Quality Research Journal of Canada.* **36(2):** 233-251.

Fox, G. A. Gilman, A. P., Peakall, D. B., and Anderka, F. W., 1978, Behavioral abnormalities of nesting Lake Ontario herring gulls, *J Wildl Manage.* **42(3):** 477-483.

Fry, D. M. and Toone, C. K., 1981, DDT-induced feminization of gull embryos, *Science.* **213:** 922-924.

Fry, D. M., Toone, C. K., Speich, S. M., and Peard, J. R., 1987, Sex ratio skew and breeding patterns of gulls: Demographic and toxicological considerations, *Studies in Avian Biology.* **10:** 26-43.

Garry, V., Harkins, M. E., Erickson, L. L., Long-Simpson, L. K., Holland, S. E., and Burroughts, B. L., 2002, Birth defects, season of conception, and sex of children born to pesticide applicators living in the Red River Valley of Minnesota, USA, *Environ Health Perspect.* **110(suppl 3):** 441-449.

Gershon, E. S. and Rieder, R. O., 1992, Major disorders of mind and brain, *Scientific American.* **Sept:** 127-133.

Gillberg, C. and Wing, L., 1999, Autism: Not an extremely rare disorder, *Acta Psychiatr Scand.* **99(6):** 399 .

Gray, L. E. Jr., Ostby, J., Monosson, E., and Kelce, W. R., 1999, Environmental antiandrogens: Low doses of the fungicide vinclozolin alter sexual differentiation of the male rat, *Toxicol Ind Health.* **15(1-2):** 48-64.

Guillette, E. A., 1999, An anthropological interpretation of endocrine disruption in children, in: Guillette, L. J. Jr. and Crain, D. A., eds. Environmental endocrine disrupters: An evolutionary perspective, Taylor and Francis, New York. pp. 322-339.

Guillette, L. J. Jr., Gross, T. S., Masson, G. R., Matter, J. M., Percival, H. F., and Woodward, A. R., 1994, Developmental abnormalities of the gonad and abnormal sex hormone concentrations in juvenile alligators from contaminated and control lakes in Florida, *Environ Health Perspect.* **102(8):** 680-688.

Gupta, C., 2000, Reproductive malformations of the male offspring following maternal exposure to estrogenic chemicals, *Proc Soc Exp Biol Med.* **224(2):** 61-68

Herman-Giddens, M. E., Slora, E. J.; Wasserman, R. C., Bourdony, C. J., Bhapkar, M. V., Koch, G. G., and Hasemeier, C. M., 1997, Secondary sexual characteristics and menses in young girls seen in office practice: A study from the Pediatric Research in Office Settings Network, *Pediatrics.* **99(4):** 505-512.

Hunt, G. L. Jr. and Hunt, M. W., 1977, Female-female pairing in western gulls (*Larus occidentalis*) in Southern California, *Science.* **196:** 1466-1476.

Jensen, T. K., Carlsen, E.; Jørgensen, N., Berthelsen, J. G., Keiding, N., Christensen, K., Petersen, J. H., Knudsen, L. B., and Skakkebaek, N. E., 2002, Poor semen quality may contribute to recent decline in fertility rates, *Hum Reprod.* **17(6):** 1437-1440.

Kaipiainen-Seppanen, O. and Savolainen, A., 2001, Changes in the incidence of juvenile rheumatoid arthritis in Finland, *Rheumatology* (Oxford). **40(8):** 928-32.

Kelce, W. R., Monosson, E., Gamcsik, M. P., Laws, S. C., and Gray, L. E. Jr., 1994, Environmental hormone disruptors: Evidence that vinclozolin developmental toxicity is mediated by antiandrogenic metabolites, *Toxicol Appl Pharmacol.* **126:** 276-285.

Kelce, W. R., Stone, C. R., Laws, S. C., Gray, L. E., Kemppainen, J. A., and Wilson, E. M., 1995, Persistent DDT metabolite p,p'-DDE is a potent androgen receptor antagonist, *Nature.* **375:** 581-585.

King, S. E. and Schottenfeld, D., 1998, The "epidemic" of breast cancer in the U.S. - determining the factors, *Oncology.* **10(4):** 453-462.

Klip, H., Verloop, J., van Gool, J. D., Koster, M. E., Burger, C. W., and van Leeuwen, F. E., 2002, Hypospadias in sons of women exposed to diethylstilbestrol *in utero*: A cohort study, *Lancet.* **359(9312):** 1102-1107.

Kristensen, F., Irgens, L. M., Andersen, A., Bye, A. S., and Sundheim, L., 1997, Birth defects among offspring of Norwegian farmers, 1967-1991, *Epidemiology.* **8(5):** 537-544.

Kubiak, T. J., Harris, H. J., Smith, L. M., Schwartz, T. R., Stalling, D. L., Trick, J. A., Sileo, L., Doucherty, D. E., and Erdman, T. C., 1989, Microcontaminants and reproductive impairment of the Forster's tern on Green Bay, Lake Michigan--1983, *Arch Environ Contam Toxicol.* **18:** 706-727.

Lary, J. M. and Paulozzi, L. J., 2001, Sex differences in the prevalence of human birth defects: A population-based study, *Teratology.* **64(5):** 237-251.

Leatherland, J. F., 1992, Endocrine and reproductive function in Great Lakes salmon, in: Colborn, T. and Clement, C., eds. Chemically-Induced Alterations in Sexual and Functional Development: The Wildlife/Human Connection. Princeton Scientific Publishing, Princeton. pp. 129-145.

Leatherland, J. F, 1998, Personal communication.

MacLachlan, J., 2000, Personal communication.

Marcus, M., Kiely, J., Xu, F. J., McGeehin, M., Jackson, R., and Sinks, T., 1998, Changing sex ratio in the United States, 1969-1995, *Fertility & Sterility.* **70(2):** 270-273.

Mastroiacovo, P., Spagnolo, A., Marni, E., Meazza, L., Bertollini, R., and Segni, G., 1988, Birth defects in the Seveso area after TCDD contamination, *J Am Med Assoc.* **259(11):** 1668-1672.

McKiernan, J. M., Goluboff, E. T., Liberson, G. L., Golden, R., and Fisch, H., 1999, Rising risk of testicular cancer by birth cohort in the United States from 1973 to 1995, *J Urol.* **162(2):** 361-363.

Mizuno, R., 2000, The male/female ratio of fetal deaths and births in Japan, *Lancet.* **356:** 738-739.

Muir, T. and Zegarac, M., 2001, Societal costs of exposure to toxic substances: Economic and health costs of four case studies that are candidates for environmental causation, *Environ Health Perspect.* **109(suppl 6):** 885-903.

National Birth Defects Prevention Network, 1997, Birth defects surveillance data from selected states, *Teratology.* **56(1-2):** 115-175.

Paulozzi, L. J., Erickson, J. D., and Jackson, R. J., 1997, Hypospadias trends in two US surveillance systems, *Pediatrics.* **100(5):** 831-834.

Peakall, D. B., Fox, G. A., Gilman, A. P., Hallett, D. J., and Norstrom, R. J., 1978, Reproductive success of herring gulls as an indicator of Great Lakes water quality, in; Afghan, B. K. and Mackay, D., eds. Hydrocarbons and Halogenated Hydrocarbons in the Aquatic Environment: McMaster University. pp. 337.

Rowland, A. S., Umbach, D. M., Stallone, L., Naftel, A. J., Bohlig, E. M., and Sandler, D. P., 2002, Prevalence of medication treatment for attention deficit-hyperactivity disorder among elementary school children in Johnston County, North Carolina, *Am J Public Health.* **92(2):** 231-234.

Schettler, T., Stein, J., Reich, F., Valenti, M., and Wallinga, D., 2000, In harm's way: Toxic threats to child development, Cambridge: Greater Boston Physicians for Social Responsibility; 2000.

Shibata, A., Ma, J., and Whittemore, A. S., 1998, Prostate cancer incidence and mortality in the United States and the United Kingdom, *J Natl Can Inst.* **90(16):** 1230-1231.

Short, P. and Colborn, T., 1999, Pesticide use in the U.S. and policy implications: A focus on herbicides, *Toxicol Ind Health.* **15(1-2):** 240-275.

Swan, S. H., Elkin, E. P., and Fenster, L., 1997, Have sperm densities declined: A reanalysis of global trend data, *Environ Health Perspect.* **105(11):** 1228-1232.

Toppari, J., Haavisto, A. M., and Alanen, M., 2002, Changes in male reproductive health and effects of endocrine disruptors in Scandinavian countries, *Cad. Saúde Pública*, Rio De Janeiro. **18(2):** 413-420.

van Maanen, J. M. S., Albering, H. J., de Kok, T. M. C. M., van Breda, S. G. J., Curfs, D. M. J., Vermeer, I. T. M., Ambergen, A. W., Wolffenbuttel, B. H. R,; Kleinjans, J. C. S., and Reeser, H. M., 2000, Does the risk of childhood diabetes mellitus require revision of the guideline values for nitrate in drinking water? *Environ Health Perspect.* **108(5):** 457-461.

Weidner, I. S., Moller, H., Jensen, T. K., and Skakkebaek, N. E., 1998, Cryptorchidism and hypospadias in sons of gardeners and farmers, *Environ Health Perspect.* **106(12):** 793-6.

Weiss, B., 1998, A risk assessment perspective on the neurobehavioral toxicity of endocrine disruptors, *Toxicol Ind Health.* **14(1-2):** 341-59..

ENDOCRINE DISRUPTION AND HYPOSPADIAS

George F. Steinhardt*

1. INTRODUCTION

In 1996 a conference was held in Europe to discuss the ramifications of a number of disturbing clinical observations relating to male reproductive health. The rate of testicular cancer was noted to be increasing in many countries (Weybridge, 1995). Additionally these same countries were noted to have increasing problems with male infertility correlated with decreasing sperm counts (Carlesen et al., 1995). Plausibly, testicular cancer and infertility could be tied to environmental agents affecting fetal testicular structure and function. That individuals could demonstrate more than one problem, i.e. cancer and infertility, suggested the existence of shared etiologic influences. Substantial evidence supported the notion that testicular function and male development can be affected by changes in the hormonal milieu of the developing fetus (Toppari et al., 1996). Concentrations of hormonally active compounds of no consequence late in fetal development can have profound long-term consequences if present during earlier critical periods of male genital development (Ema et al., 2000). From this conference came the definition of a endocrine disruptor as an exogenous substance that causes adverse health effects in an intact organism or its progeny, consequent to changes in endocrine function. A potential endocrine disruptor is a substance that possesses properties that might be expected to lead to endocrine disruption in an intact organism ((Joffe, 2001).

In addition to the observations that in utero exposure to endocrine disruptors may possibly effect fertility and cause cancer, was another group of observations concerning hormonally active chemicals affecting male genital anatomy, providing the exposure occurred during susceptible periods of development. In contrast to female genital development, which is largely independent of hormone interaction, masculinization is strictly dependent upon the elaboration of androgenic hormones from the testes with subsequent activation of hormone dependent pathways. Hormones, or hormonally active substances can disrupt this cascade of events,

* DeVos Children's Hospital, Grand Rapids, Michigan 49503

Hypospadias and Genital Development, edited by
L. Baskin, Kluwer Academic/Plenum Publishers, 2004

resulting in penile deformities (Sharpe, 2001).   Most of the early experimental
support for the changes in male phenotype attendant to endocrine disruption came
from observations of testicular abnormalities   (epididymal abnormalities and
maldescent) in male offspring of diethylstilbestrol (DES) exposed pregnant
mothers ((McLaschlan et al., 1975. Skakkebaek's Testicular Dysgenesis
Syndrome(TDS) (Skakkebaek, 2002; Skakkebaek et al., 2002; Damgaard et al.,
2002) emphasized the clinical, experimental and epidemiologic studies that
implied   a common cause behind the   observations of increased genital
abnormalities, testicular cancer, and male infertility. An environmental etiology
for TDS was further suggested by geographic variations such as that in Finland
the rates of testes cancer , cryptorchidsm, and hypospadias (Adami et al., 1994;
Paulozzi, 1999) are much higher than in Denmark and that Danes also have better
semen quality (Jensen et al., 2000; Calrsen et al., 1992).

McLachlan suggested that the estrogen-associated alterations in male genital
tract phenotype be called the Developmental Estrogenization Syndrome, and he
used this model to predict that conditions with excessive prenatal estrogen
exposure would be associated with   genital anomalies in male offspring
(McLachlan et al., 2001). The model successfully predicted increased genital
deformities and testicular cancer   in a few maternal conditions that have been
demonstrated to be associated with hyperestrogen states, ie. obesity, hyperemesis,
hypertension, and first pregnancy (Santti et al., 1994). If endocrine disruptors
pass from the maternal circulation through the placenta to effect the male genitalia,
there may be other consequences as well.   The observed decrease in the
male:female birth ratio might relate to a higher male fetal loss from   the effect of
endocrine disruption on early male development (Lary and Paulozzi, 2001; Allan
et al., 1997; Mizuno, 2000). There are documented non-genital consequences of
such endocrine disruption . For example, PCBs can profoundly effect thyroid
elaboration in the fetus, which obviously can have systemic consequences
including fetal loss (Brouwer et al., 1998; Brucker-Davis, 1998).

While definitive proof is still lacking, there is great concern, derived from
diverse sources, that normal male genital development and testicular function are
currently threatened by the ubiquitous presence of manmade endocrine disruptors.
The purpose of this chapter is to review evidence that the increased incidence of
male penile anomalies may very well derive from disturbed morphogenesis
resulting from endocrine disruption at a critical period of male genital development
(Baskin et al., 2001; Kennedy and Snyder, 1999).

2.   HYPOSPADIAS: OCCURRENCE AND ASSOCIATIONS

Hypospadias is a congenital deformity of the penis with varying severity
demonstrating characteristic deformation of the prepuce (dorsally located) and
proximal placement of the urethral meatus on the ventral penile shaft most often
associated with downward penile curvature(chordee) (Baskin et al., 1998). While
there are genetic causes of hypospadias, 70% to 80% of boys with HPS have no
genetic pattern or clear cut etiology (Boisen et al., 2001; Boehmer et al., 2001).

Brothers of affected boys have a 1 in 6 chance of having the deformity and the heritability for first degree relatives of affected boys was 57% in one study from France (Stoll et al., 1990). The genetic causes of hypospadias are assumed to be static but there are several sources that suggest that this deformity is increasing in occurrence. Multiple reports document the increased incidence of hypospadias in several western countries in the last few decades (Paulozzi, 1999). In England and Wales the rate of hypospadias increased from 7.3 per 10,000 to 16 per 10,000 (all births, not just male) between 1960 and 1980 with a subsequent leveling off (Matlai and Beral, 1985; Dolk, 1998). A recent report from the Netherlands discerned a four to six fold increase in the frequency of hypospadias in newborn boys compared to previous studies, with more severe deformities noted (Pierik, 2002). In the U.S. the rates of hypospadias have increased in all parts of the country as shown by multiple reporting systems. Both the Birth Defects Monitoring Program (BDMP) and the Metropolitan Atlanta Congenital Defects Program (MACDP) demonstrated an unequivocal doubling in the rate of hypospadias using standardized definitions and uniform reporting techniques over 25 and 30 year time periods (Paulozzi et al., 1997) The BDMP registry shows a statistically significant increase from 20.2 cases per 10,000 in 1970 to 39.7 per 10,000. While it is arguable that reporting systems and registries are more effective in picking up mild defects, the MACDP showed an increase in the occurrence of severe deformities within the population based registry. It is unlikely that severe deformities would have been missed in the early years of the program Not all studies have found an increased occurrence of hypospadias in the U.S. A recent study from New York State evaluated both the incidence and repair rates for hypospadias between 1983 and 1985 and found no observable trend in either occurrence of the deformity or surgical repairs (Choi et al., 2001). Other data from New York suggest that while the deformity is not increasing in frequency, maternal age is associated with a higher rate of hypospadias (Fisch et al., 2001). It may be that older mothers have a longer time to accumulate xenobiotic compounds with greater effect on male genital development.

Obviously there are ethnic and geographic differences in the occurrence of the genital deformity. For example, hypospadias is more common in Caucasians than African Americans (CHUNG ET AL., 1968) while the rate of hypospadias has not changed in Finland over the last 30 years (AHO ET AL., 2000). High rates of hypospadias have been noted in geographic areas noted to have impaired population fertility (mean female parity) (IRVINE, 2000). The major contention of the TDS syndrome is multiple male problems (genital deformities, testicular cancer and infertility) can result from same fetal endocrine disrupting insult. In this regard, the significant association of cryptorchidism and hypospadias is also worth noting (WEIDNER ET AL., 1999). These two deformities were observed to increase in the sons of gardeners and farmers, suggesting a relationship to prenatal exposure to occupationally related chemicals rather than a strict geographic influence (Weidner et al., 1998). A mild increase in hypospadias was seen in families living adjacent to hazardous landfills in Europe (Dolk et al., 1998). Not all endocrine disruptors are manmade organochlorine compounds. Naturally occurring phytoestrogens are present in foods, particularly those deriving from soy, a protein staple for vegetarians who avoid meat ingestion.   In a study of 7928   male births, 2.2% of the vegetarian mothers gave birth to boys with hypospadias compared to 0.6% of omnivore mothers (North and Golding, 2000). Also worth noting is the fact that the rate of hypospadias is fivefold higher with mothers who have undergone in vitro fertilization (Silver et al., 1999).   Assisted reproduction protocols require maternal administration of progesterone during early gestation .   It is hypothesized that this estrogenic stimulus accounts for the increased occurrence of the genital deformity in this group of patients. Hypospadias is also more common in low birthweight infants, suggesting a common insult affecting intrauterine growth as well as genital development (Hussain et al., 2002; Gatti et al., 2001). As monozygotic twins have the same genetic make up, the noted increase in hypospadias in the twin with the lower birth weight suggests again the influence of environmental factors affecting both intrauterine growth and genital development in the smaller of the identical twins (Fredell et al., 1998).   Taken together, these studies suggest that hypospadias is increasing in frequency, and the reason may well have do to with altered hormonal milieu affecting both fetal well-being and normal penile development.

3 . DES: MODEL FOR CONSEQUENCES OF FETAL ENDOCRINE EXPOSURE

Diethylstilbestrol is a synthetic compound that is structurally dissimilar to estrogen but nevertheless is a very potent stimulator of the estrogen receptor. The recognition that maternal administration of this compound to control nausea early in pregnancy resulted in profound consequences for the female offspring was breathtaking in the clinical ramifications.   The development of vaginal cancer in the daughters of women so treated brought into sharp focus the concept that transplacental transfer of maternal substances can have lifelong consequences for the developing fetus (Herbst et al., 1971).   As the index drug for maternal-fetal interactions, DES has been intensely scrutinized in terms of male genital development and subsequent adult reproductive function. Early case reports noted not only testicular maldescent and epididymal deformities in sons of DES treated mothers, but also sporadic cases of hypospadias (Kaplan, 1959; Henderson et al., 1976). While long term studies of sons found no impaired fertility, genital deformities were more common in DES exposed men compared to non-exposed controls (Wilcox et al., 1995). Another important observation   in this well studied cohort of men with fetal DES exposure, is that genital deformities were twice as high in those exposed before the 11th week of gestation compared to those whose mothers ingested DES after the 11th week.   While there is no ironclad evidence that maternal DES causes hypospadias in exposed male offspring, one relevant study documented a transgenerational effect of DES such that sons of women exposed in utero to DES were found to have an increased risk for hypospadias (Klip et al., 2002). In experimental models the effect of maternally administered estrogen is similar to those of DES and other synthetic estrogens both on the developing brain and on the genitalia (Arai et al., 1983).   DES and the synthetic estrogens seem to cause effects at lower doses than estrogen because they are not bound by sex hormone-binding globulin, which binds 95% of circulating estrogen, making it inactive (Sheehan and Young, 1979).

4 .ENDOCRINE DISRUPTORS: IDENTIFICATION AND IMPLICATIONS

In terms of endocrine disruption and genital development, the current focus is not on estrogen or synthetic estrogen compounds. Rather most of the concern centers around a host of man made organochlorine compounds widely distributed in industry and agriculture. These chemicals function as endocrine disrupters by mimicking estrogen (or androgen), antagonizing estrogen (or androgen) action, altering estrogen (or androgen) synthesis or modifying estrogen (or androgen) receptor activity. Regarding estrogenic activity, for a long time the only way to assess the estrogenic activity was to check *in vivo* for hyperplastic activity of uterine epithelium exposed to the compound in question (Sonnenschein and Soto, 1998). Another such assay to test for endocrine disruption effects relied on morphologic changes in testicular development in male quail embryos when the eggs yolks were injected with suspect compounds at varying concentrations (Berg et al., 1999). These assays reliably demonstrate estrogenic effects of suspected endocrine disruptors but clearly are not suitable for large scale screening of potentially harmful xenobiotic compounds. The E-SCREEN system is a more efficient and quantitative assay which measures cellular proliferation in yeast cells subsequent to estrogen stimulation (Sonnenschein and Soto, 1998). Another test for estrogen stimulation is based on estrogen induced activation of the progesterone receptor in MCF7 cells (Soto et al., 1995). The MCF7 cell proliferation assay can also test antiestrogenic action by exposing the cells to minimally stimulating doses of estradiol in addition to varying concentrations of the estrogen antagonists. With such assays a wide variety of manmade organochlorine compounds such as polychlorinated biphenyls (PCB's) have been shown to have estrogenic activity. PCB's have been used for 80 years for industrial applications as flame retardants, dielectric capacitors and transformer fluids. PCB has over 200 recognized congeners, all them now banned from production in the United States, but , as PCB's are not metabolized, they persist in the environment and are today measurably present in most people with aggregate mean concentrations for all congeners of 2.15ng/ml (Stellman et al., 1998). The exact toxicity varies for each congener; both estrogenic and anti-estrogenic action have been reported (Krishnan and Safe, 1993) DDT, DDE, dieldrin, phenylphenol and toxaphene (pesticides), and bisphenol-A (plasticizer) are other common organochlorine compounds that have demonstrated weak estrogenic activity with the E-SCREEN assay .

Many compounds affecting fetal male development have hormonal activity separate from estrogen stimulation or estrogen receptor blockade. Masculinization of the external genitalia is an active process that starts in the 7th week of gestation with expression of the SRY gene in the indifferent male gonad. This SRY gene is located on the Y chromosome and initiates testicular differentiation with testosterone elaboration, secretion of Mullerian-inhibiting substance (MIS) of the Sertoli cells, and conversion of testosterone to 5alpha-dihydrotestosterone in the genital tubercle and urogenital sinus to stimulate normal masculinazation of the external genitalia.   Genetic or mutational deficiencies of either the androgen receptor(AR) or the 5alpha-reductase enzyme clinically result in loss of normal male penile phenotype (Ahmed et al., 2000). In a similar way experimental pharmacologic interference binding of the androgen receptor (flutamide) or induction of partial agonist activity by promoting DNA binding (cyproterone acetate), can have phenotypic effects on the penis (Toppari and Skakkebaek, 1998). In parallel with estrogen assays has been the development of assays to detect antiandorgenic/androgenic activity of endocrine disruptors.   A PALM (PC-3-androgen receptor-luciferase-MMTV) cell line of prostatic origin uses bioluminescence to characterize the androgen agonist or antagonist activity of potential endocrine disruptor compounds (Sultan et al., 2001). Using such systems many endocrine disrupting androgens and anti-androgens have been identified. Examples of chemicals with antiandrogen activity include the fungicide vinclozolin, the pesticides DDT, and its metabolite DDE. In most cases, the mechanism of action for these compounds is AR receptor binding with subsequent inhibition of esponse element DNA regulatory sequences. Other endocrine disruptors demonstrate antihormone action deriving from AR binding to DNA but with failure of subsequent transcription (Kelce and Wilson, 1997).   While environmental antiandrogens have a low affinity for the AR it may be that the circulating levels of endogenous fetal androgens are low enough to have reduced competition for the AR receptor sites. For example, doses of vinclozolin at high doses in an adult male have little effect on fertility, but low dose administration to a pregnant rat is enough to induce hypospadias in the offspring (Gray et al., 1994).

Given the   complexity of environmental   health, it is not   possible to unequivocally attribute the increase in human penile deformities to fetal exposure to chemicals in the environment that have hormonal activity on *in vitro* assays However, a number of observations add plausibility to that hypothesis. Fetal exposure to endocrine disruptors is certain as shown by demonstration of a wide variety of persistent organochlorine compounds (PCB, DDT, and DDE) in human amniotic fluid (Foster et al., 2000). Does this exposure have effects?   In utero exposure to PCB's in Taiwan in 1978 resulted in 'Yu Chen' illness that, among other things resulted in male offspring with smaller peripubertal penes ((Guo et al., 1993). In another environmental catastrophe, hypospadias increased in males born to mothers living adjacent to the Seveso, Italy spill of TCDD (dioxin), a byproduct of industrial incineration (Mastroiacomo et al., 1988). Maternal exposure was estimated by determination of the concentration of TCDD in the soil. In the areas with the highest soil contamination hypospadias was not increased, but the rate of spontaneous abortion in this area was high.   The rate of hypospadias in the next lower zone of TCDD contamination was 1/54 male births.   Of course the absolute numbers are small, but more hypospadias was present in areas of significant contamination.

Like most organocholorine compounds , DDE (a DDT metabolite) is non-biodegradable, accumulates in the food chain and is present in almost everyone in the United States. As DDT is no longer produced in the United States, peak concentrations are found in persons born in the early 1960's.   This endocrine disruptor acts in part by inhibiting binding of androgen to AR, a necessary action for male phenotypic development (Ahmed et al., 2000; Baskin et al., 1997). With a nested case-control design, stored serum samples were used to study the maternal DDE level with the risk of having boys with hypospadias.   Approximately 42,000 women were studied and   serum from mothers with levels   of DDE greater 85.6ug/liter had adjusted odds ratios of 1.2 (95% confidence interval:0.9,4.0) for bearing a son with hypospadias compared to mothers with DDE levels less than 21.4ug/liter (Longnecker et al., 2002). The wide confidence interval admittedly softens the impact of this positive finding.   We must accept that it is not likely that definitive proof will be found for a causal relationship between endocrine disruptors and penile deformities from human surveys or epidemiologic studies. Human studies have too many confounding variables (Rittler and Castilla, 2002).

5.   CONSEQUENCES OF EXPERIMENTAL ENDOCRINE DISRUPTION ON PENILE DEVELOPMENT

For further support of the hypothesis that endocrine disruption results in male penile deformities, we must look to experimental models. Widespread experiments suggest that manmade chlorinated hydrocarbons have hormonal activity, and these ubiquitous compounds can effect fetal male genital development (Gray et al., 2001). Persistent organochlorine compounds clearly disturbing fetal penile development in experimental animals include the TCDD (dioxin), phthalate esters and the pesticides vinclozolin, linuron procymidone, and DDE (Gray et al, 1999). The effect of maternal DES on fetal development was the signal event that stimulated interest in this field. Following that, maternal DES ingestion has been extensively studied in terms of male genital development with experimental investigation. Normal male development includes the elaboration of Mullerian inhibiting substance (MIS) from the Sertoli cells of the fetal testes. Hypospadias is commonly associated with persistence of varying degrees of Mullerian remnants, most often demonstrated on voiding cystourethrography (Krystic et al., 2001). While prenatal exposure of male mice to DES does not dramatically change the penile morphology (though the effects on testicular form and function are extensive) such treatment results in persistence of Mullerian remnants (Newbold et al., 1987). Such persistence derives from failure of the Mullerian tissue to respond to MIS with appropriate apoptosis and subsequent involution. Lack of mesenchymal integration of apoptotic and proliferative activity also seems to be responsible for the DES induced " hypospadias" seen in fetal female mice exposed in utero to DES (Miyagawa et al., 2002).

TCDD , noted above to have some very real toxicity in the Seveso spill is formed most often as an unwanted industrial byproduct or as a result of incineration.   Like PCB, dioxin can stimulate the Aryl hydrocarbon (Ah) receptor, but its profound effect on genital anatomy (including clefting of the phallus) may be neither estrogenic nor antiestrogenic.   The genital effects of

TCDD are well described and occur at the very lowest dosages tested. At parts per trillion TCDD (dioxin) can cause hypospadias in rat male offspring following single dose to the mother (Toppari et al., 1996). Phthalate esters are widely used in industrial applications as well as consumer products such as shampoos, paints, hairsprays, cosmetics and a wide variety of medical supplies including intravenous bags and tubing. Phthalates may be the most abundant manmade environmental pollutant, and human daily intake can be measured in tens of milligrams. Correlation between semen phthalate concentration and poor sperm counts has been observed for quite some time (Murature et al., 1987) and recently has been reemphasized with recent definitive studies (Duty et al., 2003). Skakeebaek's TDS hypothesis would predict that if our fertility were adversely affected, penile (and testicular) deformities would also occur. Perhaps it is not surprising that diethylhexyl phthalate (DHEHP) causes hypospadias in fetal male rats in addition to other genital deformities (undescended testes, epididymal agenesis) (Foster et al., 2001). It is worrisome that the adverse affects occur at phthalate concentrations less than those commonly demonstrated in human tissue. Further studies emphasize that there is a critical period during fetal development where the male pup is most sensitive to maternal ingestion of the phthalate ester (Ema et al., 2000). The mechanism of phthalate endocrine disruption has yet to be fully described, but it may act by decreasing fetal testosterone levels to female levels during this critical time of penile development (Parks et al., 2000). Whatever the mechanism, the effect on the developing penis is unmistakable. The widespread and continued exposure of our population to phthalate esters is thus a source of great concern in the context of increasing rates of hypospadias in our country.

Vinclozolin is a fungicide that reproducibly causes hypospadias in experimental animals when administered to pregnant mothers, again at critical periods of gestation (Wolf et al., 2000). The endocrine disruption activity of vinclozolin derives from AR antagonism, quite different from the mechanism of phthalate disruption. By binding to AR, vinclozolin prevents maximal AR-DNA binding, thus altering androgen-dependant gene expression. The effects of vinclozolin on the developing penis are so well characterized that they serve as a model of AR antagonism when analyzing the mechanism of endocrine disruption for other xenobiotic chemicals. Procymidone, a fungicide structurally similar to vinclozolin, produces a pattern of effects in treated animal identical to vinclozolin, suggesting AR antagonism as its sole mechanism of action. On the other hand, linuron, a urea-based herbicide with weak AR antagonist activity also causes penile deformities subsequent to fetal exposure. However, the profile of other biologic effects such as testicular and epididymal abnormalities (not seen with vinclozolin) suggests that, in addition to AR antagonism, linuron also has toxicity similar to the pthalates (Gray et al., 1999). DDE also causes hypospadias in male rats exposed fetally, comparable in action but much less potent than vinclozolin and other known AR ligands. The widespread experimental proof that multiple organochlorine compounds affect penile development by a variety of mechanisms at concentrations found currently in human tissue and in amniotic fluid provides indirect evidence that the putative increase in the rate of hypospadias is more real than imagined.

Current studies are doing a great deal to describe penile development, not in classic embryologic terms, but rather in terms of molecular developmental processes. Data from the mouse suggests that urethral seam formation occurs via cellular migration, not by epithelial mesenchymal transformation or epithelial apoptosis, and that disruption of this process results in hypospadias (Baskin et al., 1999; Kurzrock et al., 1999). Similar processes occur in the rabbit, which has been used (via 5 alpha reductase inhibition) as a model for the induction of hypospadias (Kurzrock et al., 2000). In organ culture sonic hedgehog (Shh) gene is dynamically expressed during the development of mouse genital tubercle explants; other genes with importance in genital tubercle formation include patched 1 (Ptch1), bone prophogenietic protein 4 (Bmp4) Hoxd13, and fibroblast growth factor 10 (FGF10) (Haraguchi et al., 2001). It will take further work to understand the action of endocrine disruption on these specific cellular pathways.

6. SUMMARY

The complexity of human biology makes it impossible to know for certain if endocrine disruption accounts for human penile deformities. Toxicologists point out that an overall assessment of risk must include other factors in addition to exposure including absorption, metabolism, excretion, bioaccumulation and other chemical interactions (Harrison et al., 1997). Many skeptics observe lack of analytic ability to document contaminant levels during critical windows of exposure (Safe, 2000). Further, the environmental estrogens studied (DDT, PCB and bis-phenol A) are quite weak compared to the well studied potent estrogen DES which did not cause penile deformities (Joffe, 2001). While environmental estrogens may be unlikely in contributing to penile deformities, the antiandrogens (phthalates, vinclozolin and DDE) are more plausible is this regard, as maleness is critically dependent upon androgen action. Observers note that, in general, the environmental concentrations of persistent organochlorine compounds have been decreasing over the past two decades. Some feel that our current levels of exposure are too low and the potency of the anti-androgens too weak to account for any significant developmental genital effect (Williams et al., 2001). Caution and restraint are always reasonable in matters of data intrepretation. Past researchers were reassured that pthtalate esters were quite safe when they first were assessed for possible harmful effects on male fertility. Unfortunately it took different models, analyzing transgenerational effects, before it became crystal clear that these compounds can dramatically affect male genital development following experimental maternal exposure at dosages and concentrations currently present in most women. We can not now be so reassured that our male development is unaffected by any of the over 65,000 manmade organochlorine compounds on the planet. Multiple observations from diverse disciplines provide credible evidence that proliferation of xenobiotic chemicals can cause potentially disastrous unintended consequences for the male gender, and upon reflection, our species.

Acknowledgement: The author would like to thank Louis K. Huisman of the Amberg Health Sciences Library, Spectrum Health, for her invaluable assistance in preparation of this manuscript.

REFERENCES

Adami, H.O., Bergstrom, R., and Mohner, M., 1994, Testicular cancer in nine Northern European countries. *Int. J. Cancer* **59**: 33.

Ahmed, S.F., Cheng, A., Dovey, L., Hawkins, J.R., Martin, H., Rowland, J., Shimura, N., Tait, A.D., and Hughes, I.A., 2000, Phenotypic Features, androgen receptor binding and mutational analysis in 278 clinical cases reported as androgen insensitivity syndrome. *J. Clin. Endocr. Metab.* **85**: 658..

Ahmed, S.F., Cheng, A., Dovey, L, Hawkins, J.R., Martin, H., Rowland, J., Shimura, N., Tait, A.D., and Hughes, I.A., 2000, Phenotypic features, androgen receptor binding, and mutational analysis in 278 clinical cases reported as androgen insensitivity syndrome. *J. Clin. Endocrinol. Metab.* **85**: 658.

Aho, M., Koivisto, A.M., Tammela, T.L., Auvinen, A., 2000, Is the incidence of hypospadias increasing? Analysis of Finnish hospital discharge data 1970-1994. *Environ. Health Perspect.* **108**: 463.

Allan, B.B., Brant, R., Seidel, J.E., Jarrell, J.F., 1997, Declining sex ratios in Canada. *Can. Med. Assoc. J.* **156**: 37.

Arai, Y. Mori, T., Suzuki, Y., and Bern, H.A., 1983, Long-term effects of perinatal exposure to sex steroids and diethylstilbestrol on the reproductive system of male mammals. *Int. Rev. Cyto.* **84**: 235.

Baskin, L.S., Sutherland, R.S., DiSandro, M.J., Hayward, S.W., Lipshutz, J., and Cunha, G.R., 1997, The effect of testosterone on androgen receptors and human penile growth. *J. Urol.* **158**: 1113.

Baskin, L.S., Erol, A., Li, Y.W., and Cunha, G.R., 1998, Anatomical studies of hypospadias. *J. Urol.* **160**: 1108.

Baskin, L.S., Erol, A., Jegatheesan, P., Li, Y., Liu, W., Cunha, G.R., 2001, Urethral seam formation and hypospadias. *Cell Tissue Res.* **305**: 379.

Baskin, L.S., Himes, K., and Colborn, T., 2001, Hypospadias and Endocrine disruption: Is there a connection? *Environ Health Persp.* **109**: 1175.

Berg, C., Halldin, K., Fridolfsson, A.K., Brandt, I., and Brunstrom, B., 1999, The avian egg as a test system for endocrine disrupters: effects of diethylstilbestrol and ethynlestradiol on sex organ development. *The Science of the Total Environment* **233**: 57.

Boehmer, A.L., Nijman, R.J., Lammers, B.A., De Coninck, S.J., Van Hemel, J. O., Themmen, A. P., Mureau, M.A., de Jong, F. H., Brinkmann, A.O., Niermeijer M.F., and Drop, S.L., 2001, Etiological studies of severe or familial hypospadias. *J. Urol.* **165**: 1246.

Boisen, K.A., Main, K.M., Rajpert-DeMeyts,E., Skakkebaek, N.E., 2001, Are male reproductive disorders a common entity? *Ann. New York Acad. Sci.* **948**: 90.

Brouwer, A., Morse, D., Lans, M.C., Schuur, A.,G., Murk, A.J., Klasson-Wehler, E., Bergman, A., and Visser, T.J., 1998, Interactions of persistent environmental organohalogens with the thyroid hormone system: Mechanisms and possible consequences for animal and human health. *Toxic. Indust. Health* **14**: 59.

Brucker-Davis, F., 1998, Effects of environmental synthetic chemicals on thyroid function. *Thyroid* **8**: 827.

Carlesen, E., Giwercman, A., Keiding, N., and Skakkebaek, N.E., 1995, Declining Semen Quality and Increasing Incidence of Testicular Cancer: Is There a Common Cause? *Envron. Health Perspect* **103** (suppl 7): 137.

Carlsen, E., Giwercman, A., Keiding, N., and Skakkebaek, N.E., 1992, Evidence for decreasing quality of semen during past 50 years. *Br. Med. J.* **305**: 609.

Choi, J., Cooper, K.L., Hensle, T.W., and Fisch, H., 2001, Incidence and surgical repair rates of hypospadias in New York State. *Urology* **57**: 151.

Chung, C.S., Myrianthopoulos, N.C., and Yoshizaki, H., 1968, Racial and prenatal factors in major congenital malformations. *Am. J. Hum. Gen.* **20**: 44.

Damgaard, I.N., Main, K.M., Toppari, J., Skakkebaek, N.E., 2002, Impact of exposure to endocrine disrupters in utero and in childhood on adult reproduction. *Best Practice &Res Clin End. and Metab.* **16**: 289.

Dolk H., 1998, Rise in prevalence of hypospadias. *The Lancet* **351**: 770.

Dolk, H., Vrjheid, M., Armstrong, B., Abramsky, L., Bianchi, F., Garne, E., Nelen, V., Robert, E., Scott, J.E., Stone, D., Tenconi,R., 1998, Risk of congenital anomalies near hazardous-waste landfill sites in Europe: the EUROHAZCON study. *Lancet* **352**: 423.

Duty, S.M., Silva, M.J., Barr, D.B., Brock, J.W., Ryan, L., Chen, Z., Herrick, R.F., Christiani, D.C., and Hauser, R., 2003, Phtalate exposure and human semen parameters. *Epidem.* **14:** 269.

Ema, M., Miyawaki, E., Kawashima, K., 2000, Critical period for adverse effects on development of the reproductive system in male offspring of rats given di-n-butyl-phthalate during late pregnancy. *Toxicol. Lett.* **111:** 271.

Ema, M., Miyawaki, E., Kawashima, K., 2000, Critical period for adverse effects on the development of reproductive system in male offspring of rats given di-n-butyl- phthalate during late pregnancy. *Tox. Lett.* **111:** 271.

Fisch H., Golden, R.J., Libersen, G.L., Hyun, G.S., Madsen, P., New, M.I., and Hensle, T.W., 2001, Maternal age as a risk factor for hypospadias. *J. Urol.* **165:** 934.

Fredell, L., Lichtenstein, P., Pedersen, N.L., Svensson, J., and Nordenskjold, A., 1998, Hypospadias is related to birth weight in discordant monozygotic twins. *J Urol.* **160:** 2197.

Foster, W., Chan, S., Platt, L., and Hughes, S., 2000, Detection of endocrine disrupting chemicals in samples of second trimester human amniotic fluid. *J. Clin. Endocrinol. Metab.* **85:** 2954.

Foster P.M. Mylchreest, E., Gaido, K.W., Sar, M., 2001, Effects of phthalate esters on the developing reproductive tract of male rats. *Human Reprod. Update* **7:** 231.

Gatti, J.M., Kirsch, A.J., Troyer, W.A., Perez-Brayfield, M.R., Smith, E.A., and Scherz, H.C., 2001, Increased incidence of hypospadias in small-for-gestational age infants in a neonatal intensive-care unit. *BJU Int.* **87:** 548.

Gray, L.E., Ostby, J.S., Kelce, W.R., 1994, Developmental effects of an environmental antiandrogen: the fungicide vinclozolin alters sex differentiation of the male rat. *Toxicol. Appl. Pharmacol.* **129:** 46.

Gray, L.E., Wolf, C., Lambright, C., Mann, P., Price, M., Cooper, R.L., and Ostby, J., 1999, Administration of potentially antiandrogenic pesticides (procymidone, linuron, irodione, chlozolinate, p,p'-DDE, and ketoconazole) and toxic substances (dibutyl- and diethylhexyl phthalate, PCB 169, and ethane dimethane sulphonate) during sexual differentiation produces diverse profiles of reproductive malformations in the male rat. *Tox. And Ind. Health* **15:** 94.

Gray, L.E., Ostby J., Furr, J., Wolf, C.J., Lambright, C., Parks, L., Veeramachaneni, D.N., Wilson, V., Price, M., Hotchkiss, A., Orlando, E., and Guilette, L., 2001, Effects of environmental antiandrogens on reproductive development in experimental animals. *Hum. Reprod. Update* **7:** 248.

Guo, Y.L., Lai, T.J., Ju, S.H., Chen, Y.C., and Hsu, C.C., 1993, Sexual developments and biological findings in Yu Chen Children. *Organohalogen Compounds* **14:** 235.

Haraguchi, R., Mo, R., Hui, C., Motoyama, J., Makino, S., Shiroishi, T., Gaffield, W., and Yamad, G., 2001, Unique functions of Sonic hedgehog signaling during external genitalia development. *Development* **128:** 4241.

Harrison, P.T., Holmes, P, and Humfrey, C.D., 1997, Reproductive health in humans and wildlife: are adverse trends associated with environmental chemical exposure? *The Sci. of Total Environ.* **205:** 97.

Henderson, B.E., Benton, B., Cosgrove, M.., Baptista, J., Aldrich, J., Townsend, D., Hart, W., and Mack, T.M., 1976, Urogenital tract abnormalities in sons of women treated with diethylstilbestrol. *Peds.* **58:** 505.

Herbst, A.L., Ulfelder, H. and Poskanzer, D.C., 1971, Adenocarcinoma of the vagina. Association of maternal stilbestrol therapy with tumor appearance in young women. *New. Engl. J. Med.* **284:** 878.

Hussain, N., Chaghtai A., Herndon, C.D., Herson, V.C., Rosenkrantz, T.S, and McKenna P.H., 2002, Hypospadias and Early Gestation Growth restriction in Infants. *Peds.* **109:** 473.

Irvine, D.S., 2000, Male reproductive health: cause for concern? *Andrologia* **32:** 195.

Jensen, T.K., Vierula, M., and Hjollund, N.H and the Danish First Pregnancy planner study team, 2000, Semen quality among Danish and Finnish men attempting to conceive. *Eur. J. Endocrinol.* **142:** 47.

Joffe, M., 2001, Are problems with male reproductive health caused by endocrine disruption? *Occ. Envir. Med.* **58:** 281.

Joffe, M., 2001, Are problems with male reproductive health caused by endocrine disruption? *Occup. and Environ. Med.* **58:** 281.

Kaplan, N.M., 1959, Male pseudohermaphrodism. Report of a case, with observations on pathogenesis. *New Engl. J. Med.*, **261:** 641.

Kelce, W.R, Wilson, E.M., 1997, Environmental antiandrogens: Developmental effects molecular mechanisms , and clinical implications. *J. Mol. Med.* **75:** 198.

Kennedy, W.A., and Snyder, H.M., 1999, Paediatric andrology: the impact of environmental pollutants. *BJU Int.* **83:** 195.

Klip, H., Verloop, J., van Gool, J.D., Koster, M.E., Burger, C.W., van Leeuwen, F.E., 2002, Hypospadias in sons of women exposed to dielthylstilbestrol in utero: a cohort study. *The Lancet* **359:** 1102.

Krishnan, V., Safe, S., 1993, Polychlorinated biphenyls (PCB's), dibenzo-p-doxins (PCDDs), and dibenzofurans (PCDFs) as antiestrogens in MCF-7 human breast cancer cells: quantitative structure-activity relationships. *Toxicol. Appl. Pharmacol.* **120**: 55.

Krstic, Z.D., Smoljanic, Z., Micovic, Z., Vukadinovic, V., Sretenovic, A., and Varinac, D., 2001, Surgical treatment of the Mullerian Duct Remnants. *J. Ped. Surg.* **36**: 870.

Kurzrock, E.A., Baskin, L.S., and Cunha, G.R., 1999, Ontogeny of the male urethra: Theory of endodermal differentiation. *Differentiation.* **64**: 115.

Kurzrock, E.A., Jegatheesan, P., Cunha, G.R. and Baskin, L.S., 2000, Urethral development in the fetal rabbit and induction of hypospadias: A model for human development. *J. Urol.* **164**: 1786.

Lary, J.M., and Paulozzi, L.J., 2001, Sex differences in the prevalence of human birth defects: A population-based study. *Teratology* **64**: 237.

Longnecker, M.P., Klebanoff, M.A., Brock, J.W., Zhou, H., Gray, K.A., Needham, L.L., and Wilcox, A.J., 2002, Maternal serum level of 1,1-Dichloro-2,2-bis(p-chlorophenyl)ethylene and risk of cryptorchidsm, hypospadias, and polythelia among male offspring. *Am J. of Epid.* **155**: 313.

Mastroiacovo P., Spagnolo, A., Marni, E., Meazza, L., Bertollini, R., Segni, G. and Borgna-Pignatti,C., 1988, Birth defects in the Seveso area after TCDD contamination. *JAMA* **259**: 1668.

Matlai, P. and Beral, V., 1985, Trends in congenital malformations of external genitalia. *Lancet* **i**: 108.

McLachlan, J.A., Newbold, R.R., and Bullock, B., 1975, Reproductive tract lesions in male mice exposed prenatally to diethylstilbestrol. *Science,* **190**: 991.

McLachlan, J.A., Newbold, R.R., Burow, M.E., and Li, S.F., 2001, From malformations to molecular mechanisms in the male: three decades of research on endocrine disrupters. *APMIS* **109**: 263.

Miyagawa, S., Buchanan, D.L. Sato, T. Ohta, Y., Nishina, Y., and Iguchi, T., 2002, Characterization of Diethylstilbestrol-induced hypospadias in female mice. *The Anatom. Record* **266**: 43.

Mizuno, R., 2000, The male/female ratio of fetal deaths and births in Japan. *Lancet* **356**: 738.

Murature, D.A., Tang, S.Y., Steinhardt, G.F., and Dougherty, R.C., 1987, Phthalate esters and semen quality parameters. *Biomed. Environ. Mass. Spectrom.* **14**: 473.

Newbold, R.R., Bullock, B.C., and McLachlan, J.A., 1987, Mullerian remnants of male mice exposed prenatally to diethylstilbestrol. *Terat.Carcin. and Mutag.* **7**: 377.

North, K., Golding, J., 2000, A maternal vegetarian diet in pregnancy is associated with hypospadias. *BJU Int.* **85**: 107.

Parks, L.G., Ostby, J.S., Lambright, C.R., Abbott, B.D., Klinefelter, G.R., Barlow, N.J., and Gray, L.E., 2000, The plasticizer diethylhexyl phthalate induces malformations by decreasing fetal testosterone synthesis during sexual differentiation in the male rat. *Tox. Sci.* **58**: 339.

Paulozzi L.J., Erickson, J.D., and Jackson, R.J., 1997, Hypospadias trends in two U.S. surveillance systems. *Peds.* **100**: 831.

Paulozzi, L.J., 1999, International trends in rates of hypospadias and cryptorchidsm. *Environ. Health Perspec.* **107**: 297.

Paulozzi, L.J., 1999, International trends in rates of hypospadias and cryptorchidsm. *Environ.Health Prospect.* **107**: 297.

Pierik, F.H., Burdorf, A., Nijman, J.M., de Muinck S.M., Juttmann, R.E., and Weber, R.F., 2002, A high hypospadias rate in The Netherlands. *Human Reprod.* **17**: 1112.

Rittler, M., and Castilla, E.E., 2002, Endocrine disruptors and congenital anomalies. *Cad. Saude Publica.* **18**: 421.

Safe, S.H., 2000, Endocrine Disruptors and Human Health--Is there a problem? An Update. *Environ. Health. Persp.* **108**: 487.

Santti, R., Newbold, R.R., Makela S., Pylkkanen, L., and McLachlan JA., 1994, Developmental estrogenization and prostatic neoplasia. *Prostate* **24**: 67.

Sharpe, R.M., 2001, Hormones and testis development and the possible adverse effects of en- vironmental chemicals. *Toxicol. Lett* **120**: 221.

Sheehan, D.M., and Young, M., 1979, Diethylstilbestrol and estradiol binding to serum albumin and pregnancy plasma of rat and human. *Endocrinology* **104**: 1442.

Silver, R.I., Rodriguez, R., Chang, T.S., and Gearhart, J.P., 1999, In vitro fertilization is associated with an increased risk of hypospadias. *J. Urol.* **161**: 1954.

Skakkebaek, N.E., Rajpert-DeMeyts,E., and Main, K.M., 2001, Testicular dysgenesis syndrome: an increasingly common developmental disorder with environmental aspects. *Hum. Repro.* **16**: 972.

Skakkebaek, N.E., 2002, Endocrine disrupters and testicular dysgenesis syndrome. *Horm. Res.* **57(supp2)**: 43.

Sonnenschein, C., and Soto, A.M., 1998, An updated review of environmental estrogen and androgen mimics and antagonists. *J. Steroid. Biochem. Molec. Biol.* **65**: 143.

Soto, A.M., Sonnenschein, C., Chung, K.L., Fernandez, M.F., Olea, N., and Serrano, F.O., 1995, The E-SCREEN assay as a tool to identify estrogens: an update on estrogenic environmental pollutants. *Environ. Health Perspect.* **103:** 113.

Stellman, S.D., Djordjevic, M.v., Muscat, J.E., Gong, L., Bernstein, D., Citron, M.L., White, A., Kemeny, M., Busch, E., and Nafziger, A.N., 1998, Relative abundance of organochlorine pesticiedes and polychlorinated biphenyls in adipose tissue and serum of women in Long Island, New York. *Cancer Epid. Biomarkers and Prev.* **7:** 489.

Stoll,C., Alembik, Y., Roth, M.P., and Dott, B., 1990, Genetic and environmental factors in hypospadias. *J. Med. Genet.* **27:** 559.

Sultan, C., Balaguer, P., Terouanne, B., Georget, V., Paris, F., Jeandel, C., Lumbroso, S., and Nicolas, J., 2001, Environmental xenoestrogens, antiandrogens, and disorders of male sexual differentiation. *Molec. Cell. Endoc.* **178:** 99.

Toppari, J., Larsen J.C., Christiansen, P., Giwercman, A., Grandjean, P., Guillette, L.J., Jegou, B., Jensen, T.K. et al., 1996, Male reproductive health and environmental xenoestrogens. *Environ. Health Persp.* **104:** 741-803.

Toppari, J., Skakkebaek, N.E., 1998, Sexual Differentiation and environmental endocrine disruptors. *Bailliere's Clin. Endoc. Metab.* **12:** 143.

Weidner, I.S., Moller, H., Jensen, T.K., and Skakkebaek, N.E., 1998, Cryptorchidism and hypospadias in sons of gardeners and farmers. *Environ. Health Perspec.* **106:** 793.

Weidner, I.S., Moller, H., Jensen, T.K., Skakkebaek, N.E., 1999, Risk factors for cryptorchidism and hypospadias. *J. Urol.* **161:** 1606.

Weybridge, 1995, EUR 17549 European workshop on the impact of endocrine disrupters on human Health and Wildlife. Report of Proceedings. Environment and Climate Research Programme of DG XII of the European Commission.

Wilcox, A.J. Baird, D.D., Weinberg, C.R., Hornsby, P.P., and Herbst, A.L., 1995, Fertility in men exposed prenatally to diethylstilbestrol. *New Engl. J. Med.* **332:** 1411.

Williams, K., McKinnell, C., Saunders, P.T., Walker, M., Fisher, J.S., Turner, K.J., Atanassova, N., and Sharpe, R.M., 2001, Neonatal exposure to potent and environmental oestrogens and abnormalities of the male reproductive system in the rat: evidence for importance of the androgen-oestrogen balance and assessment of the relevance to man. *Human Reprod. Update* **7:** 236.

Wolf, C.J., LeBlanc, G.A., Ostby, J.S., and Gray, L.E., 2000, Characterization of the period of sensitivity to fetal male sexual development to vinclozolin. *Toxicol. Sci.* **55:** 152.

TOXICANT-INDUCED HYPOSPADIAS IN THE MALE RAT

L. Earl Gray Jr.*, Joseph Ostby, Johnathan Furr, Carmen Wolf,
Christy Lambright, Vickie Wilson, Nigel Noriega

Disclaimer. The research described in this article has been reviewed by the National Health Environmental Effects Research Laboratory, U. S. Environmental Protection Agency, and approved for publication. Approval does not signify that the contents necessarily reflect the views and policies of the Agency nor does mention of trade names or commercial products constitute endorsement or recommendation for use.

1. INTRODUCTION

Prenatal exposure to endocrine disrupting chemicals that interfere with the androgen or insulin like factor 3 signaling pathways during sexual differentiation can induce malformations of the reproductive tract of the male rodent offspring. The pattern of malformations in the male depends upon the specific mechanism of action of the toxicant, the dosage level administered and the timing of administration during pregnancy. Hypospadias occurs in male rats or mice after maternal treatment with 1). potent estrogens or estrogenic drugs, 2). drugs that inhibit 5 alpha reductase, 3). drugs, herbicides and dicarboximide and conazole fungicides that act as androgen receptor (AR) antagonists, and 4). drugs, herbicides and conazole fungicides that inhibit cytochrome P450 enzymes involved in steroid hormone synthesis. In addition, 5). several phthalate diesters including di-n-butyl phthalate (DBP), di-n-ethylhexyl phthalate (DEHP), and benzylbutyl phthalate(BBP) also induce hypospadias and by altering fetal testis Leydig cell differentiation, resulting in reduced steroid and peptide hormone production.

In the laboratory rat (Sprague-Dawley and Long Evans Hooded) the background rate of severe malformations like hypospadias, spontaneous sex reversals and agenesis of sex accessory and epididymal tissues is infinitesimally low, if it occurs at all (Figure 1).

*L. Earl Gray Jr, MD-72, Endocrinology Branch, RTD, NHEERL, ORD, USEPA, RTP, NC, 27711. Phone. 919-541-7750. fax 919-541-4017. email gray.earl@epa.gov

Hypospadias and Genital Development, edited by
L. Baskin, Kluwer Academic/Plenum Publishers, 2004

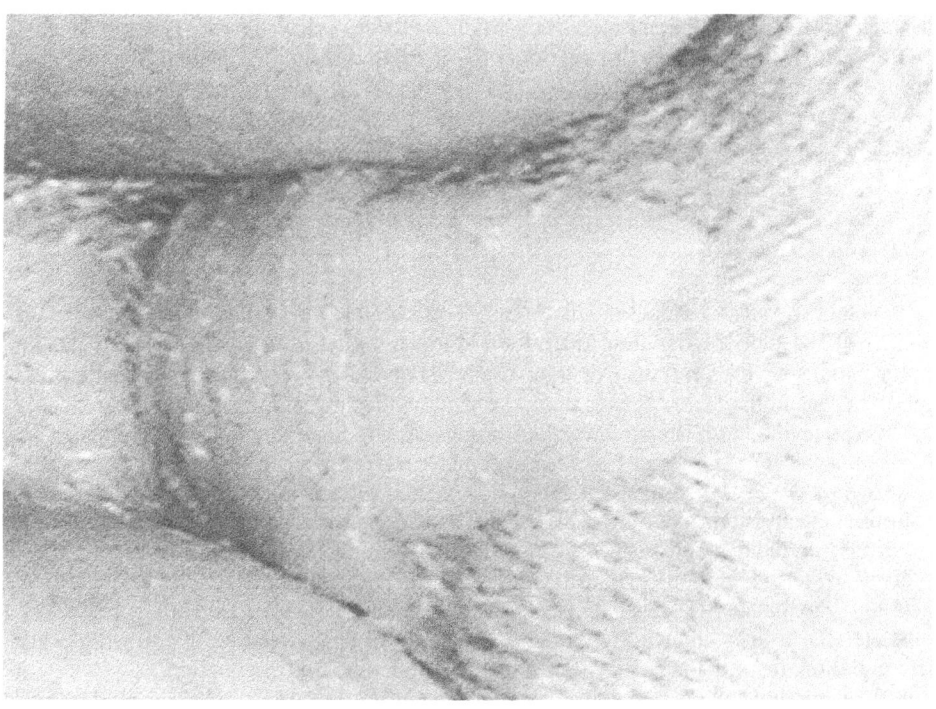

Figure 1 a, b. and c. Glans penis of an untreated adult male rats . The prepuce has been manually retracted to display the glans structure

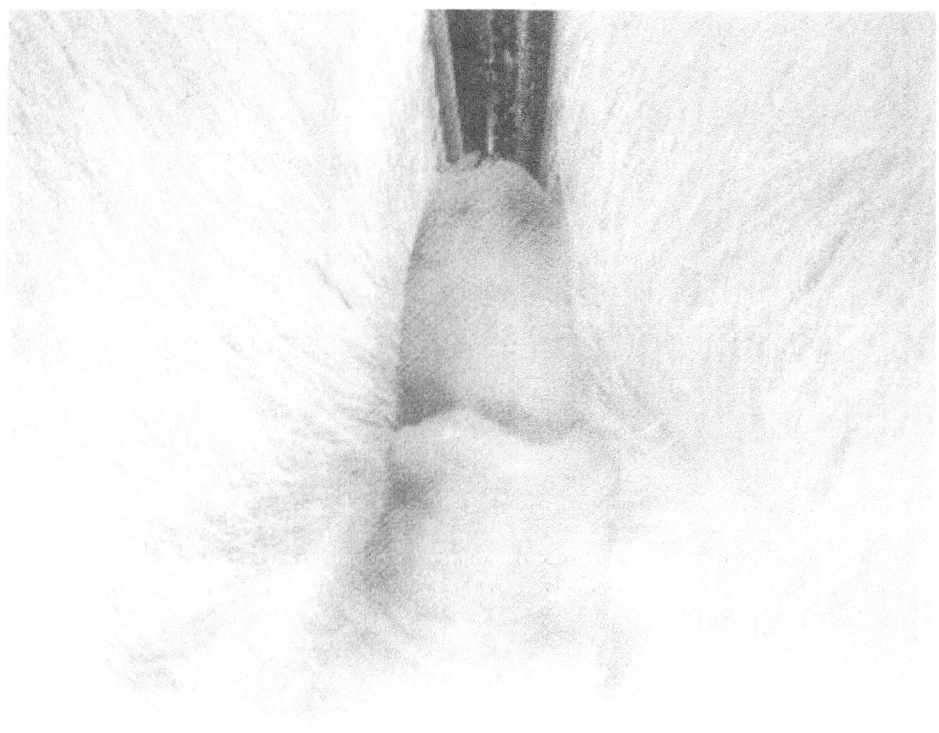

Figure 1b

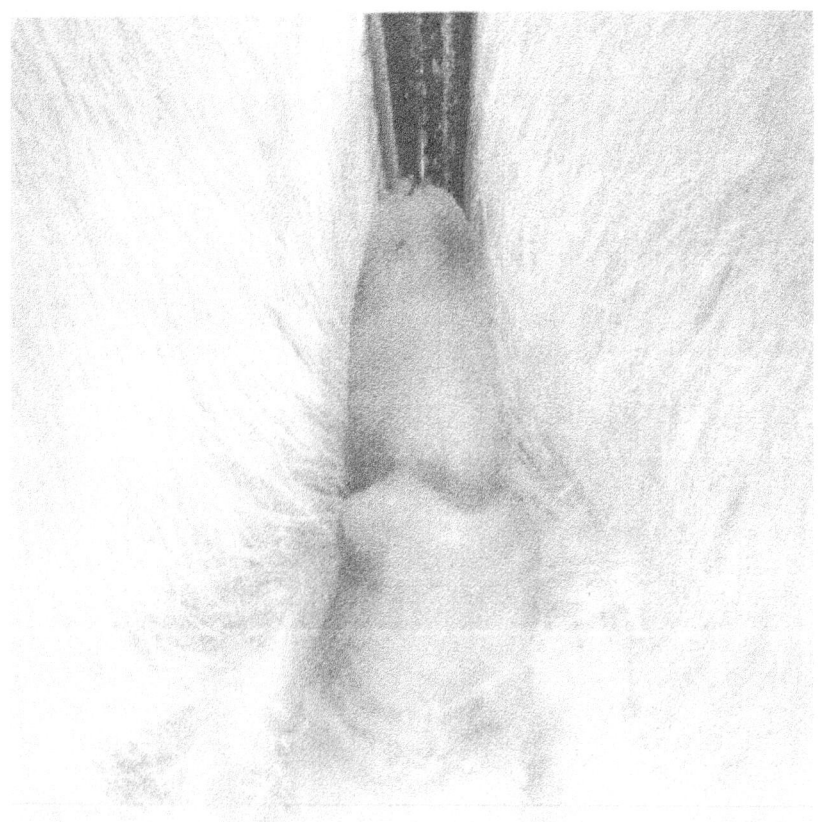

Figure 1c

In addition, when hypospadias is induced by toxicant administration during late gestation isolated hypospadias may occur in a few individuals, but typically a variety of additional effects accompany this malformation. With androgen receptor antagonists like vinclozolin and procymidone the dose-related progression of alterations (from low to high dose) includes 1). reduced anogenital distance (AGD), 2). induction of female-like areolas/nipples and reduced ventral prostate weight, 3). reduced seminal vesicle weight, 4). hypospadias (i.e. 50% at 50 mg vinclozolin/kg/day)and agenesis of the ventral prostate,5). hypospadias with a vaginal pouch and undescended testes and 6). epididymal agenesis (low rate at 200 mg vinclozolin/kg/day). In contrast, the phthalate esters and the herbicide linuron produce a different constellation of malformations from the AR antagonists procymidone and vinclozolin.

Specifically, epididymal agenesis is present at relatively high rates (i.e. 25% at 300 mg DEHP/kg/day during pregnancy) at dose levels that induce a low rate of hypospadias in (i.e. 2% at 300 mg DEHP/kg/day). With diethylstilbestrol(DES), ethinyl estradiol and 17 beta estradiol relatively high dose levels administered to the dam in the last third of pregnancy induce hypospadias and other malformations in the male offspring, while the females are affected at lower dosage levels. Of the above chemicals, only the phthalate esters and the potent estrogens reduce insulin-like factor 3 mRNA levels and cause gubernacular agenesis. The sections of this review that follow will discuss selected examples of chemically-induced hypospadias in rodents on a chemical-by-chemical basis, and will conclude with a discussion of the cumulative effects on sexual differentitation seen when pairs of toxicants are coadministered with one another.

2.   Androgen-Receptor Antagonists

Vinclozolin,   procymidone,   p,p'   DDE   and   linuron.
The pesticides vinclozolin, procymidone, p,p' DDT and p,p= DDE, prochloraz, fenitrothion and linuron are AR antagonists. Of these, only fenitrothion fails to induce male reproductive tract malformations when administered orally to the pregnant rat during sexual differentiation. Vinclozolin metabolites, M1 and M2 (Kelce et al., 1994), procymidone (Ostby et al., 1999), p,p' DDT (and metabolites) (Kelce et al., 1995) and linuron (Lambright et al., 2000: McIntyre et al., 2000) all competitively inhibit the binding of androgens to hAR and inhibit androgen-induced gene expression. Of these, it also has been demonstrated that vinclozolin, p,p' DDE (Kelce et al., 1997), and linuron (Lambright et al., 2000) alter androgen-dependent ventral prostatic gene expression in vivo. The dicarboximide fungicides, vinclozolin, and procymidone appear to display nearly identical toxicity profiles in vitro and in vivo.

The effects of these two pesticides on androgen-dependent organ weights, serum LH and testosterone, development of the male reproductive tract and in vivo gene expression profiles, measured both by real time rtPCR (Wilson et al., 2003) or microarrays (Rosen et al., 2003) are indistinguishable from one another. Interestingly, chlozolinate and iprodione, two other dicarboximide fungicides, do not act as AR antagonists at any dose in vivo or concentration in vitro. None of these pesticides or their metabolites (examined to date) appear to display significant affinity for the estrogen receptor (Kelce et al., 1995; Waller et al, 1996a), although the vinclozolin metabolite M1 binds the rat progesterone receptor, albeit with relatively low affinity (Laws et al., 1996).

Vinclozolin-treated male rat offspring display female-like anogenital distance (AGD) at birth, retained nipples, cleft phallus with hypospadias (Figure 2), undescended testes,

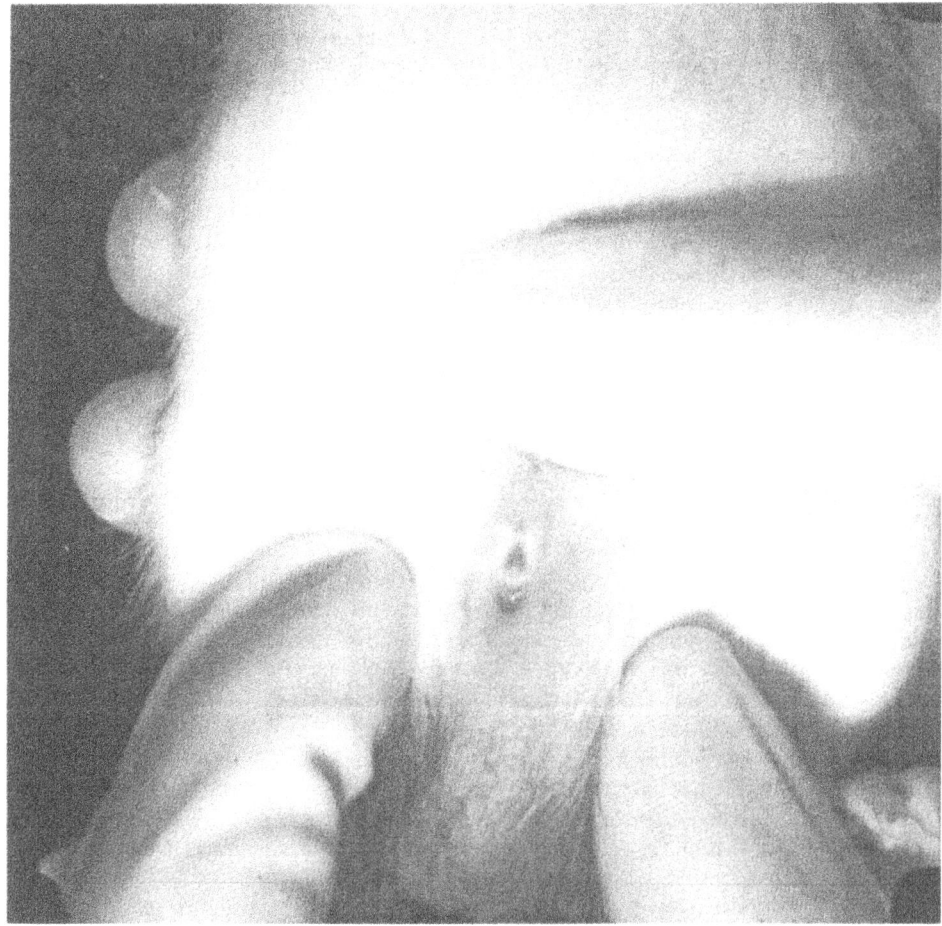

Figure 2A
These male rat offspring were exposed in utero to vinclozolin at 100 or 200 mg/kg/d by dosing the dam from gestational day 14 to day three of lactation. Figure 2A has hypospadias, with a vaginal pouch and a single descended testis. Figure 2B also has hypospadias and a vaginal pouch with two undescended testes. The left testis is in an ectopic position in the abdominal cavity. The cleft phallus of the male on the lower left has been retracted to display the vaginal pouch and os penis.

vaginal pouch, epididymal granulomas, and small to absent sex accessory glands when it is administered at 0, 3.125, 6.25, 12.5, 25, 50, 100 or 200 mg/kg/d from GD 14 to postnatal day 3 to the dam.

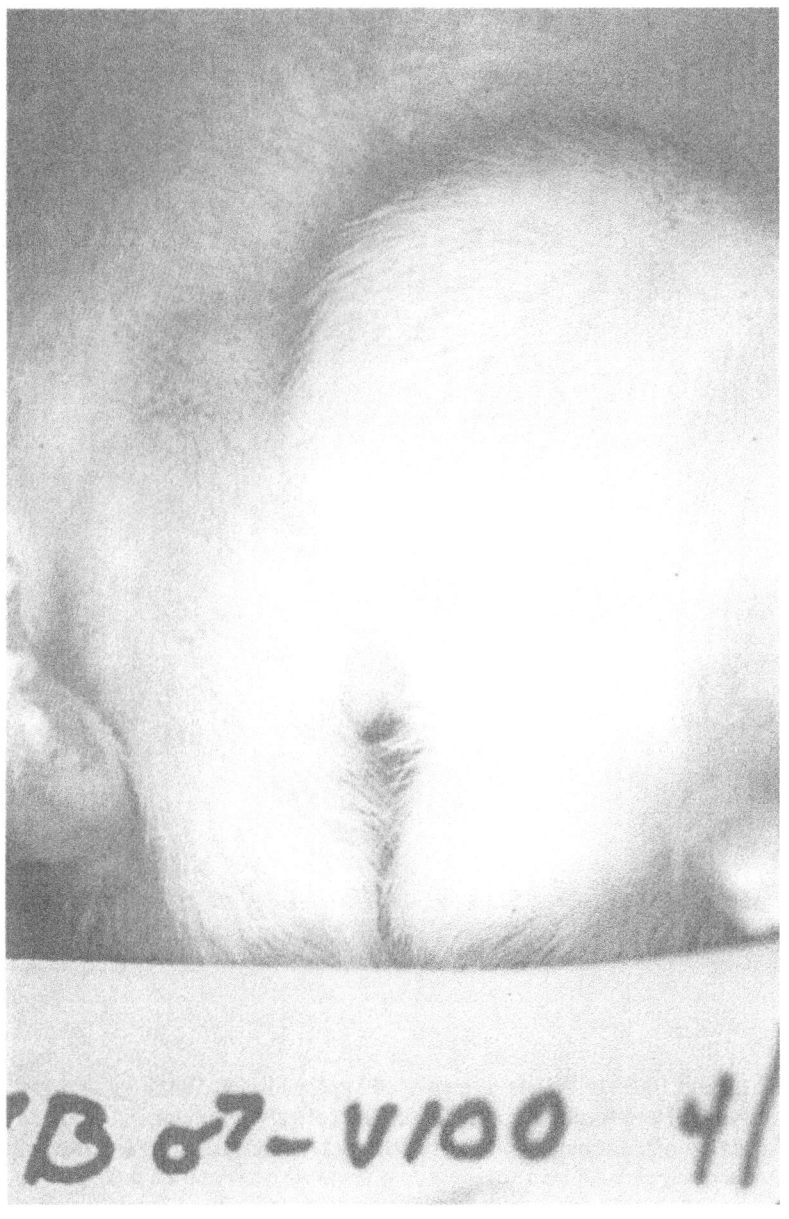

Figure 2B

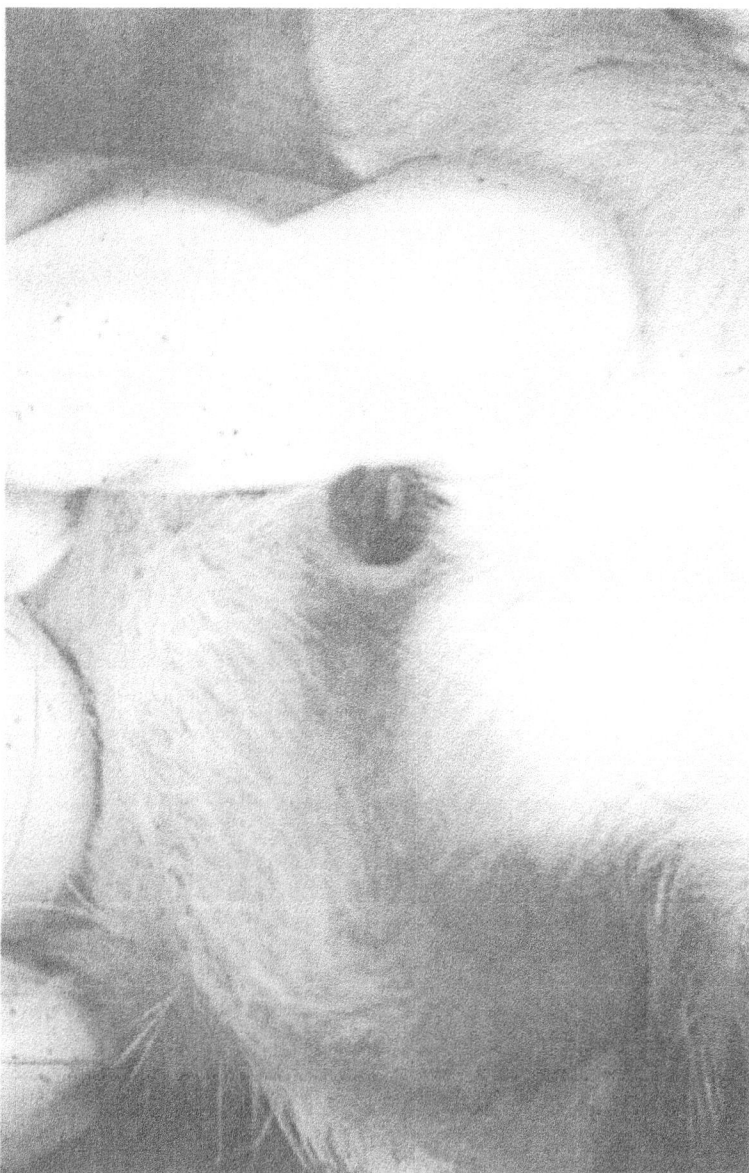

Figure 2C

Fifty percent and 100% of the males displayed hypospadias at 50 mg/kg and 100 mg/kg, respectively.   Dose-response curves for different effects of vinclozolin vary in shape and ED50 values for different androgen-dependent tissues, with AGD being significantly reduced at the lowest dosage level. Some of these dose response curves failed to display an obvious threshold (i.e. anogenital distance, induction of areolas and ventral prostate weight (Gray et al., 1999b)) and appear linear in the low dose range while the induction of hypospadias and other malformations did display apparent thresholds. Other malformations noted included males with a vaginal pouch, agenesis of the ventral prostate and ectopic, undescended testes.

Pregnant rats were dosed with by oral gavage 400 mg vinclozolin/kg/day on either GD 12-13, GD 14-15, GD 16-17, GD 18-19, or GD 20-21, or with corn oil (2.5 ml/kg) from GD 12 through GD 21 (Experiment 1). Significant effects of vinclozolin were observed in rats dosed on gestation days (GD) 14-15, GD 16-17, and GD 18-19, while the most significant effects were observed in rats treated on GD 16-17. These effects include reduced AGD; presence of areolas, nipples, and malformations of the phallus; and reduced levator ani/bulbocavernosus muscle weight. In contrast, ventral prostate weight was reduced only in the GD 18-19 group. These data indicate that the reproductive system of the fetal male rat is most sensitive to antiandrogenic effects of vinclozolin on GD 16 and 17 (Wolf et al., 2000).

3. Procymidone, chlozolinate and iprodione

When procymidone is administered by gavage on GD 14 to day 3 after birth to the dam at doses of 0, 25, 50, 100 and 200 mg/kg/day, effects were noted in all dosage groups (Ostby et al., 1999). The incidence of hypospadias was dose-related being displayed by 0, 0, 14.3, 19 and 91% of the male rat offspring at 0, 25, 50, 100 and 200 mg/kg/day, respectively(Figure 3).
Procymidone also reduced anogenital distance in male pups at all dosage levels and induced retained nipples, cleft phallus, vaginal pouch and reduced sex accessory gland size in rat

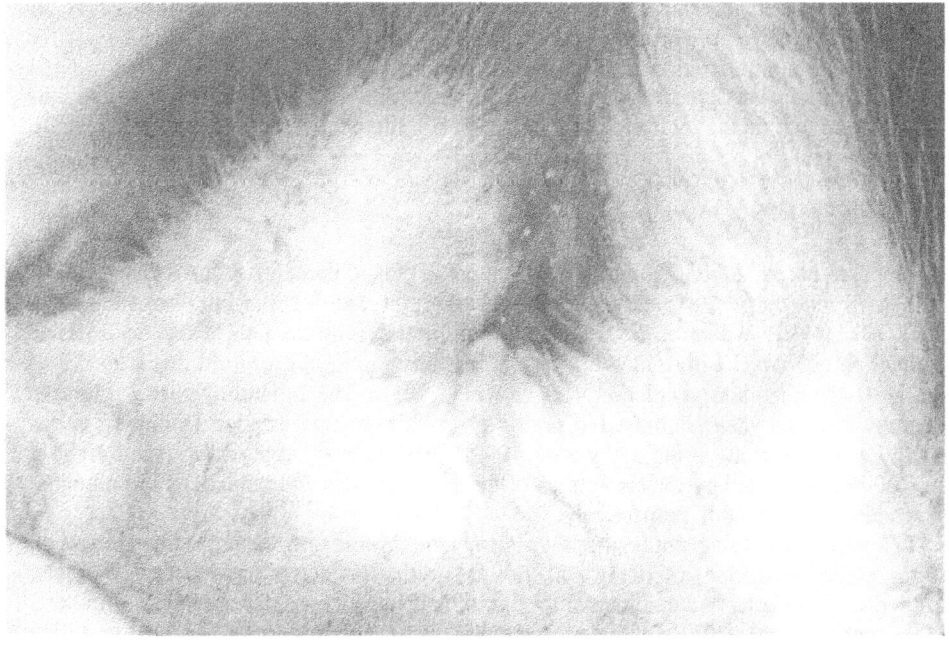

Figure 3. This male rat's dam was exposed to procymidone at 200 mg/kg/d during sexual differentiation. The male has hypospadias and an obvious vaginal pouch.

offspring and had marked effects on the histology of the dorsolateral and ventral prostatic and seminal vesicular tissues (at 50 mg/kg/day and above). The effects consisted of fibrosis, cellular infiltration, and epithelial hyperplasia (Ostby et al., 1999).

Chlozolinate and iprodione are dicarboximide fungicides, similar in structure to the antiandrogens vinclozolin and procymidone. However, when chlozolinate and iprodione were administered at 100 mg/kg/day from GD 14 to PND 3, male rat offspring were not demasculinized or feminized at this dosage level (Gray et al., 1999a).

4. Flutamide

Flutamide is a nonsteroidal androgen receptor antagonist that shares many toxic properties with the fungicides procymidone and vinclozolin, although flutamide appears to be at least ten-fold more potent in vivo than either of these pesticides. When administered in utero to the dam at 0, 6.25, 12.5, 25, or 50 mg/kg/day (po) from gestation days 12 to 21 all of the flutamide exposed male rat offspring displayed hypospadias. In addition, 55-85% of the males in the low to high dosage groups, respectively, displayed cryptorchid testes (McIntyre et al., 2001).

In utero, flutamide also exposure significantly decreased AGD on postnatal day (PND) 1 and increased areola/nipple retention in male rats on PND 13. Treated males also displayed partial or complete prostate agenesis and decreased the weights of the seminal vesicles, levator ani bulbocavernosus muscle, testes, and epididymides in a dose-related manner. Epididymal malformations were observed mainly in the highest dose group.

Imperato-McGinley et al. (1992) found that complete feminization of the genitalia occurred at 24 mg/kg/day flutamide (GD 12-21) in all male rat offspring. At 18 mg/kg/day, Wolffian ductal differentiation occurred, but seminal vesicle weight was decreased while at dosages of 100, 200, and 300 mg/kg/day, the vas deferens was absent unilaterally or bilaterally, with small remnants of epididymal head and tail present. At dosages of 24 mg/kg/day and above, the prostate was absent.

Flutamide administration over narrower time intervals (GD 13-15, 16-17, or 18-19) revealed maximal interference with testicular descent during GD 16-17; however, the authors did not comment on the incidence of hypospadias in this study (Husmann and McPhaul, 1991).

5. The Herbicide Linuron

The urea-based herbicide linuron binds rat prostatic and human AR (hAR) and inhibits DHT-hAR induced gene expression in vitro (Lambright et al., 2000; McIntyre et al., 2000; Cook et al., 1993; Waller et al., 1996b). Linuron also inhibits fetal testis testosterone production ex vivo (Lambright et al., 2003; Hotchkiss et al., in prep). In this regard, it is similar to the fungicide prochloraz (Norigea et al., 2003). The antiandrogenicity of linuron is quite apparent when administered orally by gavage during gestation (McIntyre et al., 2000; Lambright et al., 2000; Gray et al. 1999a) or in a Hershberger assay (Lambright et al., 2000) as indicated by malformations of the epididymides and reductions in androgen-dependent organ weights, respectively.

In a multigenerational study, the linuron-treated (chronic oral maternal treatment with 40 mg/kg/day) offspring (F1) (Gray et al., 1999a) sired 40% fewer pups and treated F1 males had reduced testicular and epididymal weights, and lower testicular spermatid numbers. When administered at 100 mg/kg/day to the dam by gavage from days 14-18 of gestation (Gray et al., 1999a) linuron reduced AGD in male offspring and the incidence of areolas (with and without nipples) in infants males was increased. Linuron-treatment induced epispadias (1/13 male offspring) (Figure 4) and caused agenesis of the caput and/or corpus epididymides, and atrophic, or fluid filled and flaccid testes in 56% of the male rat offspring (Gray et al., 1999a; Lambright et al., 2000; McIntyre et al. 2000).

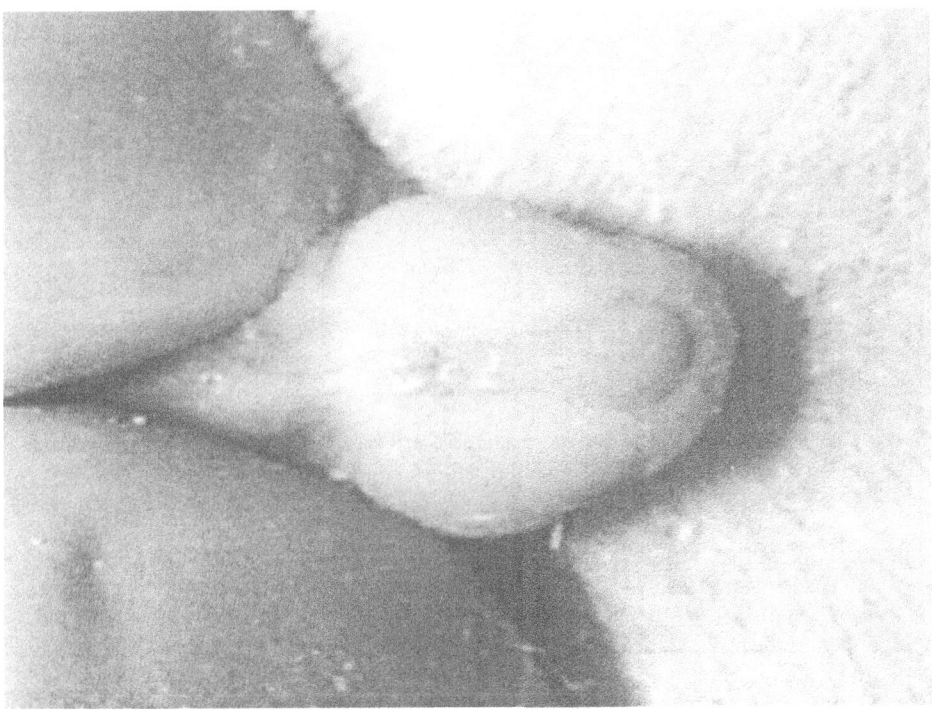

Figure 4. Epidspadias in a adult male rat exposed to the herbicide linuron during sexual differentiation. The dam was treated with linuron at 100 mg/kg/d from day 14 to 18 of pregnancy. The urethral opening is about is in the middle of the glans penis rather than at the tip.

6. The insecticide DDT and DDE: AR antagonists

Although use of DDT has been banned in some countries, some wildlife populations still display incredibly high total DDT residue levels (Elliot et al., 1994; Williams, 1999; Guillette et al. 1999) as a result of decades of former use of this persistent, bioaccumulating pesticide,. In the orchards and fields sampled by Elliot et al. (1994), birds had high tissue levels of p,p= DDE (up to 103 ppm in fat) while at Lake Apopka, FL, fat sample levels in birds were even higher.

When p,p' DDE is administered to pregnant rats at 100 mg/kg/day (days 14-18 of gestation) it reduces AGD and induces a low rate (about 2-3%) hypospadias (Figure 5), retained nipples, and reduces weights of androgen-dependent tissues in treated Long Evans Hooded (LE) and Sprague-Dawley (SD) male offspring (Gray et al., 1999a).

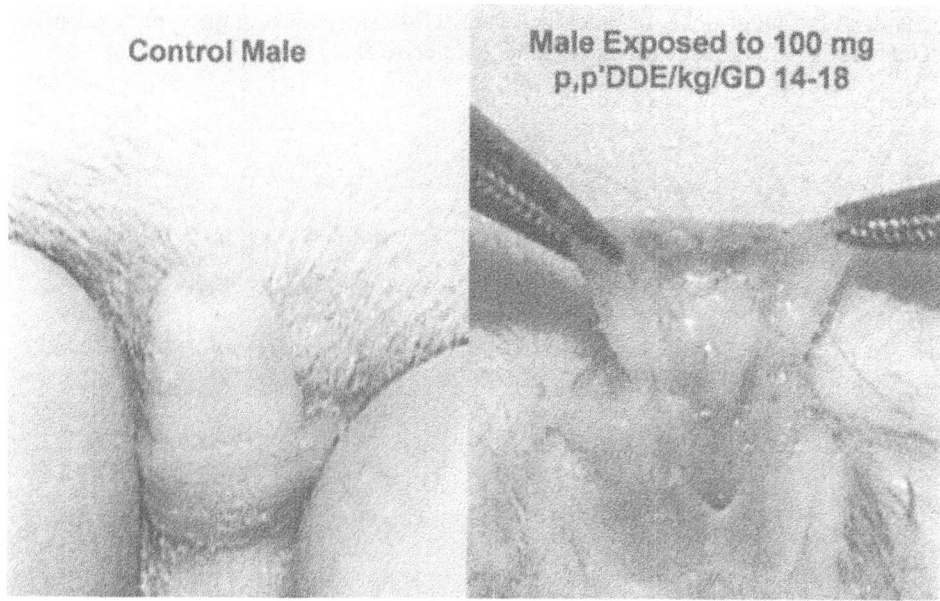

Figure 5. The normal glans penis of an untreated control male rat is shown at the left, while on the right, a male rat whose mother was treated with p,p' DDE at 100 mg/kg/d on gestational days 14-18 is shown on the right.

While these alterations were evident in both rat strains, only the SD strain displayed hypospadias and other effects of DDE were of a greater magnitude in this strain. You et al. (1998) also found that p,p'-DDE (10 and 100 mg/kg/day oral on GD 14-18) induced antiandrogenic effects on AGD and areola development in LE and SD rats and also noted an increased incidence of chronic suppurative prostatitis in treated male progeny (You et al., 1999a); not an uncommon observation for males exposed in utero to an EDC (i.e. polychlorinated biphenyls (PCBs) and procymidone). These adverse developmental effects were correlated with fetal rat tissue of p,p' DDE levels ranging from 1 to 2 µg/g on GD 21 and 10-20 µg/g on GD 20 (You et al., 1999b).

7.   Inhibitors of steroid hormone synthesis

Some classes of fungicides inhibit sterol synthesis for cell membranes and fungal growth by inhibiting cytochrome P450 enzymes, especially 14 alpha-demethylation of lanosterol in the sterol pathways. These fungicides are fairly nonspecific inhibitors of CYP450 enzymes, so the endocrine effects induced are not always limited to the reproductive system and often include affects on adrenal and liver steroid metabolism in vertebrates and ecdysteroid synthesis in invertebrates.

8.   The Fungicide Prochloraz

Prochloraz is a conazole-type fungicide that is widely used in Europe, the United Kingdom, Australia and other areas of the world, but it not licensed for use in the US. There are literally hundreds of triazole and imadizole fungicides in the conazole group that are used as pesticides and drugs. Prochloraz is the first member of the class that has been shown to act as an AR antagonist in vitro and in vivo and to alter male reproductive tract development (Noriega et al., 2003). Prochloraz is a novel EDC because it has several distinct endocrine mechanisms of action. Prochloraz inhibits P450 aromatase (Laignelet et al., 1989; Andersen et al., 2002; Vinggaard et al., 2000) and 17,20-lyase (involved in the conversion of progesterone to androgens) activities, it binds the androgen receptor (Noriega et al., 2003; Vinggaard et al., 2002) and inhibits androgen-induced transactivation in vitro (Noriega et al., 2003; Andersen et al., 2002). In MDA-kb2 cells containing endogenous androgen receptor and stably transfected with a MMTV-luc reporter, in vitro concentrations of prochloraz above 1M caused a dose-dependent inhibition of DHT-induced luc expression in the absence of cytotoxicity. It also has been reported that prochloraz inhibits MCF-7 cell proliferation and estrogen receptor-mediated gene transactivation in vitro (Andersen et al., 2002).

Maternal prochloraz exposure from GD 14-18 by oral gavage decreases testosterone while increasing progesterone production by fetal GD 18 rat testes. This, along with the AR antagonism, demasculinzes the fetal male rat and delays deliver when administered at 0, 31.25, 62.5, 125 or 250 mg/kg/day. Hypospadias was common in male offspring exposed to 125 and 250 mg/kg/day. Male offspring also displayed reduced AGD at three days of age and permanently retained nipples and altered reproductive tract tissue weights. This study is important because prochloraz was shown to induce hypospadias (Figure 6).

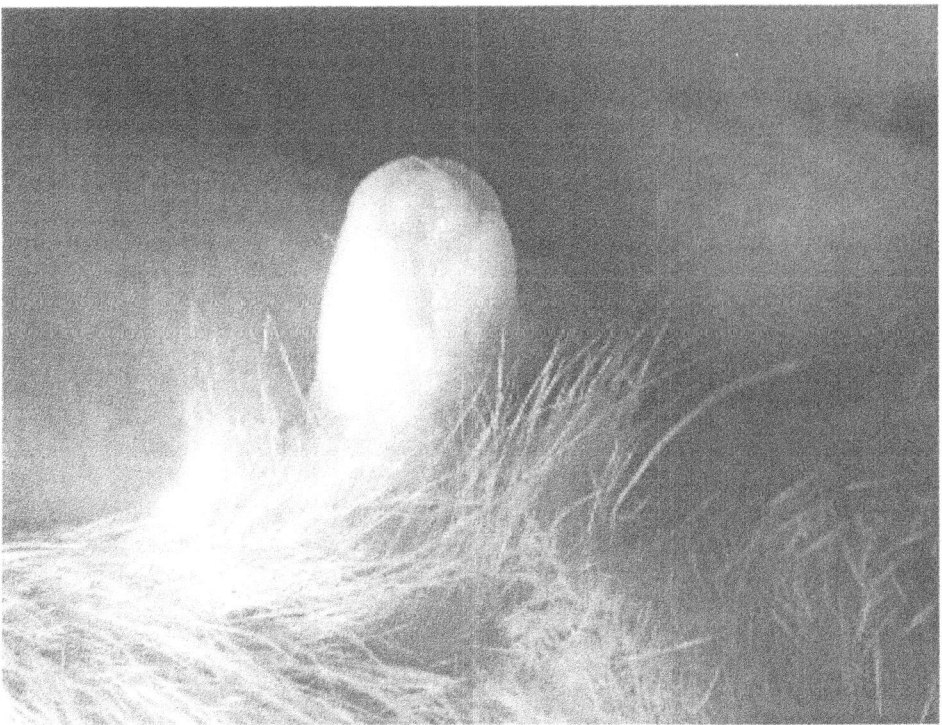

Figure 6A.
The male rats at the top with epispadias was exposed to the fungicide prochloraz during sexual differentiation at 250 mg/kg/d from gestational day 14 to 18. The male rat (6B) has hypospadias with an exposed os penis is from the same dose group. The male at the

Figure 6B

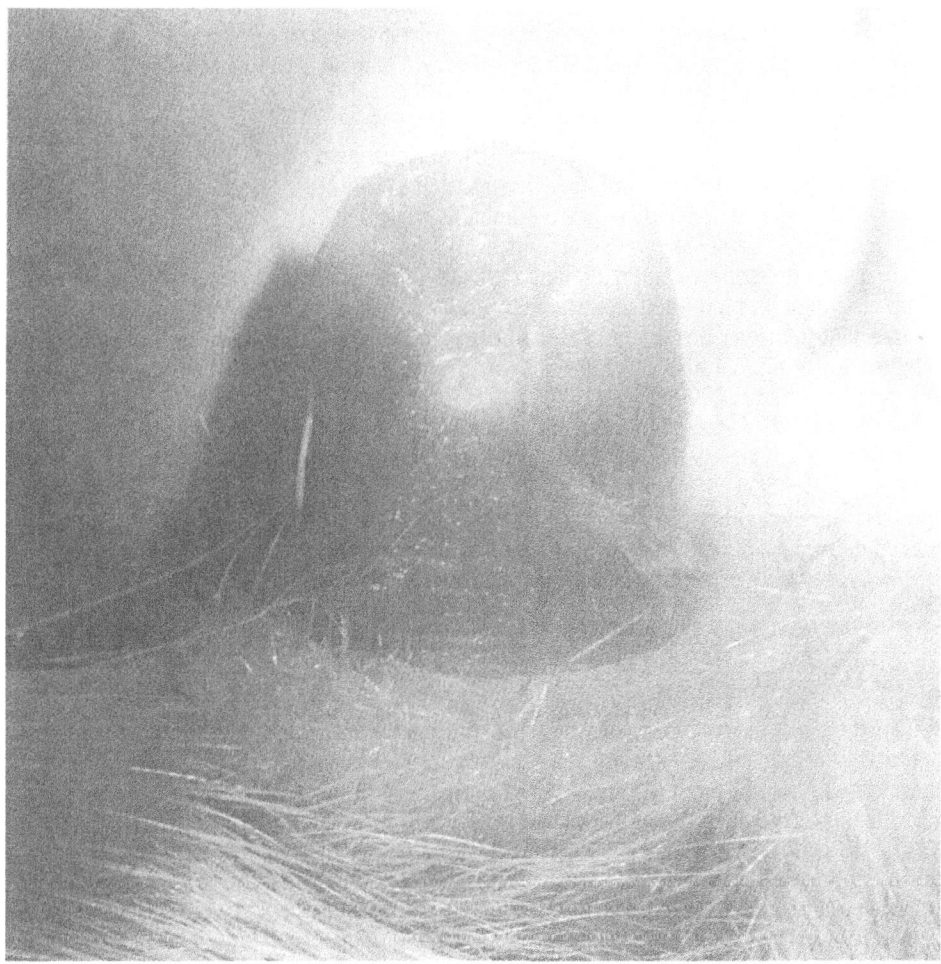

Figure 6C

Such findings may be cause for awareness regarding several other fungicides and pharmaceuticals aimed at modulating P450 enzymes.   The multiple mechanisms of action demonstrated by prochloraz have potential for inducing wide ranges of effects related to hormone production. Prochloraz=s P-450-specific activity prompts the hypothesis that it can simultaneously affect androgen, and estrogen-related physiology as well as physiology associated with other steroids. Such effects may be worthy of investigation because of the ubiquity of compounds containing imidazole rings mostly used in the preparation of antifungal pharmaceutical and pesticide agents. Imidazole itself is used to produce curing agents of epoxy resins, as well as photosensitive materials.

9. Synthetic Steroids that inhibit 17-20 lyase

Two synthetic steroid analogues inhibit steroidogenesis in a manner that appears similar to prochloraz. The first, 16beta- bromo-3beta, 17alpha- dihydroxy- 5alpha- pregnane-

11,20-dione, and the second 17 beta-ureido-1,4-androstadien-3-one irreversibly inhibit the active site of rat testicular microsomal steroid 17alpha-hydroxylase and C17-20 lyase in vitro; 24 h exposure produced potent inhibition of these testicular enzymes in vivo. When administered to pregnant rats during the critical period of male organogenesis they produced hypospadias (Goldman et al., 1976).

Ketoconazole

The antifungal drug ketoconazole inhibits various enzymes which belong to the cytochrome-P450-dependent mono-oxygenases such as side chain cleavage of cholesterol, 11 beta-hydroxylase in the adrenal, and 17 alpha-hydroxylase and C17-20 lyase in the testes. Human testicular mono-oxygenase activities in vitro are reduced by 50% from 3.1 µM ketoconazole. Schurmeyer and Nieschlag (1984) demonstrated that ketoconazole and other imidazole fungicides inhibited testosterone production in men, while Pepper et al. (1990) reported that ketoconazole was useful in the treatment of ovarian hyperandrogenism in women. Ketoconazole also has been shown to alter hepatic steroid catabolism in mice (Wilson and Leblanc, 2000).

Similar to prochloraz, administration of ketoconazole by oral gavage from GD 14-18 at 100 mg/kg/day delays delivery, causing extensive pup mortality (Gray et al., 1999a). However, in contrast to prochloraz, lower dosage levels of ketoconazole did not affect reproductive development of the male offspring. At 25 and 50 mg/kg/day, ketoconazole delayed delivery and the numbers of live pups were reduced but treated male pups did not display hypospadias or any reproductive alterations (Gray et al., 1999a).

10. Fenarimol

The fungicides prochloraz and fenarimol both inhibit the P450 enzyme aromatase, preventing the conversion of androgens to estrogens (Hirsch et al., 1987, above). When fenarimol was administered continuously by oral gavage from weaning, treated male rats, as adults, displayed altered mating behavior (failure to mount a sexually receptive female). As this behavior is dependent upon the local conversion of testosterone to estradiol in the brain of the male rat, this effect likely resulted from an inhibition in the production of estrogens in the brain (Gray and Ostby, 1998). Serum hormone levels were unaffected by fenarimol treatment but male rats from the highest dosage group of 70 mg/kg/d from weaning until necropsy at adulthood displayed reduced sex accessory gland weights and increased liver size, effects that could indicate that fenarimol was inhibiting P450 enzymes in the testis and liver. This chemical also reduced ovarian weights and caused delayed parturition in female rats. Unlike prochloraz, male offspring did not display hypospadias or any other reproductive tract malformation.

11. Inhibitors of 5 alpha reductase

The conversion of testosterone to 5 alpha-dihydrotestosterone by the enzyme 5 alpha-reductase is inhibited by finasteride and other drugs (Wier et al., 1990). In a dose response study with maternal dosing of finasteride from GD 15 to PND 21, male rat AGD was decreased by 13 and 27% on PND 1 at 0.03 and 3 mg/kg/day, respectively (Clark et al., 1990). In a second dose-response study, dose-related effects indicated that 0.1 mg/kg/day was a threshold dose for induction of hypospadias in male offspring with a 100% effect level of 100 mg/kg/day (with dosing through Day 20 of gestation). Abnormalities in external genitalia were also seen in male rats exposed in utero to finasteride, a 5 alpha-reductase inhibitor (Clark et al., 1993). When Clark et al (1993) administered 20 mg/kg/day of finasteride for successive 2-day periods during gestation, they found that the period of GD 16 to 17 was the most sensitive for inducing hypospadias, cleft prepuce, decreased anogenital distance, reduced prostate weight, and nipple formation in male offspring. This critical period is just prior to the appearance on day 18 of gestation of a midline mesenchymal plate between the urogenital sinus and the rectum in normal male fetuses. This midline plate does not appear in finasteride-exposed fetuses destined to have hypospadias as demonstrated in a previous study.

Imperato-McGinley et al., (1992) found that a dose of 25 mg/kg/day 12-21 of finasteride resulted in significant feminization of the external genitalia but that there was no further feminization of the genitalia of the male offspring at doses up to 300 mg/kg/day.

12. Phthalate Esters

The phthalate esters DBP, BBP and DEHP alter male rat sexual differentiation when administered orally to the dam during gestation by altering fetal testis endocrine function during this critical period (Gray et al., 2000; Parks et al., 2000). Maternal treatment with one of these phthalate diesters reduces fetal testis steroid and peptide hormones (Lambright et al., 2003). In contrast to flutamide, vinclozolin or other EDCs that induce hypospadias, the phthalate diesters and their monoester metabolites do not bind the androgen receptor. In addition, they fail to display any estrogenicity at any dosage level in vivo, or in vitro at relevant concentrations. DEHP has been reported to inhibit the P450 steroidogenic enzyme aromatase, but only at dosage levels (2 g/kg) an order of magnitude higher than those reported to alter male rat sexual differentiation and pregnancy maintenance.

When DEHP, BBP or DBP are administered orally to the pregnant rat during sexual differentiation, fetal testis testosterone production is reduced altering differentiation of androgen-dependent tissues and mRNA for the peptide hormone insulin-like-factor 3 is reduced, altering gubernacular development.

When administered by gavage at 750 mg/kg/day (GD 14- postnatal day 3), BBP, DBP and DEHP induce hypospadias and epididymal and gubernacular agenesis in more than 50% of the male offspring. When DBP is administered at 500 mg/kg/day over three four-day intervals including GD 8-11, 12-15 and 16-19, effects were most evident with exposure on GD 16-19 (reduced AGD, retained nipples, reduced androgen-dependent tissue weights), with only minimal effects induced by treatment on GD 12-15 (retained areolas/nipples in male infant rats). When DBP is administered to the dam for longer periods, including this critical period, hypospadias is induced at very low rates at 100, 300 mg/kg/day (GD 8 to weaning; Gray et al., 2003) and 500 mg/kg/day (Gray et al., 1999a).

Hypospadias is not the most common reproductive tract malformation in these studies. For example, when DBP is administered at 500 mg/kg/d from GD 14-postnatal day 3, about 25% of the male offspring displayed epididymal agenesis but only 6.2% of the male offspring displayed hypospadias (Gray et al., 1999b). Similarly, when DBP administered orally from GD 12-20 at 500 mg/kg/d, about 40% of the male rat offspring display epididymal agenesis but hypospadias is only present in 9% of the males (Mylchreest et al., 1998; 1999). When DBP is administered in longer-term studies, not only are the offspring affected, but pregnancy is disrupted (Gray et al., 1999b). Subchronic maternal treatment prior to and including gestation with 500 mg/kg/day induces mid-gestation whole litter loss and lower maternal progesterone. In summary, the phthalate esters DBP, BBP and DEHP induce hypospadias when administered during sexual differentiation but this is not the most common lesion. The most sensitive period of development for the induction of these malformations by the phthalate diesters is the same as that for EDCs that disrupt this process via other mechanisms of action. The incidence of these effects varies not only with timing and dose but also with the duration of exposure during gestation.

13. Estrogens and hypospadias in males
(DES and other potent estrogenic substances)

Almost thirty years ago McLachlan et al (1975) discovered that prenatal exposure to DES induced reproductive tract lesions in male mice offspring. Male offspring exposed to DES during gestation were infertile and display structural malformations including cryptorchidism, epididymal cysts and retained Mullerian ducts. Similarly, in utero administration of relatively high dosage levels of other potent estrogenic substances like estradiol and ethinyl estradiol have been shown to cause male reproductive tract lesions, including hypospadias. McLachlan et al. (2001) recently described these estrogen-associated alterations in the genital tract as the Developmental Estrogenization Syndrome. While many of the cellular and molecular mechanisms responsible for these lesions are still unknown, new evidence indicates how estrogens can induce the testicular dysgenesis syndrome in humans (testis cancer, cryptorchidism, hypospadias, low sperm counts).

These pathways include altered by estrogen treatment include inhibiton of fetal testis testosterone production, down regulation of androgen receptor protein and gene expression and reduction of insulin-like factor-3 (InsL3) production by fetal testis Leydig cells (Sharpe, 2003). However, the role of the relatively weak environmental estogens in this process remains controversial and unconfirmed. In this regard, Sharpe (2003) observed that although many new environmental oestrogens have been identified, their uniformly weak oestrogenicity excludes the possibility that they could induce the above disorders.

Hypospadias has been linked to in utero exposure to potent estrogenic drugs in humans and laboratory animals. Klip et al. (2002) reported that sons of women using DES during pregnancy (205 exposed boys out of 8934 total sons) displayed an increased prevalence rate of hypospadias. Four of the 205 (2%) exposed boys were reported to have hypospadias. In the remaining unexposed 8729 boys, only eight cases of hypospadias were reported (0.09%; prevalence ratio 21.3 [95% CI 6.5-70.1]). All cases of hypospadias were medically confirmed. The authors concluded that these findings suggest an increased risk of hypospadias in the sons of women exposed to DES in utero. Although the absolute risk of this anomaly is small, this transgenerational effect of DES warrants additional studies.

These results appear consistent with work on DES sons from twenty years earlier by Henderson et al., (1976) who surveyed exposed and unexposed male offspring by mail. A larger proportion of exposed than of unexposed boys experienced problems in passing urine (12.9% vs. 1.8%, P = 0.0003) and abnormalities of the penile urethra (4.4% vs. 0%; P = 0.017).

Another finding that warrants further investigation is the observation of a positive association of maternal vegetarian diet with hypospadias (North and Golding ,2000). Detailed information was obtained prospectively from mothers, including previous obstetric history, lifestyle and dietary practices, using questionnaires during pregnancy. Of 7928 boys born to mothers taking part in the study, 51 cases of hypospadias were identified. Mothers who were vegetarian in pregnancy had an adjusted odds ratio (OR) of 4.99 (95% confidence interval, CI, 2.10-11.88) of giving birth to a boy with hypospadias, compared with omnivores who did not supplement their diet with iron. Another significant association for hypospadias was with influenza in the first 3 months of pregnancy (adjusted OR 3.19, 95% CI 1.50-6.78). The authors concluded that AAs vegetarians have a greater exposure to phytoestrogens than do omnivores, these results support the possibility that phytoestrogens have a deleterious effect on the developing male reproductive system.

14. Animal studies of estrogens and hypospadias

While most of the studies on the developmental effects of potent estrogens and estrogenic drugs have focused on the Wolffian duct derivatives, testicular development and testis descent in males, several of these studies also have reported that relatively high dosage levels of estradiol, ethinyl estradiol, DES or other potent estrogenic chemicals in utero can induce hypospadias in male rat or mice offspring. For example, Vannier and Raynaud (1980) found that prenatal exposure to either estradiol or the synthetic estrogen RU 2858, sc GD 16-20, altered the genital tract of both male and female offspring.

Although the main effects were seen in the female offspring in the lower dose groups of RU 2858 (at 2 mg/rat/day and above females displayed marked hypospadias and permanent estrus), hypospadias was induced in male rat offspring (at 50 mg/rat/day).

In contrast to the potent estrogens, none of the studies using the weaker estrogenic pesticides and toxic substances or phytoestrogens have detected hypospadias in male rodent offspring exposed in utero. In fact, few, if any, reproductive tract malformations appear to be induced with these weakly estrogenic environmental chemicals. This does not mean that these chemicals are without serious adverse reproductive effect on the offspring, but the effects generally are more subtle than the frank malformations induced by the high affinity ER ligands.

In many studies, potent estrogens (i.e. DES or estradiol) and weaker xenoestrogens (i.e. the pesticide chlordecone or the fungal mycotoxin zearalenone) have been administered during neonatal life directly to rodent pups. Neonatal estrogen treatment can masculinize sexually dimorphic behaviors (Gray et al., 1985; 1982), and alter testis and sex accessory gland development, but treatment during this period of life is too late to induce hypospadias and other reproductive tract malformations.

Several studies which administered estrogenic pesticides or toxic substances orally, by gavage, in the diet, or in drinking water using multigenerational protocols also have failed to detect malformations including hypospadias in the male offspring. In these chronic studies, estrogenic effects were observed with estradiol (Biegel et al., 1998), DES (Odum et al., 2002), methoxychlor (Gray et al., 1989; Chapin et al., 1997; Staub et. al., 2002; You et al., 2002), nonylphenol (Hossaini, et al., 2001; de Jager et al., 1999; Chapin et al., 1999; Nagao et al., 2001), genistein (Roberts et al., 2000; Delclos et al., 2001) but not with octylphenol (Tyl et al., 1999) or bisphenol A (Tyl et al., 2002; Kwon et al., 2000; Ema et al., 2001).

While the weight-of-evidence from these multigenerational studies would suggest that the weakly estrogenic toxicants do not induce reproductive tract malformations or hypospadias in rodents when administered to the dam throughout breeding, gestation and lactation, such a broad conclusion is tempered by the observation that oral administration of 17 beta estradiol or DES, a known human and animal teratogen in such multigenerational studies failed to detect reproductive tract malformations in the offspring (Biegel et al., 1998; Odum et al., 2002). In contrast, when given to the dam at higher dosage levels for a few days late in pregnancy, DES and estradiol do induce malformations, including hypospadias in male offspring.

Odum et al., (2002) commented that they did not establish a unique role for exposures in utero or during the early neonatal period to DES. When administered chronically, estradiol and DES interfere with reproductive function in male and female rats at low doses which preclude that administration of doses that induce frank reproductive terata. On the other hand, short-term administration to the dam during sexual differentiation of the offspring can induce malformations without inducing pregnancy loss and infertility.

Taken together, the above results with estradiol and DES cause one to wonder if any of the weaker xenoestrogens also could induce reproductive tract malformations in the offspring if administered at high dosage levels only during the sensitive period of sexual differentiation. If such responses were detected they could impact on the acute risk assessments on the xenoestrogens, while the multigenerational results would likely be more relevant for chronic risk assessments.

15. TCDD and PCBs: Putative Antiandrogens

Ah receptor agonists such as 2,3,7,8-tetrachlorodibenzo-p-dioxin (TCDD), PCBs and polychlorinated dibenzofurans (PCDFs) induce developmental toxicity in humans, nonhuman primates, rodents, fish, mink, and other wildlife species (Golub et al., 1991); effects are expressed at ppt concentrations. 2,3,7,8 TCDD, and other toxicants bind to a cellular steroid hormone-like receptor, termed the Ah receptor, and alters many hormone levels, growth factors, and/or their receptors, as well as hormone synthesis (Birnbaum, 1994) but do not bind the androgen receptor.   In utero exposure to very low doses of 2,3,7,8 TCDD produces infertility in rodent progeny (Khera and Ruddick, 1973; Murray et al., 1979; Gray et al., 1995; Mably et al., 1992 a,b,c; Bjerke et al., 1994 a,b,c). A single dose of TCDD ranging from 50 ng/kg to 2 g/kg during sex differentiation of the rat or hamster results in a number of reproductive alterations in male and female progeny (Wolf et al., 1999; Gray and Ostby, 1995; Gray et al. 1997 a,b). The alterations include delayed puberty, reduced ejaculated and epididymal sperm numbers, and reductions in the size of the ventral prostate, seminal vesicle, and testis.

TCDD-treated males in the above studies did not display hypospadias, reduced AGD or areolas, although epididymal agenesis is induced when the dams are treated with a higher dosage level of TCDD (3 µg/kg on GD 15, Wilker et al. 1996). In female offspring, in utero treatment with TCDD induced clefting of the clitoris with a mild degree of hypospadias in females and a permanent "thread" of tissue across the opening of the vagina of the progeny(Gray et al., 1997b). Female progeny, treated earlier in gestation (day 8) with TCDD displayed reduced fecundity, a high incidence of constant estrus, and cystic endometrial hyperplasia at middle-age. Female hamster offspring, from mothers treated with 2 g/kg TCDD on day 11 of gestation, also display clitoral clefting and reduced fertility as a result of several functional reproductive problems.

Additional studies with rats indicated that development of the reproductive system is severely altered when fetal rat TCDD concentrations reach 50 ppt (Hurst et al., 1996; 1997). The two lowest dosages (0.2 and 0.05 µg/kg on GD 15) produced fetal concentrations of 5 and 13 ppt, which lowered sperm counts in male and induced reproductive tract anomalies in female offspring (Gray et al., 1997a,b). The PCB congener 169 is an Ah receptor agonist which is about 0.001 times the potency of TCDD. PCB 169 treatment at 1.8 mg/kg on GD 8 alters reproductive development of LE hooded male and female rats in a manner nearly identical to TCDD (Gray et al., 1999a).

Although the overall profile of effects in the rats exposed in utero to these Ah receptor agonists bears some resemblance to that seen in animals exposed to antiandrogens like vinclozolin or the phthalates, the fact that Ah agonists fail to reduce AGD or induce areolas, retained nipples, or male hypospadias suggests that TCDD and the dioxin-like PCBs may be affecting the developing reproductive tract via alternative pathways from the AR.

16. Cumulative effects of toxicants on the induction of hypospadias

Within the last decade,   several classes of pesticides and toxic substances have been shown to disrupt differentiation of the male rat reproductive tract by interfering with the synthesis or action of fetal Leydig cell steroid and peptide hormones. As discussed in detail earlier, vinclozolin and procymidone are pesticides that induce malformations by acting as AR antagonists, whereas, some phthalate esters induce malformations by inhibiting fetal Leydig cell hormone production.

The pesticides linuron (L) and prochloraz act both as AR antagonists and by inhibiting steroid hormone production. Typically, the potential risk of these chemicals to humans and wildlife is assessed on a chemical-by-chemical basis without consideration of concurrent exposures to other endocrine disrupting chemicals (EDCs). However, in 1996 the Food Quality Protection Act mandated that the US Environmental Protection Agency (EPA) consider the cumulative risk from multiple chemical exposures that acted via a common mechanism. EPA is now in the process of evaluating how it is going to assess the cumulative effects of exposure to antiandrogenic pesticides.

The studies described herein begin to provide a framework for the Agency in this endeavor. We are executing relatively simple mixture studies to demonstrate what types of EDCs produce cumulative reproductive toxicity when administered during sexual differentiation of the male reproductive tract. In these studies, pairs of "antiandrogenic" chemicals were coadministered on days 14-18 of gestation with one another or individually using sub- or near threshold dosage levels equivalent to about 1/2 the ED50 for causing hypospadias (Gray et al., 2001; 2003).

Studies combined 1). two AR antagonists vinclozolin plus procymidone; each at 50 mg/kg/day), 2). two phthalate esters DBP and BBP, each at 500 mg/kg/day), 3). a phthalate ester plus an AR antagonist (DBP (500 mg/kg/day) plus procymidone(50 mg/kg/day)) and 4). linuron (75 mg/kg/day) plus BBP (500 mg/kg/day). Dose response studies have shown that the relative potency for induction of hypospadias of DBP and BBP is about ten-fold less than than of vinclozolin or procymidone (relative potency factors (RPF) for hypospadias).

We expected that by themselves vinclozolin, procymidone, linuron, DBP and BBP would not induce hypospadias or any other malformations at these dose levels when given on GD 14-18, but, if the reproductive toxicity was cumulative, mixing two together would be equivalent to doubling the dose of one toxicant and this "dose" would induce hypospadias in about 50% of the males. Reproductive organ weights and malformations of other tissues should also display cumulative responses to the mixtures. All of the results to date indicate that the effects are dose-additive but they are not synergistic. This would indicate that the "antiandrogens" included in the "common mechanism group" to be evaluated in a cumulative risk assessment should be broadly rather than narrowly defined to include both AR antagonists and inhibitors of fetal hormone production.

We also found that the relative potency factor (RPF) of each chemical varied predictably from endpoint to endpoint, based upon the specific mechanism of action. In contrast to the RPFs for hypospadias, where vinclozolin = procymidone = 1 and DBP = BBP = 0.1, the RPFs for gubernacular agenesis are DBP = BBP = 1 whereas linuron, vinclozolin and procymidone = 0. The RPFs vary from tissue to tissue because all the above toxicants disrupt the androgen pathway but only the phthalate esters inhibit Leydig cell insl3 synthesis. Gubernacular agenesis was displayed in the BBP+DBP group, but not in the L+BBP or DBP+P groups. Our results indicate that chemicals that alter differentiation of the reproductive system during the same period of development will produce cumulative, dose-additive effects. Our ability to predict these effects is enhanced by specific knowledge about the mechanism of toxicity of each chemical.

REFERENCES

Andersen, H. R., Vinggaard, A. M., Rasmussen, T. H., Gjermandsen, I. M. and Bonefeld-Jorgensen, E. C., 2002. Effects of currently used pesticides in assays for estrogenicity, androgenicity, and aromatase activity in vitro, *Toxicol Appl Pharmacol.***179**: 1-12.

Biegel, L.B., Flaws, J.A., Hirshfield, A.N., O'Connor, J.C., Elliott, G.S., Ladics, G.S., Silbergeld, E.K., Van Pelt, C.S., Hurtt, M.E., Cook, J.C., Frame, S.R., 1998. 90-day feeding and one-generation reproduction study in Crl:CD BR rats with 17 beta-estradiol, *Toxicol Sci.* **44(2)**:116-42.

Birnbaum, L. S., 1994. Endocrine effects of prenatal exposure to PCBs, Dioxins, and other Xenobiotics: Implications for Policy and Research, *Environ. Health Perspectives.* **102**: 676-679.

Bjerke, D.L., Brown, T.J., MacLusky, N.J, Hochberg, R.B., and Peterson, R. E. ,1994a. Partial demasculinization and feminization of sex behavior in male rats by in utero and lactational exposure of male rats to 2,3,7,8 tetrachlorodibenzo-p-dioxin (TCDD) is not associated with alterations in estrogen receptor binding or volumes of sexually dimorphic brain nuclei, *Toxicol Appl Pharm.* **127**: 258-267.

Bjerke, D.L., Sommer, R.J., Moore, R.W., and Peterson, R.E ,1994b. Effects of in utero and lactational exposure on responsiveness of the male rat reproductive system to testosterone stimulation in adulthood, *Toxicol Appl Pharm.* **127**: 250-257.

Bjerke, D.L. and Peterson, R. E., 1994c. Reproductive toxicity of 2,3,7,8-tetrachlorodibenzo-p-dioxin in male rats: Different effects of in utero versus lactational exposure, *Toxicol Appl Pharm.* **127**: 241-249.

Chapin, R.E., Harris, M.W., Davis, B.J., Ward, S.M., Wilson, R.E., Mauney, M.A., Lockhart, A.C., Smialowicz, R.J., Moser, V.C., Burka, L.T., Collins, B.J., 1997. The effects of perinatal/juvenile methoxychlor exposure on adult rat nervous, immune, and reproductive system function, *Fundam Appl Toxicol.* **40(1)**: 138-57.

Chapin, R.E., Delaney, J., Wang, Y., Lanning, L., Davis, B., Collins, B., Mintz, N., Wolfe, G,. 1999. The effects of 4-nonylphenol in rats: a multigeneration reproduction study, *Toxicol Sci.* **52(1)**:80-91.

Clark, R.L., C.A. Anderson, S. Prahalada, Y. M. Leonard, J. L. Stevens and A.M. Hoberman., 1990. 5-alpha reductase inhibitor-induced congenital abnormalities in male rat external genitalia, *Teratology.* 41,5, A 544.

Clark, R.L., Anderson, C.A., Prahalada, S., Robertson, R.T., Lochry, E.A., Leonard, Y.M., Stevens, J.L., Hoberman, A.M., 1993. Critical developmental periods for effects on male rat genitalia induced by finasteride, a 5 alpha-reductase inhibitor, *Toxicol Appl Pharmacol.* **119(1)**: 34-40.

Cook, J.C., Mullin, L.S., Frame, S.R., Biegel, L.B., 1993. Investigation of a mechanism for Leydig cell tumorigenesis by linuron in rats, *Toxicol Appl Pharmacol.* **119**: 195-204.

Delclos, K.B., Bucci, T.J., Lomax, L.G., Latendresse, J.R., Warbritton, A., Weis, C.C., Newbold, R.R., 2001. Effects of dietary genistein exposure during development on male and female CD (Sprague-Dawley) rats, *Reprod Toxicol.* **15(6)**: 647-63.

de Jager, C., Bornman, M.S., Oosthuizen, J.M., 1999. The effect of p-nonylphenol on the fertility potential of male rats after gestational, lactational and direct exposure, *Andrologia.* **31(2)**: 107-13.

Ema, M., Fujii, S., Furukawa, M., Kiguchi, M., Ikka, T., Harazono, A., 2001. Rat two-generation reproductive toxicity study of bisphenol A, *Reprod Toxicol.* **15(5)**: 505-23.

Goldman, A.S., Eavey, R.D., Baker, M.K., 1976. Production of male pseudohermaphroditism in rats by two new inhibitors of steroid 17alpha-hydroxylase and C 17-20 lyase, *Endocrinol.* **71(3)**, 289-97.

Gray, L.E.,Jr., 1982. Neonatal chlordecone exposure alters behavioral sex differentiation in female hamsters, *Neurotoxicology.* **3(2)**: 67-80.

Gray, L.E.Jr., Ferrell, J.M., and Ostby, J.S., 1985. Alteration of behavioral sex differentiation by exposure to estrogenic compounds during a critical neonatal period: Effects of Zearalenone, Methoxychlor, and Estradiol in hamsters, *Toxicology and Applied Pharmacology.* **80 (1)**: 127-136.

Gray, L.E. Jr., Ostby, J., Ferrell, J., Rehnberg, G., Linder, R., Cooper, R., Goldman, J., Slott, V., Laskey, J, 1989. A dose-response analysis of methoxychlor-induced alterations of reproductive development and function in the rat, *Fundam Appl Toxicol.* **12(1)**: 92-108.

Gray, L.E., Kelce, W.R., Monosson, E., Ostby, J.S., and Birnbaum, L.S.,1995. Exposure to TCDD during development permanently alters reproductive function in male LE rats and Hamsters: Reduced ejaculated and epididymal sperm numbers and sex accessory gland weights in offspring with normal androgenic status, *Toxicol and Appl Pharmacol.* **131**: 108-118.

Gray, L.E. Jr., and Ostby, J.S., 1995. In utero 2,3,7,8 Tetrachlorodibenzo-p-dioxin(TCDD) Alters Reproductive Morphology and Function in Female Rat Offspring, *Toxicol Appl Pharm.* **133**: 285-294.

Gray, L.E. Jr, Ostby, J.S., Kelce, W.R., 1997a. A dose-response analysis of the reproductive effects of a single gestational dose of 2,3,7,8 tetrachlorodibenzo-p-dioxin (TCDD) in male Long Evans hooded rat offspring, *Toxicol Appl Pharmacol.* **146**: 11-20.

Gray, L.E., Wolf, C., Mann, P., Ostby, J.S., 1997b. In utero exposure to low doses of 2,3,7,8-tetrachlorodibenzo-p-dioxin alters reproductive development of female Long Evans hooded rat offspring, *Toxicol Appl Pharmacol.* **146(2)**: 237-244.

Gray, L.E. Jr and Ostby, J., 1998. Effects of pesticides and toxic substances on behavioral and morphological reproductive development: Endocrine versus nonendocrine mechanisms, *Toxicol Industrial Health.* **14**: 159-184.

Gray, L.E. Jr, Ostby, J., Monosson, E., Kelce, W.R., 1999a. Environmental antiandrogens: low doses of the fungicide vinclozolin alter sexual differentiation of the male rat, *Toxicol Ind Health* **15 (1-2)**: 48-64.

Gray, L.E. Jr., Wolf, C., Lambright, C., Mann, P., Price, M., Cooper, R.L., Ostby, J.,1999b. Administration of potentially antiandrogenic pesticides (procymidone, linuron, iprodione, chlozolinate, p,p'-DDE, and ketoconazole) and toxic substances (dibutyl- and diethylhexyl phthalate, PCB 169, and ethane dimethane sulphonate) during sexual differentiation produces diverse profiles of reproductive malformations in the male rat, *J Toxicol Ind Health.* **15(1-2)**: 94-118.

Gray, L.E. Jr, Ostby, J., Furr, J., Price, M., Veeramachaneni, D.N.R., Parks, L., 2000. Perinatal exposure to the phthalates DEHP, BBP and DINP but not DEP, DMP or DOTP alters sexual differentiation of the male rat, *Toxicol Sci.* **58**: 350-365.

Gray, L.E. Jr, Ostby, J., Furr, J., Wolf, C.J., Lambright, C., Parks, L., Veeramachaneni, D.N.R., Wilson, V., Price, M., Hotchkiss, A., Orlando, E., Guillette, L., 2001. Effects of environmental antiandrogens on reproductive development in experimental animals, *Human Reproduction Update.* **7(3)**: 248-64.

Gray, L.E. Jr., Ostby, J., Furr, J., Wolf, C., Hotchkiss, A., Parks, L., Price, M., Lambright, C. and Wilson, V., 2003. Cumulative Developmental Effects of Endocrine Disrupters: Synergy or Additivity? *Biology of Reproduction, Suppl.*

Guillette, L., Brock, J., Rooney, A., Woodward, A., 1999. Serum concentrations of various environmental contaminants and their relationship to sex steroid concentrations and phallus size in juvenile male alligators, *Arch Environ Contam Toxicol.* **36**: 447-455.

Henderson, B.E., Benton, B., Cosgrove, M., Baptista, J., Aldrich, J., Townsend, D., Hart, W., Mack, T.M., 1976. Urogenital tract abnormalities in sons of women treated with diethylstilbestrol, *Pediatrics.* **58(4)**: 505-7.

Hirsch, K.S.,m Adams, E.R., Hoffman, D.G., Markham, J.K., and Owen, N.V., 1987. Studies to elucidate the mechanism of fenarimol-induced infertility in the male rat, *Toxicology and Applied Pharmacology.* **86**: 391-399.

Hossaini, A., Dalgaard, M., Vinggaard, A.M., Frandsen, H., Larsen, J.J., 2001. In utero reproductive study in rats exposed to nonylphenol, *Reprod Toxicol.* **15(5)**: 537-43.

Hurst, C., DeVito, M., Abbott, B., and Birnbaum L., 1996. 2,3,7,8 tetrachlorodibenzo-p-dioxin (TCDD) in pregnant rats: Distribution to maternal and fetal tissues, *The Toxicologist.* **30**: 198.

Hurst, C., DeVito, M., Abbott, B., and Birnbaum, L., 1997. Dose-response of (TCDD) distribution in pregnant Long Evans rats, *The Toxicologist.* **36**: 257.

Husmann, D.A., McPhaul, M.J., 1991. Time-specific androgen blockade with flutamide inhibits testicular descent in the rat, *Endocrinology.* **129(3)**: 1409-16.

Imperato-McGinley, J., Sanchez, R.S., Spencer, J.R., Yee, B. and Vaughan, E.D., 1992. Comparison of the effects of the 5 alpha-reductase inhibitor finasteride and the antiandrogen flutamide on prostate and genital differentiation: Dose-response studies, *Endocrinology.* **131**: 1149-1156.

Kelce, W.R., Monosson, E., Gamcsik, M.P., Laws, S., Gray, L.E. Jr., 1994. Environmental hormone disruptors: Evidence that vinclozolin developmental toxicity is mediated by antiandrogenic metabolites, *Toxicol Appl Pharmacol.* **126**: 275-285.

Kelce, W.R., Stone, C.R., Laws, S.C., Gray, L.E., Kemppainen, J.A., Wilson, E.M., 1995. The persistent DDT metabolite p,p' DDE is a potent androgen receptor antagonist, *Nature.* **375**: 581-585.

Kelce W., Lambright, C., Gray, L. and Roberts, K., 1997. Vinclozolin and p,p' DDE alter androgen-dependent gene expression: In vivo confirmation of an androgen receptor mediated mechanism, *Toxicol Appl Pharm.* **142**: 192-200.

Khera, K.S. and Ruddick, J.A., 1973. Polychlorodibenzo-p-dioxins: Perinatal effects and the dominant lethal test in Wistar rats. In: Chlorodioxins-origin and fate (E.H. Blair, Ed.) pp. 70-84. *American. Chem. Soc.*, Washington, D.C.

Klip, H., Verloop, J., van Gool, J.D., Koster, M.E., Burger, C.W., van Leeuwen, F.E., 2002. Hypospadias in sons of women exposed to diethylstilbestrol in utero: a cohort study, *Lancet.* 2002 Mar 30;**359(9312)**:1081-2.

Kwon, S., Stedman, D.B., Elswick, B.A., Cattley, R.C., Welsch, F., 2000. Pubertal development and reproductive functions of Crl:CD BR Sprague-Dawley rats exposed to bisphenol A during prenatal and postnatal development, *Toxicol Sci.* **55(2)**: 399-406.

Laignelet, L., Narbonne, J. F., Lhuguenot, J. C. and Riviere, J. L., 1989. Induction and inhibition of rat liver cytochrome(s) P-450 by an imidazole fungicide (prochloraz), *Toxicology.* **59**: 271-84.

Lambright, C., Ostby, J., Bobseine, K., Wilson, V., Hotchkiss, A.K. and Gray, L.E. Jr., 2000. Cellular and molecular mechanisms of action of linuron: An antiandrogenic herbicide that produces reproductive malformations in male rats, *Toxicology Sciences.* **56:** 389-399.

Lambright, C., Wilson, V., Furr, J., Wolf, C., Noriega, N. and Gray, L.E., 2003. Effects of endocrine disrupting chemicals on fetal testes hormone production, *The Toxicologist* **72:** S1, A 1323 p 272.

Laws, S.C., Carey, S.A., Kelce, W.R., Cooper, R.L., Gray, L.E., 1996. Vinclozolin does not alter progesterone receptor (PR) function in vivo despite inhibition of PR binding by its metabolites in vitro, *Toxicology.* **112(3):** 173-82.

Mably T., Moore. R.W., Peterson, R.E., 1992a. In utero and lactational exposure of male rats to 2,3,7,8-tetrachlorodibenzo-p-dioxin. 1. Effects on Androgenic status, *Toxicol Appl Pharmacol.* **114:** 97-107.

Mably, T., Bjerke, D.L., Moore, R.W., Gendron-Fitzpatrick, A., Peterson, R.E., 1992c. In utero and lactational exposure of male rats to 2,3,7,8-tetrachlorodibenzo-p-dioxin. 3. Effects on spermatogenesis and reproductive capability, *Toxicol Appl Pharmacol.* **114:** 118-126.

Mably, T., Moore, R.W., Goy, R.W, Peterson, R.E., 1992b. In utero and lactational exposure of male rats to 2,3,7,8-tetrachlorodibenzo-p-dioxin. 2. Effects on sexual behavior and the regulation of luteinizing hormone secretion in adulthood, *Toxicol Appl Pharmacol.* **114:** 108-117.

McIntyre, B.S., Barlow, N.J., Wallace, D.G., Maness, S.C., Gaido, K.W., Foster, P.M., 2000. Effects of in utero exposure to linuron on androgen-dependent reproductive development in the male Crl:CD(SD)BR rat, *Toxicol Appl Pharmacol.* 1;**167(2):** 87-99.

McIntyre, B.S, Barlow, N.J., Foster, P.M., 2001. Androgen-mediated development in male rat offspring exposed to flutamide in utero: permanence and correlation of early postnatal changes in anogenital distance and nipple retention with malformations in androgen-dependent tissues, *Toxicol Sci.* **62(2):** 236-249.

McLachlan, J.A., Newbold, R.R., Bullock, B., 1975. Reproductive tract lesions in male mice exposed prenatally to diethylstilbestrol, *Science.* **190(4218):** 991-2.

McLachlan, J.A., Newbold, R.R., Burow, M.E., Li, S.F., 2001. From malformations to molecular mechanisms in the male: three decades of research on endocrine disrupters, *APMIS.* **109(4):** 263-72.

Murray, F.J., Smith, F.A., Nitschke, K.D., Humiston, C.G., Kociba, R.J., and Schwetz, B.A., 1979. Three-generation reproduction study in rats given 2,3,7,8-tetrachlorodibenzo-p-dioxin (TCDD) in the diet, *Toxicol Appl Pharmacol.* **50:** 241-252.

Mylchreest, E., Cattley, R., Foster, P.M.D., 1998. Male reproductive tract malformations in rats following gestational and lactational exposure to DBP: An antiandrogenic mechanism, *Toxicol Sciences.* **43:** 47-60.

Mylchreest, E., Sar, M., Cattley, R.C. and Foster, P.M., 1999. Disruption of androgen-regulated male reproductive development by DBP during late gestation in rats is different from flutamide, *Toxicol Appl Pharmacol.* **1556(2):** 81-95.

Nagao, T., Wada, K., Marumo, H., Yoshimura, S., Ono, H., 2001. Reproductive effects of nonylphenol in rats after gavage administration: a two-generation study, *Reprod Toxicol.* **15(3):** 293-315.

Noriega, N., Gray, L.E., Ostby, J., Lambright, C., and Wilson, V., 2003. Prenatal exposure to the fungicide prochloraz alters the onset of parturition in the dam and sexual differentiation in male rat offspring, *The Toxicologist.* **72:** S1, A 1375, p 283.

North, K., Golding, J., 2000. A maternal vegetarian diet in pregnancy is associated with hypospadias. The ALSPAC Study Team. Avon Longitudinal Study of Pregnancy and Childhood., *BJU Int.* **85(1):** 107-13.

Odum, J., Lefevre, P.A., Tinwell, H., Van Miller, J.P., Joiner, R.L., Chapin, R.E., Wallis, N.T., Ashby, J., 2002. Comparison of the developmental and reproductive toxicity of diethylstilbestrol administered to rats in utero, lactationally, preweaning, or postweaning, *Toxicol Sci.* **68(1):** 147-63.

Ostby, J., Kelce, W.R., Lambright, C., Wolf, C.J., Mann, P., Gray, L.E. Jr., 1999. The fungicide procymidone alters sexual differentiation in the male rat by acting as an androgen-receptor antagonist in vivo and in vitro, *Toxicol. Ind. Health.* **15:** 80-93.

Parks, L.G., Ostby, J.S., Lambright, C.R., Abbott, B.D., Klinefelter, G.R, Barlow, N.J., Gray, L.E. Jr., 2000. The plasticizer diethylhexyl phthalate induces malformations by decreasing fetal testosterone synthesis during sexual differentiation in the male rat, *Toxicol. Sci.* **58:** 339-349.

Pepper, G., Brenner, S., Gabrilove, J., 1990. Ketoconazole use in the treatment of ovarian hyperandrogenism, *Fertil Steril.* **54:** 38-444.

Roberts, D., Veeramachaneni, D.N., Schlaff, W.D., Awoniyi, C.A., 2000. Effects of chronic dietary exposure to genistein, a phytoestrogen, during various stages of development on reproductive hormones and spermatogenesis in rats, *Endocrine.* **13(3):** 281-6.

Rosen, M., Wilson, V., Schmid, J. and Gray, L.E., 2003. Gene array analysis of the ventral prostate in rats exposed to either vinclozolin or procymidone, *The Toxicologist.* **72:** S1 A454, p 94.

Schurmeyer, T., Nieschlag, E., 1984. Effect of ketoconazole and other imidazole fungicides on testosterone biosynthesis, *Acta Endocrinologica.* **105:** 275-280.

Sharpe, R.M., 2003. The 'oestrogen hypothesis'- where do we stand now? *Int J Androl.* **26(1):** 2-15.

Staub, C., Hardy, V.B., Chapin, R.E., Harris, M.W., Johnson, L., 2002. The hidden effect of estrogenic/antiandrogenic methoxychlor on spermatogenesis, *Toxicol Appl Pharmacol.* **180(2):** 129-35.

Tyl, R.W., Myers, C.B., Marr, M.C., Thomas, B.F., Keimowitz, A.R., Brine, D.R., Veselica, M.M., Fail, P.A., Chang, T.Y., Seely, J.C., Joiner, R.L., Butala, J.H., Dimond, S.S., Cagen, S.Z, Shiotsuka, R.N., Stropp, G.D., Waechter, J.M., 2002. Three-generation reproductive toxicity study of dietary bisphenol A in CD Sprague-Dawley rats, *Toxicol Sci.* **68(1):** 121-46.

Tyl, R.W., Myers, C.B., Marr, M.C., Brine, D.R., Fail, P.A., Seely, J.C., Van Miller, J.P., 1999. Two-generation reproduction study with para-tert-octylphenol in rats, *Regul Toxicol Pharmacol.* **30(2 Pt 1):** 81-95.

Vannier, B. and Raynaud, J.P., 1980. Long-term effects of prenatal oestrogen treatment on genital morphology and reproductive function in the rat, *Journal of Reproduction and Fertility.* **59:** 43-49.

Vinggaard, A. M., Hnida, C., Breinholt, V. and Larsen, J. C., 2000. Screening of selected pesticides for inhibition of CYP19 aromatase activity in vitro, *Toxicol In Vitro.* **14:** 227-34.

Vinggaard, A. M., Nellemann, C., Dalgaard, M., Jorgensen, E. B. and Andersen, H. R., 2002. Antiandrogenic effects in vitro and in vivo of the fungicide prochloraz, *Toxicol Sci.* **69:** 344-53.

Waller, C.L., Oprea, T.I., Chae, K., Park, H.K., Korach, K.S., Laws, S.C., Wiese, T.E., Kelce, W.R., Gray, L.E. Jr., 1996a. Ligand-based identification of environmental estrogens, *Chem Res Toxicol.* **9(8):** 1240-8.

Waller, C, Juma, B, Gray, L. E., and Kelce, W., 1996b. Three dimensional Quantitative structure activity relationships for androgen receptor ligands, *Toxicol Appl Pharm.* **137:** 219-227.

Weir, P.J., Conner, M. and Johnson, C.M., 1990. Abnormal development of male urogenital sinus derivatives produced by a 5alpha-reductase inhibitor, *Teratology.* **41:** 5, A 599.

Williams, T., 1999. Lessons from Lake Apopka,. *Audubon.* **101(4):** 64-72.

Wilker, C., Johnson, L. and Safe, S., 1996. Effects of developmental exposure to indole-3-carbinol or 2,3,7,8-tetrachlorodibenzo-p-dioxin on reproductive potential of male rat offspring, *Toxicol Appl Pharmacol.* **141:** 68-75.

Wilson, V.S. and LeBlanc, G.A., 2000. The contribution of hepatic inactivation of testosterone to the lowering of serum testosterone levels by ketoconazole, *Toxicol Sci.* **54:** 125-137.

Wilson, V.S., Wood, C., Held, G., Lambright, C., Ostby, J., Furr, J. and Gray, L,E., 2003. Comparison of the effects of two AR antagonnists on tissue weights and hormone levels in male rats on expression of three androgen-dependent genes in the ventral prostate, *The Toxicologist.* **72:** S1, A639 p 131.

Wilson, V.S., Lambright, C., Furr, J., Ostby, J., Wood, C., Held, G., Gray, L.E. Jr. (in prep). Phthalate ester-induced gubernacular ligament lesions are associated with reduced Insl3 gene expression in the fetal rat testis during sexual differentiation.

Wolf, C.J., Ostby, J.S., Gray, L.E. Jr., 1999. Gestational exposure to 2,3,7,8-tetrachlorodibenzo-p-dioxin (TCDD) severely alters reproductive function of female hamster offspring, *Toxicol Sci.* **51(2):** 259-64.

Wolf, C.J., LeBlanc, G.A., Ostby, J.S., Gray, L.E. Jr., 2000. Characterization of the period of sensitivity of fetal male sexual development to vinclozolin, *Toxicol Sci.* **55(1):** 152-61.

You, L., Casanova, M., Archibeque-Engle, S., Sar, M., Fan, L., d'A Heck, H., 1998. Impaired male sexual development in perinatal Sprague-Dawley and Longs-Evans Hooded rats exposed in utero and lactationally to p,p'-DDE, *Toxicol Sciences.* **45:** 162-173.

You, L., Brenneman, K.A., Heck, H., 1999a In utero exposure to antiandrogens alters the responsiveness of the prostate to p,p'-DDE in adult rats and may induce prostatic inflammation, *Toxicol Appl Pharmacol.* **161(3):** 258-66.

You, L., Gazi, E., Archibewque-Engle, S., Casonova, M., Conolly and d' A Heck, H. 1999b. Transplacental and lactational transfer ofg p,p' DDE in Sprague-Dawley rats, *Toxicol Appl Pharm* **157:** 134-144.

You, L., Casanova ,M, Bartolucci, E.J., Fryczynski, M.W., Dorman ,D.C., Everitt, J.I., Gaido, K.W., Ross, S.M., Heck, H., 2002. Combined effects of dietary phytoestrogen and synthetic endocrine-active compound on reproductive development in Sprague-Dawley rats: genistein and methoxychlor, *Toxicol Sci.* **66(1):** 91-104.

MASCULINIZATION OF FEMALE MAMMALS: LESSONS FROM NATURE

Ned J. Place* and Stephen E. Glickman
Department of Psychology
University of California, Berkeley 94720-1650

1. INTRODUCTION

Conventional understanding of mammalian sexual differentiation, first proposed by Alfred Jost in the late 1940's and early 1950's, requires the presence of androgens acting during a critical stage of fetal (or neonatal) life for normal development of the masculine penis and scrotum from embryonic primordia (Jost, 1953; Wilson et al., 1981). Postnatally, a marked acceleration in growth of the penis is typically associated with increased secretion of androgen during puberty (at least in primates).

But, the preceding generalizations are based on the study of a relatively small number of species. There is a substantial cluster of mammal species in which all females display some masculinization of the external genitalia. In many ways, this natural masculinization of the female external genitalia is reminiscent of several human pathological conditions. The degree of virilization varies across species much as it does in several disorders of sexual development. For example, a cohort of girls born with congenital adrenal hyperplasia (CAH), owing to a deficiency in the steroidogenic enzyme 21-hydroxyase, may present with ambiguous genitalia that can vary from essentially feminine to almost completely masculine. Across species we witness the same degree of variability, from the isolated clitoromegaly of bonobos and spider monkeys (Wislocki, 1936), to the hypospadic penile clitoris of various lemurs (Hill, 1958), to the extreme masculinization of female spotted hyenas (Matthews, 1939). This continuum of naturally masculinized female mammals may represent a set of useful animals models, whereby studies of females may provide insights into the ontogeny of male disorders of sexual differentiation.

The existence of such "masculine" clitoral development could be reconciled with the conventional theory, if: (a) androgens were found to circulate during fetal life with the requisite timing, and (b) interruption of androgenic activity during such critical periods resulted in a clitoris of reduced size, that was not traversed by a urethral canal. Moreover, since females in all of the preceding species, save for the spotted hyena, retain an external vaginal opening (not a scrotum), delicacy of timing for circulating androgens, or some

---

*Spotted Hyena Project, Department of Psychology, University of California, Berkeley, 3210 Tolman #1650, Berkeley, CA 94720-1650, Ph (510) 643-5154, Fax (510) 642-8321, e: mail , ned@socrates.berkeley.edu

Hypospadias and Genital Development, edited by
L. Baskin, Kluwer Academic/Plenum Publishers, 2004

odd distribution of androgen receptors may be required to account for selective "masculinization." To date, there is no direct evidence bearing on clitoral development in species characterized by natural masculinization. Although that can be attributed to gaps in the research literature, it is also possible that novel non-androgenic mechanisms are at work. The latter will not be found if they are not sought.

The present chapter is focused on two examples of natural masculinization in female mammals: (1) moles, which display a penile clitoris, but retain a distal vagina separate from the urethra, and (2) the spotted hyena, which is the only extant mammalian species with both a penile clitoris and a pseudoscrotum, but no distal vagina. Development of the female external genitalia in eutherian mammals has generally been considered a passive event, occurring by default in the absence of potent androgens, specifically 5α–dihydrotestosterone (DHT). A corollary to this assumption is that any masculinization of the female genitalia must require an alternate source of androgen. In each case, our research trail begins with a search for naturally circulating androgens in a female mammal. Our observations in moles and spotted hyenas raise important questions with regards to the conventional theory of sexual differentiation.

2. MOLES

2.1. Background

The layman rarely cares if the mole destroying his garden or lawn is male or female, and he probably could not tell if he did. Most, if not all species of mole belonging to the mammalian family Talpidae are difficult to sex, especially outside of the breeding season. In several of the species studied to date, females have a peniform clitoris, which is traversed by the urethra through most or all of its length. The vagina is also completely occluded prior to sexual maturity and during the nonbreeding season. In males, testes are intra-abdominal throughout the year, thus depriving individuals of another important clue for sexing moles. Scientific investigators frequently rely on internal inspection of the reproductive organs to definitively sex moles. Upon such inspections, some interesting findings have been described that add further to this unusual story.

The European mole (*Talpa europaea*) and several other related species in the genus *Talpa* are remarkable in that all females carry "ovotestes", bipolar gonads that display discrete features of both ovaries and testes. This nomenclature has been considered by some to be inaccurate, as the "ovotestes" do not exhibit any spermatogenic function, thus warranting the simple designation "ovaries" (Beolchini et al., 2000. However, these ovaries are unusual and do contain a testis-like region with interstitial tissue and interspersed medullary cords (Jiménez et al., 1993; Sanchez et al., 1996). Ovarian follicles are limited to a distinct pole in the gonad, while the ovarian interstitial gland (OIG) is found at the opposite pole. Additionally, the two poles vary seasonally in size and endocrine activity (Matthews, 1935; Deansesly, 1966; Whitworth et al., 1999). The follicular pole enlarges at sexual maturity and at the beginning of subsequent breeding seasons, while the interstitial gland regresses; the arrangement reverses at the end of the breeding season. Testosterone, principally secreted from the OIG, increases in females outside of the breeding season to levels that are comparable to that of males (Whitworth et al., 1999). Interestingly, this peculiar gonadal arrangement is present at birth (Gorman and Stone, 1990), and is found in pre- and post-pubertal females (Jiménez et al., 1996). The timing of gonadal differentiation in *Talpa* has not been well studied, but Matthews (1935) noted differentiation of both regions of the female gonad *in utero*. Matthews also noted the abnormal development of epididymides adjacent to the OIG, which suggested to him that testosterone is secreted by the fetal ovary, and may provide a mechanism for masculinization of the female's external genitalia.

2.2. Comparative Studies

However, there is considerable variation in ovarian morphology among mole species not belonging to the genus *Talpa*. Moles of North America can also be very difficult to sex, owing to the same problems described above for the Old World moles. Mossman and Duke (1973) tabulated and described the ovarian morphology of many mammal species, including several from the family Talpidae, which includes both the Old and New World moles. Depending on the species, the latter may have bipolar ovaries very similar to those of the *Talpa* spp. (e.g. star-nosed moles, *Condylura cristata*), conventional ovaries lacking an OIG (e.g. eastern moles, *Scalopus aquaticus*), or ovaries that are intermediate between the two extremes (e.g. shrew moles, *Neurotrichus gibbsii*). Given these findings, Rubenstein et al. (2004) recently set out to determine if the masculinization of female genitalia in several species of North American moles is consistently correlated with ovarian morphology, i.e. is a peniform clitoris ever found in species lacking an OIG. If so, an alternate source of androgen, or an alternate mechanism may be required to explain female masculinization in some species of mole. Additionally, Rubenstein et al. questioned whether the presence of ovotestes or a peniform clitoris is associated with a reduction in other indices of sexual dimorphism in moles, e.g. penile/clitoral length, anogenital distance, and body size.

In order to answer the questions above, Rubenstein et al. (2004) examined ovarian and clitoral anatomy in four species of North American moles, broad-footed, coast, star-nosed and shrew moles (*Scapanus latimanus, S. orarius, C. cristata*, and *N. gibbsii*, respectively). Specifically, they focused on the presence/absence of an OIG and the existence of a urethra that traverses the clitoris to its tip. This comparative approach set out to determine if a perfect cross-species correlation exists between the presence of an OIG and existence of a peniform clitoris. If so, these findings would be compatible with Matthews' suggestion that secretion of androgens by the ovaries is responsible for the masculinized clitoral anatomy of various moles.

Interestingly, Rubenstein et al. (2004) found that masculinization of the female gonad and external genitalia can occur in isolation of one another. All four species of mole were confirmed to have a peniform clitoris, but the ovaries of the broad-footed and coast moles (*Scapanus latimanus* and *S. orarius*) were unremarkable. Their ovaries showed no distinct polarity, as follicles were equally distributed throughout the cortex. Additionally, the *Scapanus* ovaries lacked a discrete interstitial gland, even when females were captured outside of the breeding season.

Moreover, the masculinization of the female external genitalia appears to be extensive, at least in the broad-footed mole, as the phallic length and anogenital distance AGD) of females did not differ significantly from that of males. These findings are in contrast to those reported for the European mole, in which both phallic length and AGD are significantly larger in males. However, this sexual dimorphism in the external genitalia may simply reflect the strong sex difference in body size in European moles

(Gorman and Stone, 1990). In fact, male European moles are so much larger than females that investigators can often sex individuals based on size alone. The same cannot be said about the broad-footed mole. Even though mean body size is sexually dimorphic in broad-footed moles, considerable overlap exists between the sexes, and the average male body length (165 ± 7 mm) is only about 4% larger than that of the average female (158 ± 7 mm) (Rubenstein et al., 2004). That being said, the sexual dimorphism in the genitalia of *T. europaea* might well have an androgen-dependent component after all. Phallic length and AGD are known to be larger in males than females even at birth, however this difference disappeared when Godet (1946) treated pregnant European moles with testosterone proprionate.

The mechanism behind the masculinization of the external genitalia in female moles remains a mystery. Does the presence of a peniform clitoris, traversed to its tip by the urethra, in species with and without so-called "ovotestes" mean we need to look for an androgen-independent mechanism to explain the natural masculinization of female moles? Research involving the spotted hyena suggests we might. However, we need to continue to look for alternate sources of testosterone, be it from the maternal or fetal compartment. Neither the maternal or fetal ovary needs to express an "ovotestis"-like phenotype to produce significant amounts of androgens, and theca-interstitial cells need not conglomerate into discrete interstitial glands to carry out this function. Clearly, additional research needs to be completed to further understand the "intersexuality" of these elusive and interesting animals.

3. SPOTTED HYENAS

3.1. "Masculinization" of Females

Spotted hyenas are members of the family Hyaenidae, a lineage of carnivores that split from ancestral viverrids 25 - 30 million years ago. At this time, there are four extant species in the family: striped hyenas (*Hyaena hyaena*), brown hyaenas (*Parahyaena brunnea*), spotted hyenas (*Crocuta crocuta*) and a small termite-eating outlier, the aardwolf (*Proteles cristatus*). Among the female hyenids, only spotted hyenas are characterized by exceptional masculinization of the external genitalia, and their internal urogenital anatomy follows the typical mammalian pattern.

Spotted hyenas are also the largest and most social of the hyenids, living in multi-male, multi-female "clans" and, when hunting in large groups, capable of killing zebra and cape buffalo. Within these groups, every adult female resident totally dominates every adult male immigrant, both at kills and in the course of other social interactions (Kruuk, 1972; Frank 1986a,b; Mills 1990). When viewed in combination with their "masculine" genitalia, this female behavioral dominance has provoked a search for naturally circulating androgens. Such androgens would have had to be present during prenatal life, in order to account for the genital masculinization which is complete at birth.

3.2. Androgens in Female and Male Spotted Hyenas

Testosterone and dihydrotestosterone appear to be the primary androgens circulating in adult male hyenas, while androstenedione, of ovarian origin, is the primary androgen circulating in female spotted hyenas (Glickman et al., 1992; Lindeque et al., 1986). Through a set of convergent observations on fetal steroid concentrations, and placental metabolism *in vivo* (Licht et al., 1992), and detailed study of placental metabolism in vitro (Yalcinkaya et al., 1993), it has become clear that androstenedione secreted by the maternal ovary is converted to estrogen and testosterone by the placenta and transferred to the developing fetus. This process has been observed as early as day 35 of gestation, and may well begin at a still earlier date (Licht et al., 1998). It is also the case, that formation of the fetal clitoris, with a completely enclosed central urogenital canal, has been observed in a 36 day fetus, prior to differentiation of the fetal ovaries (Licht et al., 1998). This latter observation raised the possibility that formation of the clitoris of the female spotted hyena is, either driven via a non-traditional androgenic source (e.g., the placenta, or the fetal adrenal), or that it is formed by a non-androgenic mechanism. Experiments designed to test that possibility are described below.

3.3. Some Essential Genital Morphology in Relation to Behavior

As previously noted, the genital swellings of the female spotted hyena have fused to form a pseudoscrotum instead of an external vagina. The genital tubercle has hypertrophied, and is superficially similar to the penis of the male. From the first months of life, female and male hyenas display full erections during meeting ceremonies. In such "ceremonies," a pair of hyenas typically stand side-by-side, head-to-tail, and subject the ano-genital region of the partner to careful olfactory and (sometimes) gustatory inspection. Hyena etiquette requires that the subordinate hyena make the initial offering of the erect phallus for inspection by the dominant animal (East et al., 1993).

There are clear and significant differences in phallic morphology, as is appropriate, given the very different tasks that these organs have evolved to fulfill (Neaves et al., 1980; Frank et al., 1990). The penis of the male is long and thin, with a small opening at the narrow tip of the glans. This urogenital meatus has little, if any, elasticity. But, the small meatus is adequate for urination and delivery of seminal fluid during mating. The angular shape of the glans facilitates entry into the prepuce and/or clitoral meatus during mating.

The shaft of the clitoris is shorter and thicker than the male penis and, at the tip of the blunt glans, the urogenital meatus is much larger and more elastic than that of the male (Glickman et al., 1992). The size and elasticity of the shaft and urogenital meatus permit receipt of the male penis during mating, and expansion of the organ during delivery of the fetus. However, at the time of the first birth, the meatus cannot stretch sufficiently to permit delivery of a 1.5 kg fetus and the urogenital meatus has to tear before the infant hyena can emerge (Frank and Glickman 1994). This is a reproductively costly arrangement, resulting in stillbirths as much as 60% of the time (Frank et al., 1995). Once torn, by the passage of the first fetus, subsequent births through the clitoris are uneventful.

Cunha et al. (in press) have identified a set of internal sex differences in phallic morphology that facilitate the diverse burdens placed on the clitoris and the penis. For example, the anterior placement of the clitoris on the abdomen, relative to the more usual location of an external vagina in close proximity to the anus, requires that the elongate male "flip" his semi-erect organ against the abdomen of the female as he "searches" for the opening of the clitoral prepuce. In both females and males cross-hatched collagen fibers provide more-then-usual firmness of structure. This would facilitate the flipping of the male penis, and may assist in stabilizing the clitoris for receipt of the male. In addition, the male urethra is centrally located within the shaft of the penis, surrounded by erectile tissue and a thick tunica, promoting firm erectile capability. In the female, the urogenital canal is located more ventrally within the shaft of the clitoris, and is not surrounded by an unyielding tunica, permitting appropriate expansion during mating and parturition (see Neaves et al., 1981).

3.4. Normal Postnatal Development

As observed above, the essential sexually differentiated features of the external genitalia are present at birth (Cunha et al., in press). Precise measurements are impossible in infant hyenas until 2 - 3 months of age, due to fusion of the prepuce and the glans. But, several features of postnatal development should be noted. First, sex differences in the size and elasticity of the urogenital meatus are obvious within the first six months of life. Second, we have been unable to find any evidence of a pubertal growth spurt. In fact, 80% - 90% of genital growth occurs within the first year of life, well before puberty (Glickman et al., 1998). The latter is typically achieved at roughly 18 - 24 months of age in males, and 30 - 36 months of age in females (Kruuk, 1972; Glickman et al., 1992).

3.5. Effects of Pre-Pubertal Gonadectomy on Growth of the External Genitalia

We have observed the effects of pre-pubertal gonadectomy on growth of the external genitalia, employing a comprehensive set of measures. Castration, or ovariectomy, in spotted hyenas reduced circulating testosterone concentrations to barely measureable levels, and led to a reduction of approximately 90% in plasma concentrations of andros-tenedione (Glickman *et al.*, 1992). However, removal of the testes or ovaries at 4 - 7 months of age had little effect on erect (i.e., "stretched") length of the adult penis or clitoris (Glickman et al., 1992), or on the density of "spines" found on the glans of both females and males (Glickman et al., 1998). Such "spines" are found in many mammalian species, and are usually highly sensitive to the reduction in circulating androgens that typically follows castration (Beach and Levinson, 1950; Sachs et al., 1984). In addition, although mating behavior was absent, full erectile function was displayed during meeting ceremonies, in a male hyena more than 10 years after pre-pubertal castration. The apparently normal growth, and maintenance of spine densities, observed in these pre-pubertally castrated/ovariectomized hyenas, once again raised the possibility that some non-androgenic mechanism might be at work.

To be sure, there were clear effects of prepubertal ovariectomy on clitoral morpho l-ogy. The glans was noticeably thinner in ovarectomized subjects, and the urogenital meatus of the female lost its elasticity, responding much as the meatus of the male when gently distended. These effects were largely reversed by systemic estrogen administration (Glickman et al., 1998).

A final note: given the exceptionally early growth of the external genitalia of the spotted hyena, it may be the case that gonadectomy at 4 - 7 months of age was already too late. We have some morphological data available from two male hyenas, one castrated at 30 days of age, and the other at 37 days of age. There may be effects on stretched length of the adult penis, as well as effects on shaft diameter. We need more data, and these initial observations suggest that still earlier gonadectomies are warranted. Also, considering the effects of estrogens on diameter of the glans in ovariectomized females, and noting the effects of an aromatase inhibitor on penile development *in utero* (as described below), it is clear that further experiments on pre- and post-natal effects of estrogens are required. Although there is little circulating estrogen in male hyenas, the possibility of local conversion in target tissues needs to be explored.

3.6. Effects of Prenatal Androgen Blockade

When one encounters a female mammal that is so thoroughly masculinized as the spotted hyena, you have to wonder if the process by which masculinization has occurred can be reversed. The scientific literature is replete with papers describing the effects of prenatal androgen treatment on female anatomy, physiology, and behavior, however the spotted hyena requires the obverse. As such, exhaustive efforts have been made to treat pregnant spotted hyenas with a variety of drugs, which can be generally categorized as "anti-androgens", with the goal of producing a female spotted hyena with the external genitalia of a more typical female mammal. These experiments have been complicated by the fact that such treatments in spotted hyenas are regularly associated with poor pregnancy outcomes (stillbirths and early neonatal deaths). However, a handful of anti-androgen treated animals have survived to adulthood, and even the stillborn cubs have provided useful information.

Pregnant spotted hyenas were treated with several different regimens, yet not a single female has been born with a distal vagina, and the clitoris has always been prominent. These traits persisted in female offspring whether mothers were treated with an androgen receptor blocker alone (cyproterone or flutamide) or in combination with an inhibitor of 5α-reductase (finasteride + flutamide). Even when the latter regimen was initiated well before the anticipated time of sexual differentiation (as early as 10 days post-mating), newborn females continued to display a large clitoris and the absence of an external vaginal opening.

This is not to say that there was no effect of the treatment on genital morphology. In fact, the effects of the prenatal anti-androgens are structurally and functionally significant. In anti-androgen treated males (AA-males) the penis became "feminized", in that it took on many of the characteristics of the clitoris. The penis of AA-males is shorter and thicker than that of control males (C-males), with dimensions more akin to those seen in C-females. Similarly, the diameter and elasticity of the urethral meatus is enlarged in AA-males, again approximating that of C-females (Drea et al., 1998). Finally, the morphology of the bulbocavernosus (BC) and ichiocavernosus muscle, and the number of spinal motoneurons (Onuf's nucleus) that innervate them are sexually dimorphic in spotted hyenas, as they are in other mammals. Both the number of motoneurons in Onuf's nucleus and the morpholgy of the BC muscle were "feminized" in AA-males (Forger et al., 1996).

Although the clitoris of AA-females is quite prominent, the "feminine" characteristics are more exaggerated in this group. The clitoris is even broader and shorter than that of C-females, and the urogenital meatus is much more elastic (Drea et al., 1998). From a functional point-of-view, these modest changes in phallic morphology have had robust effects. The greater meatal elasticity of AA-females seems to have negated some of the high costs associated with the delivery of a fetus through a long and tortuous birth canal. Indeed, the first cub born to a C-female has only a 33% chance of being born alive, while the rate of live births in our four AA-females has been 100%. Conversely, the changes seen in males may have rendered some of our AA-males functionally sterile (Drea et al., 2002). In the most profoundly affected AA-males the penis appears to be too short to reach the ventrally located urogenital opening. The increased thickness of the erect glans in AA-males may also make insertion problematic, especially when courting virginal females.

3.7. Effects of Prenatal Aromatase Inhibition

After several years of experiments involving prenatal treatment with anti-androgens we have recently begun to expand our repertoire by manipulating maternal and fetal hormones in new ways. We started a pilot program using the potent aromatase inhibitor Letrozole (Femura®, Novartis) for the following reasons: 1. A deficiency in placental aromatase activity has been associated with female pseudohermaphroditism in humans (Conte et al., 1994). However, Licht et al. (1992) and Yalcinkaya et al. (1993) suggested that aromatase activity in the spotted hyena placenta is relatively low as compared to that of 17β-hydroxysteroid dehydrogenase. Thus, most of the androstenedione derived from the maternal ovary is preferentially converted to testosterone rather than estrogen as it crosses the placenta. We queried whether we would induce further masculinization of the female genitalia by blocking aromatase. 2. We were also curious to know if the intense neonatal aggression seen between littermates is influenced by prenatal androgens, and if so, are those effects mediated by local aromatization of androgens into estrogens in critical areas of the brain. As circulating maternal androgens and estrogens are at their highest levels during late gestation (Licht et al., 1992), our investigations with aromatase inhibition started here.

Initially, we used the potent third generation aromatase inhibitor   Letrozole during the last 30 d of a twin pregnancy (average length of gestation = 110 d). Both cubs were males and only one was born alive, so we were unable to determine if the genitalia of a female spotted hyena was further masculinized by this treatment. However, our routine measurements of the surviving cub's phallus revealed an interesting effect. While the overall shape and contour of the phallus was unmistakably male, the elasticity of the urethral meatus was significantly greater than untreated males. The absolute values of the stretched meatal diameter were intermediate between untreated males and AA-males (recall the latter are very similar to untreated females).

This preliminary finding piqued our curiosity such that we treated a second pregnancy for a longer duration, encompassing the last 60 d of gestation. This pregnancy resulted in the birth of a single female cub with very unusual genitalia. The glans clitoridis appeared blunted, but more importantly, the urethral meatus was displaced, with part of the opening located along the ventral surface of the glans (Figure 1). For want of a better term, this appears to be a case of clitoral hypospadias.

4. SUMMARY

Although varying degrees of genital masculinization are a reasonably common phenomenon in the world of female mammals, the majority of such variation has not been investigated. In this chapter we have described research on the "masculinized" genitalia of moles and hyenas. Such research raises intriguing possibilities regarding the coordinated role that androgens, estrogens and peptide hormones (e.g., relaxin) might play, at different stages of sexual differentiation and development, in preparing genital tissues for their functional roles in reproduction. Such studies also suggest that non-androgenic mechanisms need to be considered.

Arnold (1996) and Carruth et al. (2002) have recently presented the argu   ment for broadening our view of sexual differentiation of brain and behavior, emphasizing direct genetic effects. A similar view has been presented for the Tammar wallaby, where formation of a scrotum, or a pouch, is a direct consequence of the presence/absence of two X chromosomes (Pask and Renfree, 2001). Although our research on moles and hyenas has not yet yielded such definitive results, the research reviewed in this chapter calls attention to processes that could well operate in other mammals, including humans.

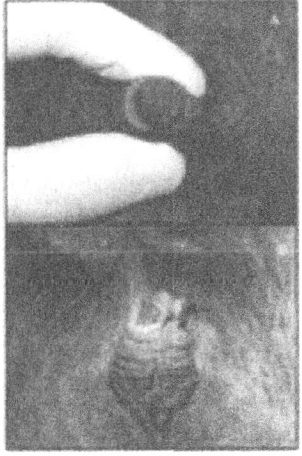

Figure 1. Coronal (A) and ventral (B) views of glans and urogential meatus of a 4 month old female spotted hyena, treated with an aromatase inhibitor (Letrozole) during the last 60 days of gestation. The urethral meatus is displaced, simulating a hypospadias-like condition, with part of the opening located along the ventral surface of the glans (arrows). In untreated hyenas the meatus would not extend onto the ventral surface of the glans.

REFERENCES

Arnold, A.P., 1996, Genetically triggered sexual differentiation of brain and behavior. *Hormones and Behavior*, **30**: 495-505.

Beach, F.A. and Levinson, 1950, Effects of androgen on the glans penis and mating behavior of castrated rats. *Journal of Experimental Zoology*, **114**: 159-168.

Beolchini, F., Rebecchi, L., Capanna, E., and Bertolani, R., 2000, Female gonad of moles, genus *Talpa* (Insectivora, Mammalia): ovary or ovotestis?. *Journal of Experimental Zoology*, **286**: 745-754.

Carruth, L. L., Reisert, I., and Arnold, A. P., 2002, Sex chromosome genes directly affect brain differentiation. *Nature Neuroscience*, **5**: 933-934.

Conte, F. A., Grumbach, M. M., Ito, F., Fisher, C. R., and Simpson, E. R., 1994, A syndrome of female pseudohermaphroditism, hypergonadotropic hypogonadism, and mulitcystic ovaries associated with missense mutations in the gene encoding aromatase (P450arom). *Journal of Clinical Endocrinology and Metabolism* , **78**: 1287-1292.

Cunha, G. R., Wang ,Y., Place, N. J., Lui, W., Baskin, L. S., and Glickman, S.E., in press, The urogenital system of the female spotted hyena (*Crocuta crocuta*): A functional histological study. *Journal of Morphology* .

Deanesly, R., 1966, Observations on reproduction in the mole *Talpa europaea*. In *Comparative biology of reproduction in mammals*, **Vol 15** pp 387-402 Ed IW Rowlands. Academic Press, London.

Drea, C. M., Weldele, M. L., Forger, N. G., Coscia, E. M., Frank, L. G., Licht, P., and Glickman, S. E., 1998, Androgens and masculinization of genitalia in the spotted hyaena (*Crocuta crocuta*). 2. effects of prenatal anti-androgens. *Journal of Reproduction and Fertility*, **113**: 117-127.

Drea, C. M., Place, N. J., Weldele, M. L., Coscia, E. M., Licht , P., and Glickman, S. E., 2002, Exposure to naturally circulating androgens in foetal life incurs direct reproductive costs in female spotted hyaenas, but is prerequisite for male mating. *Proceedings of the Royal Society, London B*, **269**: 1981-1987.

East, M. L., Hofer, H. and Wickler, W., 1993, The erect "penis" is a flag of submission in a female-dominated society: greetings in Serengeti spotted hyaenas. *Behavioral Ecology and Sociobiology*, **33**: 335-370.

Forger, N. G., Frank, L. G., Breedlove, S. M., and Glickman, S. E .,1996, Sexual dimorphism of perineal muscles and motoneurons in spotted hyenas. *The Journal of Comparative Neurology*, **375**: 333-343.

Frank, L. G., 1986a, Social organisation of the spotted hyaena (*Crocuta crocuta*). I. Demography *Animal Behaviour*, **34**: 1500-1509.

Frank, L. G., 1986b, Social organisation of the spotted hyaena (*Crocuta crocuta*). II. Dominance and reproduction *Animal Behaviour*, **34**: 1510-1527.

Frank, L. G. and Glickman, S. E., 1994, Giving birth through a penile clitoris: Parturition and dystocia in the spotted hyaena (*Crocuta crocuta*). *Journal of Zoology, London*, **234**: 659-665.

Frank, L. G., Glickman, S. E., and Powch, I., 1990, Sexual dimorphism in the spotted hyaena (*Crocuta crocuta*). *Journal of Zoology, London*, **221**: 308-313.

Frank, L. G., Weldele, M. L., and Glickman, S.E., 1995 Masculinization costs in hyaenas. *Nature*, **377**: 584-585.

Glickman, S. E., Frank, L. G., Pavgi, S., and Licht, P., 1992, Hormonal correlates of 'masculinization' in female spotted hyaenas (*Crocuta crocuta*). I. Infancy to sexual maturity. *Journal of Reproduction and Fertility*, **95**: 451-462.

Glickman. S. E., Coscia, E. M., Frank, L. G., Licht ,P., Weldele, M. L., and Drea, C. M., 1998, Androgens and masculinization of genitalia in the spotted hyaena (*Crocuta crocuta*). 3. Effects of juvenile gonadectomy. *Journal of Reproduction and Fertility*, **113**: 129-135.

Gorman, M. L. and Stone, R. D., 1990, *The natural history of moles*. Comstock Publishing Associates, Ithaca.

Godet, R., 1946, Biologie experimentale - modifications de l'organogenèse des voies urogénitales des embryons de taupes (*Talpa europæa* L), par action du propionate de testostérone. *Comptes Rendus Hebdomadaires Des Seances De L Academie Des Sciences*, **222**: 1526-1527.

Hill, W. C. O., 1958, External genitalia. *Primatologia*, **3**: 630-704.

Jiménez, R., Burgos, M., Sanchez, A., Sinclair, A. H., Alacron, F. J., Marin, J. J., Ortega, E., and Diaz de la Guardia, R., 1993, Fertile females of the mole *Talpa occidentalis* are phenotypic intersexes with ovotestes. *Development*, **118**: 1303-1311.

Jiménez, R., Alacron, F. J., Sanchez, A., Burgos, M., and Diaz de la Guardia, R., 1996, Ovotestis variability in young and adult females of the mole *Talpa occidentalis* (Insectivora, Mammalia). *Journal of Experimental Zoology*, **274**: 130-137.

Jost, A., 1953, Problems of fetal endocrinology: the gonadal and hypophyseal hormones. *Recent Progress in Hormone Research,* **8:** 379-418.

Kruuk, H., 1972, *The spotted hyena.* University of Chicago Press, Chicago.

Licht, P., Frank, L. G., Pavgi, S., Yalcinkaya, T. M., Siiteri, P. K., and Glickman, S. E., 1992, Hormonal correlates of 'masculinization' in female spotted hyaena (*Crocuta crocuta*). 2. Maternal and fetal steroids. *Journal of Reproduction and Fertility*, **95:** 463-474.

Licht, P., Hayes, T., Tsai, P., Cunha, G., Kim, H., Golbus, M., Hayward, S., Martin, M. C,, Jaffe, R. B., and Glickman ,S. E., 1998, Androgens and masculinization of genitalia in the spotted hyaena (*Crocuta crocuta*). 1. Urogenital morphology and placental androgen production during fetal life. *Journal of Reproduction and Fertility,* **113:** 105-116.

Lindeque, M., Skinner, J. D., and Millar, R. P., 1986, Adrenal and gonadal contribution to circulating androgens in spotted hyaenas (*Crocuta crocuta*) as revealed by LHRH, hCG, and ACTH stimulation. *Journal of Reproduction and Fertility,* **78:** 211-217.

Matthews, L. H., 1935, The oestrous cycle and intersexuality in the female mole (*Talpa europaea* Linn.). *Proceedings of the Zoological Society, London Series 2,* **1935:** 347-383.

Matthews, L. H., 1939, Reproduction of the spotted hyaena (*Crocuta crocuta* Erxleben). *Philosophical Transactions of the Royal Society, London Ser B,* **230:** 1-78.

Mills, M. G. L., 1990, *Kalahari hyaenas: Comparative behavioural ecology of two species,* Unwin-Hyman, London.

Mossman, H. W. and Duke, K. L., 1973, *Comparative morphology of the mammalian ovary,* University of Wisconsin Press, Madison.

Neaves, W. B., Griffin, J. E., and Wilson, J. D.,1980, Sexual dimorphism of the phallus in spotted hyaena (*Crocuta crocuta*). *Journal of Reproduction and Fertility*, **59:** 509-513.

Pask, A., and Renfree, M. B. 2001, Sex determining genes and sexual differentiation in a marsupial *Journal of Experimental Zoology,* **290:** 586-596.

Rubenstein, N., Cunha, G., Wang, Y. Z., Campbell, K., Conley, A. J., Catania, K. C., Glickman, S. E., and Place, N. J., 2004, Variation in ovarian morphology in four species of New World moles with a peniform clitoris. *Reproduction* (in press).

Sachs, B. D., Glater, G. B., and O'Hanlon, J. K., 1984, Morphology of the erect glans penis in rats under various gonadal conditions *Anatomical Record* 2,**10:** 45-52.

Sanchez, A., Bullejos, M., Burgos, M., Hera, C., Stomatopoulos, C., Diaz de la Guardia, R., and Jimenez, R., 1996, Females of four mole species of the genus *Talpa* (Insectivora, Mammalia) are true hermaphrodites with ovotestes. *Molecular Reproduction and Development,* **44:** 289-294.

Whitworth, D. J., Licht, P., Racey ,P. A. and Glickman, S. E., 1999, Testis-like steroidogenesis in the ovotestis of the European mole, *Talpa europaea. Biology of Reproduction* , **60:** 413-418.

Wilson, J. D., George, F. W., and Griffin, J. E., 1981, The hormonal control of sexual development. *Science,* **211:** 1278-1284.

Wislocki, G. B.,1936, The external genitalia of the simian primates. *Human Biology,* **8:** 309-347.

Yalcinkaya, T. M., Siiteri ,P. K., Vigne, J-L., Licht ,P., Pavgi, S., Frank, L. G. and Glickman, S.E., 1993, A mechanism for virilization of female spotted hyenas *in utero. Science,* **260:** 1929-1931.

INDEX

Adrenocorticotropic hormone stimulation,
  36, 39
Affected relative pair analysis, 76
Allele sharing, 75-76
Allelic variant, 37-40
5 α reductase,
    inhibition of, 232-233
5 α reductase gene, 75
5 α-reductase type 2, 48-56, 68
    expression of, 54-55
    schematic of structure, 53
    significance of mutations on
            hypospadias, 48-52, 54
Ambiguous genitalia, 4, 243
Amenorrhea, 38
Androgen binding, 8, 11
Androgen insensitivity, 34, 39, 74, 142
Androgen receptor gene, 75
Androgen receptor, 31, 56-59, 94-95, 125
    genetic mutations, 34
    role in genital tubercle development,
            125-126, 127-132, 142-143,
Androgen resistance, 56
Androgen signaling, 154-155
Androgen-receptor antagonists, 221-225
Androgens, 13, 31-34, 46, 95, 124-142,
            160, 174, 243, 246, 249
Andromedins, 155
Androstenedione, 39, 47, 246
Anencephaly, 26
Anogenital distance, 105, 195, 221
Anti-androgens, 189, 195, 208, 236, 249
Apical ectodermal ridge, 150, 160
Aromatase inhibition, 250
Assisted reproduction, 60, 206
Autosomal dominant inheritance, 74, 79
Autosomal recessive inheritance,63
Axial patterning, 153-154

17 β hydroxysteroid dehydrogenase, 38-39

Birth control, 60
Birth Defects Monitoring Program
    (BDMP), 9, 205
Birth defects, 25-26, 31-32
Bone morphogenetic proteins, 154

Castration, 248
Cesarean delivery, 67
Chlozolinate, 225-226
Clitoromegaly, 184-185
Cloaca, 88, 145, 162-163
Cloacal membrane, 31, 87-89, 117, 162
Competitive inhibition, 40
Congenital adrenal hyperplasia, 174, 243
Congenital anomalies, 3
Consanguinity, 74
Cryptochidism, 34-36, 47, 205
Cumulative risk, 237
Cytochrome P450, 217

DDT, 221, 227
Developmental estrogenization syndrome,
    204, 234
Dexamethasone, 69
1,1-dichloro-2,2bis(pchlorophenyl)ethylene
    (DDE), 195
Diethylhexyl phthalates, 210
Diethylstilbestrol (DES), 197, 204, 206-
Dihydrotestosterone, 31-33, 46, 94-95, 246
Dioxin, 208, 210, 236
Drash syndrome, 74

Ectodermal ingrowth theory, 7
Endocrine disruption, 104
    background, 189-198
        human scenario, 192
        latent activity, 190
        wildlife scenario, 191
    definition of, 190
    environmental agents on, 196

The manufacturer's authorised representative in the EU is Springer
Nature Customer Service Centre GmbH, Europaplatz 3, 69115 Heidelberg,
Germany. If you have any concerns regarding our products, please
contact ProductSafety@springernature.com

Printed and bound by CPI Group (UK) Ltd, Croydon, CR0 4YY

30/04/2026

02100146-0003